Pembroke 14th May
2015

So pleased to meet you and
in hope of future exchanges

[illegible signature]

Information Infrastructure(s)

Information Infrastructure(s): Boundaries, Ecologies, Multiplicity

Edited by

Alessandro Mongili and Giuseppina Pellegrino

Foreword by Geoffrey C. Bowker

CAMBRIDGE SCHOLARS PUBLISHING

Information Infrastructure(s):
Boundaries, Ecologies, Multiplicity,
Edited by Alessandro Mongili and Giuseppina Pellegrino

This book first published 2014

Cambridge Scholars Publishing

12 Back Chapman Street, Newcastle upon Tyne, NE6 2XX, UK

British Library Cataloguing in Publication Data
A catalogue record for this book is available from the British Library

ISBN (10): 1-4438-6655-5, ISBN (13): 978-1-4438-6655-2

In memory of Leigh Star

TABLE OF CONTENTS

List of Figures and Tables

Figures

Tables

FOREWORD

THE INFRASTRUCTURAL IMAGINATION

GEOFFREY C. BOWKER

In a classic from 1959, pragmatist sociologist C. Wright Mills wrote of the pressing need in the face of constant change and strife to develop a sociological imagination through which: "men and women hope to grasp what is going on in the world, and to understand what is happening in themselves as minute points of the intersections of biography and history within society" (7)[1]. There is no reference to computers in his tome. Actually, I found that out at the touch of a button by deploying an infrastructure he could not have imagined – the inescapable, highly problematic, deeply rich, politically wrought infrastructure of Google Books. How could Mills have foreseen this consummation not so devoutly to be wished - it was a commonplace at the time that a mere handful of computers would meet the world's needs. The millennial generation (and indeed the new millennialists such as Ray Kurzweil) cannot picture a world without the web and it is hard for them to imagine ways of forming personal, intellectual and social relations without passing through an information infrastructure.

Bruno Latour used to say that society was about the size of a pumpkin, since whenever speakers referred to "the social", their standard gesture was to bring their opened hands together in a forceful movement that left a virtual space between them about that size. It was his way of saying that society is a vacuous concept. However, I can think of no gesture and few stories that easily summon information infrastructures. And yet they are not vacuous. They are invisible, they melt away, they are what you use in order to do something. We don't think about the road when we drive our cars (except, as Star and Ruhleder would point out, when there are potholes – then we curse the local council). We all too rarely think about the ways in which our social, cultural and political values are braided into the

[1] Wright Mills, Charles. 1959. *The Sociological Imagination*. Oxford: Oxford University Press.

wires, coded into the applications and built into the databases which are so much a part of our daily lives. We are all today not citizens pure and simple - we are data citizens; and our data doubles (uncanny doppelgangers) are an integral part of our lives.

Mills' biography, history and society could all seemingly be experienced without a technological substrate. For my own life, my "minute point of intersection", I know that I spend more time with my computers than with humans - dearly as I occasionally love the latter. I talk more often and more deeply with distant rather than proximate interlocutors. And it's not just about the words and such - I feel an ache in my fingers if I have been away from the keyboard for too long … Yet this mediation through computers is indeed hard to grasp. The perspectives of design, ecology, mobility and multiplicity developed in the four sections below are a rich way in.

It is fitting that contributors to this volume, inspired by STS Italia, include contributions from four other countries. Just as it is increasingly difficult to imagine our infrastructures today stopping at any one country's borders, so is it clear that theorizing infrastructure is a distributed task. The geographical as well as the analytical range of this work is most impressive.

Susan Leigh Star sadly and suddenly died in 2010. Her next book would have been entitled "Infrastructural Poetics". The contributions to this volume would have been richly interwoven into that text. Her and my time in Sardinia sipping Prosecco on balconies; talking with Alessandro in a polyglot mix of Italian, French and English with a touch of Russian; exploring the traces of Nuragic civilization; and meeting the scholars of STS Italia was a transformative part of our lives – even if the internet connection wasn't so hot. I speak for her, and join my own voice, in thanking Alessandro and Giuseppina for bringing this collection together. I speak for many future readers in saying that if you want to exercise infrastructural imagination, this is a marvelous place to start.

Long Beach, 23 July, 2014

CONTRIBUTORS

Editors

Alessandro Mongili is a Professor of Sociology at the University of Padua. He is also affiliated with the Padova Science, Technology and Innovation Studies center (PaSTIS). His books include *La chute de l'U.R.S.S. et la recherche scientifique* (L'Harmattan 1998) and *Tecnologia e società* (Carocci 2007). His current research interests and publications focus on the design and development practices in computing and on the relevance of computing in standardization of minority languages.

Giuseppina Pellegrino is a Lecturer in Sociology of Communication at the University of Calabria (Italy). Her core research interests intersect STS and Mobilities, as well as Media, Gender and Organization Studies. A visiting fellow at the University of Edinburgh, Lancaster University, IAS-STS (Graz) and TU Darmstadt, she was appointed a Fulbright Distinguished Lecturer in Italian Studies at the University of Pittsburgh in 2013.

Authors

Michela Cozza is a sociologist and research fellow at the Department of Information Engineering and Computer Science at the University of Trento. In recent years, she has been teaching Sociology of Organisation and Sociology of Innovation. Her main research interests concern the processes of organizational change and technological innovation.

Stefano Crabu is a sociologist with a PhD in Social Science, and a research collaborator at the Department of Philosophy, Sociology, Education & Applied Psychology (FISPPA) at the University of Padua. His main work revolves around technoscientific innovation in biomedicine.

Jérôme Denis is an Assistant Professor in Sociology at Telecom ParisTech. Articulating a concern for writing practices and STS, he studies the production and maintenance of different kinds of information infrastructures in such domains as administrative work, transportation systems, IT security, open data and collaborative cartography. He is the co-founder of Scriptopolis, a collective scientific blog about writing practices (www.scriptopolis.fr/en).

Paolo Giardullo is a PhD Fellow in Sociology of cultural phenomena and normative processes at the University of Urbino "Carlo Bo". His main research areas are the public representation of technoscience, specifically, communication and perception of risk related to food and environmental issues. Recently, he has approached Internet Studies from an epistemological and methodological point of view.

Simona Isabella, PhD, is a Research Fellow at the University of Calabria (Italy). Her main research interests concern the sociology of communication and in particular her studies focus on Computer Mediated Communication, mobile phone and mobile Internet, and Social Networking both from a theoretical and methodological point of view.

Stefan Klein is a Professor of Interorganizational Systems at the University of Münster. His current research areas are Information Infrastructures, Network Economy, Information Management, Social Media and the transformation of work. He has published widely and is a member of the editorial board of several international IS journals.

Gian Paolo Lazzer is a PhD student in Sociology and social research at the University of Verona. His PhD research project concerns sociology of money and the aim of his dissertation is to analyse immigrants' monetary networks in the Veneto Area (Italy), in particular credit and consumption practices.

Giuseppe Lugano works in Brussels as Science Officer for COST, the longest-running European framework for cooperation in science and technology. He holds a MSc degree in Computer Science and a PhD in Cognitive Science. His research focuses on the conceptual design of technologies and services for community and sustainable living.

Francesco Miele is a researcher at the Bruno Kessler Foundation. His research interests include academic spin-offs, high tech inter-organizational networks and organizing processes in healthcare. He has also worked on childcare services, gender and carework and occupational health.

Oana Mitrea works in the field of Sociology of Technology with a focus on innovation in IT, mobility, and transportation. She completed her doctoral studies at the University of Bucharest and TU Darmstadt and is currently a member of the "Transportation Informatics Group" at the Alpen-Adria University of Klagenfurt, Graz, and Vienna.

Federico Neresini teaches Science, Technology and Society and Sociology of Innovation at the University of Padua (Italy). His research has focused on biotechnology and nanotechnology, and he has also recently addressed the relationship between Big Data and scientific research activities, as well as the implications of Big Data for the social sciences.

Giacomo Poderi obtained his PhD in Sociology and Social Research at the University of Trento with specialization in Information System and Organizations. Currently he is a post-doc researcher at the Department of Information Engineering and Computer Science (University of Trento).

David Pontille is a researcher at the French National Center for Scientific Research/Centre de Sociologie de l'Innovation. His work focuses on writing technologies in different workplaces: scientific authorship and research evaluation tools, the manufacture of deeds in bailiffs' offices, the production of biomedical databases, and the design and maintenance of subway wayfinding systems. He is the co-founder of Scriptopolis, a collective scientific blog about writing practices (www.scriptopolis.fr/en).

Stefan Schellhammer is a faculty member of the Department of Information Systems at the University of Münster. In his PhD he studied electronic ordering systems in the pharmaceutical distribution industry. His present research interests include studying the emergence of information infrastructures and the transformation of the modern workplace.

Mauro Turrini is a Marie Curie researcher at the CETCOPRA (Centre for the study of techniques, knowledge and practices) at the University of Paris 1 "Panthéon-Sorbonne". He has edited a book on *Biocapital* (ombre corte 2011) and has published articles in several journals, including *Etnografia e Ricerca Qualitativa*, *Human Studies*, *Sociologia del Lavoro*, *Spontaneous Generations*, *Studi Culturali*, *Tecnoscienza*, and *Visible*.

Assunta Viteritti has been Lecturer in Sociology of Education and Knowledge, Organization and Learning at La Sapienza University of Rome since 2002. Her main research interests are STS and Learning, Laboratory studies, Education, Organization and Professional Skills.

THE BOUNDARIES OF INFORMATION INFRASTRUCTURES: AN INTRODUCTION

ALESSANDRO MONGILI
AND GIUSEPPINA PELLEGRINO

1. Where this book comes from

Reflections and research which have converged and are summarized in these pages started a few years ago and – like the topics of this book – circulated across multiple locations.

The "material motive" was provided by a series of panels at different Italian and European conferences, organized and co-chaired by both of us.

First, at the EASST conference "Practicing Science and Technology, Performing the Social" organized and hosted by STS Italia in Trento (September 2010) we held a track on "Performative Infrastructures, Multiple Mobilities", which put together our respective interest in Information Infrastructures and Mobility Studies.

Two years later in Rovigo, at the 4th STS Italia Conference "Emerging Technologies, Social Worlds" (June 2012) the track on "Information Infrastructures: from Standardization to Multiple Uses" aimed at focusing "on diversity of infrastructure uses, which constrain local contexts, and on torqued infrastructures, emerging from situated activities. In particular, new expertise, power users, development of open-ended information systems and professional communities able to manipulate such systems, shifted the interest towards torquing and reversibility of infrastructures, as well as advanced professional uses, such as design of technology and architecture, scientific and experimental work, medical and diagnostic activities"[2].

However, all of these joint adventures would not have taken place without the STS Italia founding conference in Cagliari (June 2006) which hosted Leigh Star and Geof Bowker's joint keynote speech. It was the first

[2] http://www.stsitalia.org/conferences/STSITALIA_2012/STS_Track3.pdf

time they had come to Italy and also the first time we had the opportunity to listen to them and approach their vision of Boundary Objects (henceforth BOs) and Information Infrastructures (henceforth IIs) in depth. Alessandro Mongili, who organized and hosted that conference as first President of STS Italia – a newly constituted network of Italian scholars in Science and Technology Studies (STS)[3] - brought them to us as result of an intellectual interest for both Leigh's and Geof's work. Their lecture in Cagliari was centered on the presentation of their ecological approach to BOs and IIs and particular emphasis was given to the consequences of classification and standards in terms of exclusion, looking at residuality in IIs and the myriad of "orphans" generated by discretization and categorical work.

Since then, Leigh and Geof have embraced and supported the emergent scientific community of Italian STS scholars, including Italy, in their continental contacts already developed in France and in Nordic countries[4].

As a follow up to that encounter in Cagliari, Giuseppina Pellegrino interviewed both Geof and Leigh on the topic of "Information Society and Inequalities" (Pellegrino 2006), focusing on issues widespread in the public debate, such as the digital divide and information overload. These themes were critically analyzed by Leigh and Geof in light of their approach to IIs and residuality. This is based on a critique towards concepts such as stabilization and principal agent, and a narrative problematizing diversity and orphans of infrastructure through an ethics of ambiguity, conceived as a constant, respectful concern for multiplicity[5].

Lecturing during a seminar series promoted by the PhD program directed by Dominique Vinck in Grenoble in 2007, Leigh Star resumed on the concept of *boundary object* in all its sociological and STS density[6]. In her intervention, entitled *The History and Boundary of Boundary Objects*, she focused on the "roots" of BOs and the "Limits and Common

---

[3] www.stsitalia.org

[4] One of the resources which fertilized the Italian field and prepared for the arrival of Leigh and Geof was the work carried out by Silvia Gherardi and her group in Trento on organizations, technology and practice, which developed around the doctoral program on "Information Systems and Organizations".

[5] On the invisibilization produced by the systematic exclusion inherent to standardization, the article entitled "Enacting Silence" was published in 2007 (Star and Bowker 2007).

[6]www.cluster-gospi.fr/IMG/pdf/Livret_Seminaire_BO-BW.pdf. This series of lectures produced different reflections on the concept of intermediary and boundary objects, published later in a special issue of the *Revue d'Anthropologie des Connaissances* (2009, 3/1). Leigh Star's intervention was publisehd also in English (Star 2010).

Misconceptions Concerning Boundary Objects". In this lecture, Leigh summarized some interesting ideas, useful for understanding how BOs and IIs were created to deal with specific sociological and STS problems, which were both theoretical and empirical. As "roots" of BOs, Leigh pointed out critical issues in Symbolic Interactionist Theory, especially those linked to Social Worlds, and all the anomalies and "triangulation" created to examine social intersections among and between them. Furthermore, she emphasized the question of "Borderlands as a Resource", linked to the idea that BOs are "enmeshed" in standards and classification. This constitutive character of the BOs (i.e. their consistency in standard and classifications) is unfortunately overlooked in many cases, and consequently confused or overlapped with any kind of intermediary object.

Star also asserted the grounded character of this concept, valid "at a medium level of scale", and "rarely useful on a global or mythic level". The use of BOs is situated in a specific history, both material and intellectual. To understand BOs' uses, articulation and design, "interpretive flexibility" is not enough. Beside semiotic issues, there are other dimensions, often linked to their materiality. According to Star, BOs are impossible to understand "without understanding of infrastructure, information needs, standards and classification". She underscored very often the "middle-range" and the "grounded" character of her theoretical production, and also in Grenoble she noted that "the concept of boundary objects is grounded, resting in the principles of grounded theory: constant comparison, iterative returns to the data, theoretical sampling, and awareness of the level of abstraction" (Star 2007). This approach also influenced many of this book's contributors, and not only because of the circulation of texts and ideas, but also because of direct contact with her. With great generosity, which was so typical of her, Leigh accepted an invitation to visit the University of Cagliari, in Sardinia, where a Grounded Theory Workshop was set up in December 2007, directed to PhD students and sociologists coming from many Italian universities. She taught some methods of coding, and discussed her approach to Grounded Theory at a roundtable with a wide range of Italian scholars specialized in research methods.

All the events remembered above, jointly with the readings, writings and exchanges they promoted, paved the road for this book to come out, as a contribution to the ongoing and fascinating debate on the role of IIs and BOs in our lives. On the other hand, many contributors dealt with the same issues during their fieldwork, or during their work on the data coming from their research, and at that point they met the ecological approach as a resource for their analysis. So, this book is placed at a crossroad of various paths and genealogies, all of them connected with the problem of the intersection among different levels of scale throughout devices and networks.

2. Information infrastructures

Interest in infrastructures and focus on concepts like convergence, intersections, cooperative and distributed activities are two streams that have characterized social sciences since the mid '90s.

Actually infrastructures - and in particular IIs - allow, facilitate, mediate, saturate and influence our material and immaterial surroundings. Furthermore, they are often shaped and intertwined with networks of relations and distributed agency. They even enable the existence of networks, being in turn produced by them, often in a distributed way. Infrastructures are not static and immobile in time and space: they need maintenance and repair, which become an important aspect of their use as well.

Looking for a definition of information infrastructure (II) means facing a phenomenon which displays different characters in time, spaces, uses, design, and scaling. IIs are not mere information artifacts, although they keep a genealogical relation with many information artifacts, or at least they can include some information artifacts, if we consider an information artifact as a "wide array of tools, systems, interfaces and devices for storing, tracking, displaying and retrieving information" (Star, Bowker and Neumann 2003). The main difference between any information artifact and an II is provided by its relational character towards organized practices (Jewett and Kling 1991; Star and Ruhleder 1996, 113). Some information artifacts become infrastructures when under certain events or circumstances they "converge with human behavior to form a (…) whole", Star and Lampland (2009, 20) assert, defining it as "the nesting character of infrastructures". So, IIs carry information out in different spaces, but also in different times, creating an important texture not only for a basic exchange of information or data, but also for work, mobility, leisure, and many activities of everyday life. From a mere technical viewpoint, an II is composed of a list of numbers, technical specifications and hidden mechanisms, more precisely by standards, wires and settings (Star 1999, 377, 379), "incorporating specific elements into the system, to link them together in a specific way, and to codify interactions by a (…) set of rules and procedures" (Turner *et al.* 2006, 91).

The relational characters of IIs were pointed out in a seminal scheme included in Susan Leigh Star and Karen Ruhleder's essay (1996, 113).

Recently, another attempt at general definition was provided by Eric Monteiro *et al.* (2013, 576-578). IIs were defined by Star and Ruhleder according to their embeddedness (an infrastructure is "sunk" into other material, technical or social structures), transparency (it is taken for granted by their users, transparent in both temporal and spatial dimension) and reach or scope (in both spatial and temporal terms). Interest in the convergence between infrastructure and human behavior led to the highlighting of two other aspects: the learning of IIs as part of membership in some social world, and the link with conventions of practice already widespread. This last characteristic is mirrored by the fact that an infrastructure "does not grow" *ex novo*, but is built on an installed base and has to be linked and interoperable with its elements. Finally, its transparency is threatened by accidents, and so it becomes visible upon breakdown. Star and Ruhleder's scheme describes the relational existence of infrastructures paying great attention to other entities and to events wherein the infrastructures exist. In the "working definition" provided by Monteiro *et al.* (2013) the importance of an extended design perspective is emphasized, "to capture how workplace technologies can be shaped across multiple contexts and over extended periods of time" (p. 576). This accentuated chronotopic stance enhances the ecological argument which underscores the fact that IIs provide a set for distributed activities (Bowker, Timmermans and Star 1995; Bowker, Baker *et al.* 2010). The importance of IIs' design is so equated to their implementation and use:

> "As a working definition, IIs are characterized by openness to number and types of users (no fixed notion of 'user'), interconnections of numerous modules/systems (i.e. multiplicity of purposes, agendas, strategies), dynamically evolving portfolios of (an ecosystem of) systems and shaped by an installed base of existing systems and practices (thus restricting the scope of design, as traditionally conceived). IIs are also typically stretched across space and time: they are shaped and used across many different locales and endure over long periods (decades rather than years)." (Monteiro *et al.* 2013, 576).

metad Infrastructuring is, therefore, an ongoing process of creating, implementing and using infrastructures (Karasti and Baker 2004). So, the role of design and other originating practices in IIs becomes continuous rather than discrete: a large array of actors intervene continuously in changing elements and in fixing them, as a constitutive part of infrastructuring (Monteiro *et al.* 2013, 59).

IIs do not exist but *occur where and when* a series of tensions (between local and global, today's requirements and tomorrow's users, research and development; between project and originating practices; implementation and maintenance/repair; individual and community; but also identities and practices, planned and emergent courses of action) are resolved (Star and Ruhleder 1996, 114; Bietz *et al.* 2010, 249). Infrastructuring is precisely the collective practice which manages all these tensions. It is difficult to ascribe it to some specific actor: maybe not by chance different scholars refer to this collective dimension using metaphors such as *web of computing* (Kling and Scacchi 1982) or *web of users and developers* (Millerand *et al.* 2010). Design, development, use, maintenance, repair, implementation are different practices and, due to the ongoing character of infrastructuring, Bietz *et al.* (2010, 249) use the metaphor of *web of infrastructuring* in order to describe this process where no one can control infrastructures and there is only a visible part (the user-at-terminal), whereas all the overwhelming practices of assemblage and articulation are invisible (Bowker and Star 1999, 319; Star 1999, 386-387).

Studying IIs is difficult work for ethnographers and sociologists. Designers, developers, and other originators often share an institutionalized discourse, which Leigh Star called "master narrative", that does not problematize diversity (Star 1999, 384). Some local ideological forms are very strong and do not help researchers to understand the multifaceted and relational character of this phenomenon. So, spatially, the participation of a lot of people and terminals distributed in very different places across a local/global axis is underestimated and, temporally, the continuous maintenance, repair, design and change confront the mythopoietic "eternity dream" of these infrastructures (Edwards *et al.* 2009, 370-371; Star 1999, 383). Studying data diversity, Geoffrey Bowker shifted the focus of his analysis from changes in infrastructural components to changes in infrastructural relations (Bowker 1994). Throughout this *infrastructural inversion*, he opened a new possibility to deliver scholars from master narratives and institutionalized discourses, which usually hide the relational side of IIs, mainly for professional purposes, and to focus on the "whole enterprise", following an ecological viewpoint (Star and Griesemer 1989, 389). This change drove to construct the question of *scaling up* (Bowker, Star and Neumann 2003) as an object of analysis crucial for this phenomenon, in both spatial and temporal terms (Edwards *et al.* 2009, 370), but also with reference to lack of integrity and absence of defined boundaries in IIs. In fact, different systems of different scales crosscut and intersect, making these infrastructures fragile and dependent on chronotopic contingencies (Mackenzie 2003, 396; Star 1999, 387). Scaling IIs up is not only their condition of existence, but also a necessary heuristic tool in order to describe and analyze them. As Ribes and Finholt (2009, 376) asserted, scaling "mirrors participants' activities" in infrastructuring. These scaling dimensions can be articulated in three different spheres: enactment, organizing and institutionalization. Enacting an infrastructure is a *technological venture*, i.e. "seeking to deploy durable resources to support work, automate (…) tasks, and enable collaboration". Organizing refers mainly to these organizational arrangements that make long term projects possible: work, organization, repair and maintenance and all those activities useful for fitting infrastructures with people. Institutionalization of infrastructures refers to those activities directed to link them to a "wider longer term goal beyond those of the project team" (Monteiro *et al.* 2013, 581).

If IIs are relational, the kind of relation they establish is crucial to discuss. At large, we can consider two main kinds of relations in order to define IIs. Firstly, the *convergence* with human behavior and social practices and secondly, the fact that IIs take on "transparency by plugging into other infrastructures and tools in a standardized fashion" (Star and Ruhleder 1996, 113; Star 1999, 381): the *embodiment of standards*.

Convergence is a concept which expresses "the double process by which information artifacts and social worlds are fitted into each other and come together" (Star, Bowker and Neumann 2003). Social world, in turn, is a concept often employed to identify groups of people sharing conventions, language, technologies and practices (Star, Bowker and Neumann 2003). This term can express the fact that in many cases humans connect to infrastructures as groups or communities of practice, rather than individuals. In each of them the same infrastructure can be instrumentalized differently and segmented in use, following different chronotopic contingencies (Lave and Wenger 1991; Wenger 1998). This difference in instrumentalization and use among different groups converging at the same II, underscores its intersectional character and the fact that it is not possible to find a corresponding stable organization, but a diverse "constellation of practices" (Wenger 1988, 241). Convergence is necessarily fragile and unstable, and its main form of articulation is *learning* as continuous participants' practice (Lave 1998) linked to any form of relation and of explicit and tacit practices: use, implementation, maintenance, repair, and design (Wenger 1998, 47; *see also* Haythornthwaite 2006, 1080). Following many scholars, learning is also a necessary part of membership in a community of practice, because technology constitutes both the workplace and the ecology of everyday activities. Becoming a member of a social world or of a community of practice is inherently intertwined with learning (Bowker and Star 1999; Lave and Wenger 1991; Star 1999, 381).

This complex existence of IIs at different scales and in different chronotopes, situates them in different positions towards humans and their lives. The reality of IIs can be considered as variable in social, technical, material and symbolic terms (Latour 1992, 144; Mackenzie 2003, 96). However, many scholars consider instead the importance of those very practices which enact IIs in front of concrete human lives, thanks to specific ontologizing politics (Mol 2002, viii, 44). As Ribes and Bowker (2009, 202) asserted, "participants often refer to ontologies as a technology, tool, or software". In computing, however, the term *ontology* is common. It is "a *description* (like a formal specification of a program) of the concepts and relationships that can exist for an agent or a community of agents" (Gruber 1993, 199). Making those "descriptions" real is a very dense and important process. Suitably, Bowker *et al.* (2010, 109; *see also* Ribes and Bowker 2009, 215) underscore a double move into this process, *building ontology* and *ontology work*, corresponding to two different aspects of ontologizing practices. With *building ontology* they understand "gathering domain knowledge, formalizing this knowledge into a machine computable format, and encoding it into machine languages", i.e. assembling an ontology becoming an infrastructure to support the users' practices in a transparent way; while for *ontology work* they understand an "act of distribution – taking knowledge out of a closed community of practice and allowing for its reuse and reshaping by others in different fields", i.e. the articulation work of an ontology. The stabilizing strength of this ontological side of infrastructuring is far from being diminished, and it is surely important for the institutionalization of infrastructures and for their "Long Now" (Ribes and Finholt 2009, 377).

Participants get an active position in infrastructuring, not only because they have to take into account the conventional dimension and take on an "unprecedented reflexivity" (Cambrosio *et al.* 2006), but also because they do not have to merely implement and use IIs, but also be involved in "extended design" practices of maintenance and repair (Monteiro *et al.* 2013, 576). These practices are linked to the particular relevance of data exchange and management and the saturation of information infrastructures as a whole. "Any interface between groups and organizations, as well as between machines – assert Edwards *et al.* (2011, 670) – is a point of resistance where data can be garbled, misinterpreted, or lost". Convergence is far from being a perfect "mechanism" and people and machines do not plug into each other in a neutral way. They propose the concept of *data friction* in order to describe this imperfect operability and functioning of IIs across such different groups and systems. We can observe two different and recursive responses to this lack of automatic correspondence between heterogeneous elements in II's convergences: *metadata development* as a process linked to data culture and their complex evolution; and *standardization*.

Metadata are often described as "data of data". They respond to certain classifying principles widespread in computing, producing classes of data, or *taxa*, describing and representing files or other objects, libraries and other data intrinsic to these entities. Metadata refer both to a single file (format, creation date, title, etc.) and to the software and hardware requirements useful for managing extrinsic data (Borgman 2000, 70-80; Bowker 2005, 116; Bowker and Star 1999, 39-40). This inescapable tool in databases is widely used to order the heterogeneity of data circulating in IIs, their chronotopic dispersion, and data disorder. However, there are no unique or standardized metadata schemes. Even if they are very diverse, they do not introduce an element of ordering and organization in data complexity, which are often inconsistent and contradictory. Metadata are "almost standards" (Edwards *et al.* 2011, 684) and are submitted to a process of standardization, in order to improve the data intensive research and use, a crucial activity in infrastructuring. Metadata standardization follows two main streams exemplified by the concepts of *project*, a site where all the datasets related to multiple field campaigns and investigators are associated, and *dataset*, a series of tables recording sets of related measurements with specified units (Karasti *et al.* 2010, 393). However, metadata schemes and their standardization are an unstable phenomenon and can be more correctly interpreted as a process of ordering in data culture (Edwards *et al.* 2011, 685).

Standards are an inescapable passage point in II's architecture and work. Their importance is at the same level as software and hardware in both IIs and computing (Karasti *et al.*, 2010, 386). As Millerand and Bowker assert (2009, 150) they permit interconnection of systems and the flow of data; their existence is possible because many other standards are already in use, and often transparent in extant IIs; they allow the growth and cultivation of shared IIs and collaborative platforms.

Leigh Star and Martha Lampland (2009, 5 *et passim*) characterized standards in this manner:

- They are nested inside one another.
- They are unevenly distributed across the sociocultural landscape.
- They are relative to communities of practice; that is, one person's well-fitting standard may be another's impossible nightmare.
- They are increasingly linked to and integrated with one another across many organizations, nations, and technical systems.
- They codify, embody, or prescribe ethics and values, often with

great consequences for individuals (consider standardized testing in schools, for example).

Standards are points in a process of standardization which is similar to other processes creating networks as well as social and socio-material links across the world. Particularly, standardization is similar to quantification and formal representation as a form of thought and production of a digital, discrete image of social, material, technoscientific as well as emotional and bodily processes (Star and Lampland 2009, 9). Standards are understood as the operational dimension of classifications, and their character of shared regulations is especially useful for producing abstract or concrete objects.

However, standardization leads to the articulation of "the same" technology elsewhere, allowing in particular the embeddedness of an II within other IIs (Monteiro *et al.* 2013, 576). So, this process can be differentiated from quantification and formal representation because of its effects on lives and "reality". It is true that standardization "stabilizes knowledge, freezes action, deletes outliers and residuals, and facilitates use" (Star and Lampland 2009, 13). Nevertheless, it is not completely coherent with empirical research to think of it as an enactment.

Standards are very crucial when they are used to connect different machines or different devices, and in this case they are often legally regulated. Legal regulation of standards improves their social strength as a mandatory way to do something, making change and opposition to standards themselves difficult to carry out.

Relationships between users, machines, designers, standards and design converge towards IIs and IIs in turn mediate and saturate relationships. The interest for a hybrid form of socio-technical alignments, a non-reductionist stance and a viewpoint centered on the collective activity more than on the analysis of the elements and entities involved in infrastructuring is essential for analyzing the enterprise as a whole. However, we think that IIs are not only a "nest" for different boundary objects, but they are in themselves a boundary phenomenon, especially as a constitutive element of the ecology of distributed hi-tech workplaces (Star 1999, 379). To define them as "information" infrastructures underscores only a descriptive character of them, i.e. to carry information out. But defining them as "boundary" infrastructures opens up to the intersectional aspect of their existence, probably the one which is sociologically denser, and their importance in building a texture across dispersed locales, in strengthening mobility, and characterizing contemporary society as a world of networks and social intersections.

Let us discuss the concept of *boundary object* before returning to the idea of *infrastructure-as-boundary*.

3. Note on the Boundary Object. BOs in infrastructures, infrastructures-as-boundary

The concept of *boundary object* is the hallmark of Susan Leigh Star's research program. Her intellectual path is a merging between STS and the tradition of Symbolic Interactionism: a sociology of coordinated action and social intersections strongly influenced by Georg Simmel's classical theories and by the pragmatist philosophy. This path makes STS blossom focusing on the development of cooperative work in absence of consensus, the coexistence of multiple translation processes - whose global coherence constituted an important research theme - and the exclusion and marginality created by those social and technoscientific processes (Star and Griesemer 1989, 388; Star 2010, 604; Trompette and Vinck 2009, 6)[7]. The idea of a *boundary object* was a useful concept to be introduced in Science and Technology Studies in order to explain why in several cases different groups work together without reaching any consensus about the use or the meaning of a device that connects them. Star underscored a dimension often disregarded in STS, i.e. the weakly structured articulation of devices, the lack of stability among different components, and the openness of socio-technical processes to some forms of stabilization, not necessarily corresponding to an overall stability of a device, socio-technical network or alignment (Star 2010, 602).

Studying a Californian Museum and its system of classification and gathering of different materials for its collection, Star and Griesemer (1989) faced the heterogeneity of social worlds involved in this work and studied the mechanisms of their intersection and articulation. The authors were interested in how they created common action, although distributed,

[7] It is difficult to summarize briefly the sources of Leigh Star's thought. She was a sociologist, a feminist, a poet with many roots, and many leaves. The importance of Symbolic Interactionism, and especially of the work of Robert Park and Everett Hughes, was crucial, though mediated by Anselm Strauss (her PhD supervisor) and Howard Becker, both Leigh's mentors. Their influence is clear in her sociological lexicon (world, conventions, rules, ecology, intersection, etc.) and has been reconstructed in several papers (Trompette and Vinck 2009, 5-8 *et passim*; Clarke 2010; Bowker 2010). The knowledge of Georg Simmel's sociology seems very important in Star's work. His concepts of *sozialer Kreis* (social circle), *Wechselwirkung* (mutual action), the Stranger "which today comes, and tomorrow remains", the excluded, the "orphan", and the "tragic" character of fate seem to resound permanently under Leigh's utterances.

without losing diversity (Trompette and Vinck 2009, 5, 8). They provided a definition of an elementary unit of those intersection mechanisms, which they called "boundary object" in order to explain how Museum workers managed both diversity and cooperation:

> "Boundary objects are objects which are both plastic enough to adapt to local needs and the constraints of several parties employing them, yet robust enough to maintain a common identity across sites. They are weakly structured in common use, and become strongly structured in individual site use. These objects may be abstract or concrete. They have different meanings in different social worlds but their structure is common enough to more than one world to make them recognizable, a means of translation. The creation and management of boundary objects is a key process in developing and maintaining coherence across intersecting social worlds" (Star and Griesemer 1989, 392).

This concept has had a long fortune not only in STS, but also in Organization and Management Studies, and elsewhere[8]. It focuses on the material aspect of social intersection, elsewhere conceptualized by Vinck (2009, 53) as *intermediary object*, as well as the data density of those coordination tools, named by Latour as *centers of calculation* (Latour 1987). Leigh Star clarified how she did not necessarily understand as "object" a material piece of something, but that BOs are objects "something people act toward and with", and their "materiality" derives from action, not from "thing"-ness (Star 2010, 603). The multi-layered constitution of this concept is maybe the reason for its own robustness and generativity in terms of its possibility to be scaled, re-inscribed or localized. Leigh Star herself individuated later three components in BOs: *a)* their interpretive flexibility; *b)* the structure of informatics and work process needs and arrangements and *c)* the dynamics between their ill-structured and well-structured uses (Star 2010, 601).

The use of the concept of interpretive flexibility can surprise, because of the differences between Star's ecological approach and the SCOT approach to STS. Interpretive flexibility is indeed a concept used by Wiebe Bijker in order to describe the different meanings which an artifact assumes among pertinent social groups involved in its use (Bijker 1995). Flexibility here is placed mainly at a semiotic level, and is viewed as a

---

[8] A comprehensive review of the "conceptual progeny" of Star and Griesemer's concept of boundary object is provided by Trompette and Vinck (2009, 3, 15). Particularly important are the concepts of boundary work developed by Thomas Gieryn, boundary shifting by Michel Callon, boundary organization and boundary spanning.

transitional period before the stabilization of an artifact. Differently, Leigh Star points out the *permanent* flexibility of BOs in managing both diversity and cooperation, accompanied by a shared structure composed of classification, standards and norms, which is invisible, "transparent" and not subjected usually to an interpretive flexibility. So BOs contain an interpretive flexibility, but act also as a "robust" coordination of shared action: they are "stuff of action", not simply an artifact (Star and Griesemer 1989, 388; Star 2010, 601, 603; Trompette and Vinck 2009, 6). During the Grenoble Conference (2007), Star asserted that "interpretive flexibility without understanding of infrastructure, information needs, standards and classification is a misunderstanding" (Star 2007; Trompette and Vinck 2009, 16): in fact, standards and BOs are "inextricably related" because standards allow BOs to work as coordination tools and to provide "robustness" both to the alignments and the distribution of action and knowledge across different worlds (Star 2010, 607; Trompette and Vinck 2009, 3, 17). The concept of BO refers permanently to both coordination and articulation. Articulation is an important concept because it allows the focusing of the technoscientific enterprise not only as a process of translation managed by some heterogenous engineer, but as a collective practice where any actor "contribute[s] to translate simultaneously in order to work together" (Star and Griesemer 1989, 388). This concept opens not only to the role of users but also to the analysis of designers' work. Other authors emphasized that technicians' work is not "what their managers believe it to be" (Orr 1996); that plans are characterized by vagueness and inherent indeterminacy (Suchman 1987); and that the role of users in socio-technical processes is far from marginal (Oudshoorn and Pinch 2003). Drawing on Bateson's analysis, Leigh Star asserted that the relationship between designers and users is a "double bind" (Star 2010, 610), namely a quasi-structural link, and a permanent lack of consensus among them despite their collaboration in use and manipulation of the same devices. Articulation is just the way in which BOs are negotiated, manipulated, activated, used, designed, learned and entangled in different social worlds or communities of practice.

BOs, as said above, are not necessarily material pieces of something, as intermediary objects are. They do not need any "thing"-ness, even if they can be tangible. They can be directories, classifications, maps, design, standardized methods but, following Leigh Star (1989, 37), they can be in turn classified in some types: repositories, idealtypes, terrain with coincident boundaries, and forms and labels. This last type, in particular, is a method of common communication across dispersed workgroups and it produces standardized indexes, i.e. a robust nucleus of data which makes a

BO an "immutable mobile" essential to coordinate a dispersed (and sometimes distributed) network of action (Star 1989, 50; Millerand and Bowker 2009, 162). This organizational level of coordinating different worlds without any interpretive consensus is considered the most useful aspect of BOs (Star 2010, 602), and it also appears useful to the re-inscription of BOs in different coalescences at different levels of scale.

Classification and its development in standards are the grammatical units of this reinscription of BOs in more complex aggregations. But standards, because of their normative/ordinative action, are also involved in originating marginality, and creating "Others" (Star 2010, 604). Standards, as formalized classifications, have a claim of completeness toward the processes they want to represent (Bowker and Star 1999, 13-16). Enacting a standard to an extant process means not only increasing circulation and common action, but also leaving some actor uncategorized, forcing similarity for different entities, producing invisibility or silence on people or things not included by extant categories (Trompette and Vinck 2009, 17). These phenomena were also analyzed with reference to IIs, and it is not a fortuity, because Leigh Star's analysis, especially in her common work with Geoffrey Bowker, was led to consider BOs not as insulated phenomena, but necessarily growing into multiple BOs and into a system of BOs, up to becoming a "boundary infrastructure" (Bowker and Star 1999; Star 2010).

The relationship between BOs and infrastructures, and especially IIs, is mainly observed as a process of plugging them into each other in a standardized fashion (Star and Ruhleder 1996, 113), i.e. throughout the common sharing of standards which enable the nesting of BOs into IIs, and the link between different BOs. In fact, standards are an inescapable element of BOs but also "the key to infrastructures" (Karasti *et al.* 2010, 383). This focus on standard and data culture in BOs and IIs has maybe shadowed the overall boundary character of both phenomena, especially in constituting a frontier but also a bridge for different experiences situated in different temporalities, spaces and memberships. It is maybe necessary to develop a sociological analysis of infrastructures as boundary and not only as coordination tools, and scaling to the infrastructural level two important considerations in Leigh Star's work on BOs. First, she pointed out how the relationship between the *well*-structured and the *ill*-structured was the field where any standardization attempts to produce a standardized object or system. Then, is this scalable to infrastructure or is it a residual effect at that level? Secondly, she focused on the production of residual categories, groups of "others" or outsiders, and the emergence of new BOs as a result of articulation work (Star 2010, 614). These deep considerations are linked to the intersectional work both of BOs and of IIs with their standardized nature and their push toward coordination of heterogeneous entities in a common enterprise. Then, as an extension, how do IIs generalize or limit the marginalization resulting from any BO? How do they express their overall *boundary* character toward the different experiences they translate simultaneously? What is the relationship between BOs and IIs? Being re-inscribed in an II, does a BO act as a mere articulation of a bigger infrastructure? Is it absorbed in a mono-directed, univocal world of coherent standards or does it maintain a certain autonomy? And what does it mean in practice? From a more empirical point of view, can we apply to an infrastructure the categories of coordination and articulation work used to deal with BOs?

Leigh Star asserted that cooperative work was more interesting than stabilization and blackboxing to her (Star 2010, 613), and this is also the line of research shared by many of the authors of this book. It is an important change in STS and particularly in the most widespread approaches. Maybe there is a "revolution" linked to revisiting ANT through an ecological perspective (Trompette and Vinck 2009, 6). Surely Star opened a new direction, and with her theoretical work succeeded in overtaking a sort of *naturalis historia* of artifact and scientific facts "evolution", also widespread in SCOT, which used to describe a process as going from a principal actor to a "stabilization of facts" as a dominant pattern in technoscience (Mongili 2008). The stabilization of facts has been shown in its fragility, as an uncertain dynamics compared to the massive cooperation and networking when actors work together through simultaneous translation (Star and Giesemer 1989, 388; Fujimura 1992). The collective dimension of Star's analysis challenged many ideas which constituted the innovation process as anchored to certain fixed points, such as the heterogeneous engineer and his entrepreneurship, the enrollment of actors and the creation of obligatory passage points. De-metaphorizing the analysis, she underscored not only the multiple character of translations, but also their uncertain direction and result.

4. Boundaries, ecologies, multiplicity

Three categories constitute the *foci* of our approach to IIs, namely boundaries, ecologies, and multiplicity, which represent interlinked dimensions of how infrastructures are shaped by, and give shape to, a complex texture of hybrid social worlds, material and immaterial arrays, spatio-temporal arrangements and alignments, as well as ecological and situated locales.

In what follows, lines of connection and continuity among boundaries, ecologies and multiplicity will be emphasized, as well as their consequences for the study of IIs, so focusing on specific dimensions of the literature on BOs and IIs presented above.

Besides constituting an attribution and an emergent process transforming objects and infrastructures, the concept of boundary as such deserves a closer analysis. First and foremost, boundaries are necessary for learning and for knowledge to occur: being essentially an act of distinction (Cella 2006), tracing boundaries is a social practice continuously performed by individuals and groups. Boundaries are "produced", mediated and negotiated by technologies, artifacts, tools, cultural forms, conventions and rituals (see Gieryn 1983): hybrid arrays which can take the form of *ad hoc* temporary networks, or consolidate and "sink" into institutions and infrastructures. Official organizations establishing classifications and standards (see Bowker and Star 1999) are a typical example of how acts of distinction can be delegated to professional communities of experts. But if we think of physical, material boundaries (e.g. geographical and political borders) we can also realize how imbued with unexpected consequences and full of contradictions the establishment of boundaries can be.

Indeed, distinction is not a neutral act at all: it can be used to orient the direction of the evolution of knowledge, science and technology: just imposing an "artificial purification" on the confusing, garbling and messy reality, Western society and *episteme* defined themselves as *modern* (Latour 1993).

Since it enables a distinction, a boundary is a limit: it allows understanding of - but it also generates - differences and similarities, divergence and convergence, conflict and consensus. On the other hand, similarity and consensus on it is what constitutes, at minimum, an institution (Douglas 1986). However, while it is a limit, a boundary is not exactly the same than a frontier, or a barrier. As in one of the etymological roots of the term "communication" (from the Latin *moenia*, the city walls, see Giaccardi 2005) the boundary (constituted by walls, namely communication) can be used either to welcome or to reject a stranger - and other-ness. This constitutive ambivalence is resembled by the mixture of stability and flexibility, similarity and diversity which characterize BOs (Star and Griesemer 1989). Boundaries can be differently porous, depending on how resistant (or flexible) the distinction they establish is, and can change at different scales, following oscillations and patterns of re-adjustment which keep contradiction and ambivalence wide alive in spite of grand narratives of change (as in the case of taxonomy's shifting boundaries in the era of DNA barcoding; see Waterton, Ellis and Wynne

2013).

Whereas a frontier usually needs a series of institutional arrangements to be crossed over, a boundary can be crossed over through the reciprocal adjustment or ongoing negotiation between human and non-human participants involved in or distinguished by the boundary itself. A boundary, then, does need consensus to be set up, whereas a frontier (or a barrier) does not[9]. In order to be crossed over or overcome, the latter instead involves a unilateral violation or infringement of a rule. Furthermore, in the shape of BOs, boundaries allow cooperation saving decisive resources which are provided by the stable core that constitutes the BO: its taken-for-grantedness, obviousness and shared naturalization, therefore, virtually annihilate difference (and boundaries), making both of them available to develop other crucial processes.

However, a boundary can be not only *crossed over*, but also *taken over*: when not anymore adequate, or when needed, boundaries can be re-negotiated and this is what happens through interoperability in infrastructures, or interconnections of systems (Monteiro *et al.* 2013). Under such circumstances, consistency and coherence of IIs gets blurred and *scaling up* (Bowker, Star and Neumann 2003) affects integrity, recognizability and effectiveness of existing and current boundaries. Indeed, interoperability is the way to de-problematize and naturalize the crossing over of boundaries in IIs, and to make their saturation transparent.

The degree of definition of boundaries, however, also affects the possibility and ease of taking them over. Flexibility and plasticity of boundaries is the requisite for allowing the existence of working (boundary) objects and infrastructures (as boundary): neither frontiers, nor walls or barriers, but boundaries are necessary to this scope.

Also because of the constitutive flexibility of boundaries, the emergence of BOs and IIs, and the way they work and enable multiple, simultaneous translation and coordination, does resemble an ecology. Now, the term 'ecology' can be a metaphor for describing certain phenomena in terms of ecosystems where fairly diverse keystones and other species co-evolve in localities over time, and this is also how Information Technology can be studied and conceived of (see Nardi and O'Day 1999).

---

[9] It is interesting to note that the concept of boundary object is translated as *objet-frontière* in French (cf. Trompette and Vinck 2009) and the concept of boundary at large refers to a constellation of meanings and translations, and among them, the terms limit, border and frontier are used depending on the context. In Italian, Alessandro Mongili translated it as *oggetto liminare*, from the Latin root *limen* (Mongili 2007).

However, when talking of IIs and BOs, ecology is not only a metaphorical but also a materially appropriate attribute to theorize, envision and study the life of BOs and IIs. Infrastructures are ecologically textured and, therefore, fragile since their interconnections, links and interdependencies expose them to vulnerability and damage, as it happens in other types of ecosystems (see Star and Strauss 2009; Monteiro *et al.* 2013).

Not by chance, one of the explicit sources of inspiration of Star's work, besides those discussed in section 3, note 7, is Gregory Bateson's theory of "an ecology of mind". Bateson's work provided a clear and evocative argument for Star and Ruhleder's seminal 1996 article on infrastructure, already quoted in this introduction and in many chapters of this book. In fact, "Steps towards an ecology of infrastructure" echoes Bateson's "Steps to an ecology of Mind" (1972, 1987) and not simply in the title, as Bateson's concept of "deutero-learning" informs the analysis carried out by the two authors on large information systems.

And just the year before, Star's influential edited collection came out, putting "Ecology" in the main title (Star 1995)."Ecologies of Knowledge" constitutes a cornerstone in that ecological approach which had already emerged as epistemological stance abandoning primacy and uniqueness in favour of multiplicity of viewpoints (Star and Griesemer 1989).

Therefore, "ecological" means not only translating some of the characteristics of ecosystems to science and artifacts as systems (e.g. infrastructures), through an analogical procedure, but especially posing some questions rather than others, and exerting a specific type of sight on what is defined as "reality".

As Star put it in her introduction to "Ecologies of Knowledge", "by ecological we mean refusing social/natural and social/technical dichotomies and inventing systematic and dialectical units of analysis" (Star 1995, 2). As a result, an ecological space is defined by four categorical axes: "continuity *versus* discontinuity; pluralism *versus* elitism; work practice *versus* reified theory; relativity *versus* absolutism" (Star 1995, 14).

Looking at phenomena as continuous, plural, practice-oriented and relative, means assuming ecology as an epistemological landscape. From such a location, following Bateson we can agree that posing

> "(…) the question of whether a computer thinks. (…) What "thinks" and engages in "trial and error" is the man *plus* the computer *plus* the environment. And the lines between man, computer, and environment are purely artificial, fictitious lines. They are lines across the pathways along which information or difference is transmitted. They are not boundaries of

> the thinking system. What thinks is the total system which engages in trial and error, which is man plus environment" (Bateson 1987, 488, *original emphasis*).

Such a statement goes back to the question of boundaries in/of ecologies (or ecosystems) but it is also a pioneering anticipation (elaborated at the end of the '60s) of the stance which will later constitute Latour's technoscience as part of the ANT perspective.

In this respect, ecology implies "an inversion of everyday thought" as a result of "emphasis on use and consequences, not antecedents" (Star 1995, 21). Such an inversion opens "to a very political but nonreductionist and nonpositivist account of (…) multiple moral orders that, taken ecologically, constitute what we think of as 'societies'." (Star 1995, 21).

All this is especially multiplied and amplified in the case of large scale IIs, where diverse and interconnected (sometimes scattered) ecologies are at work and intersect, in a material and discursive texture. And this goes straight to the issue of how coherence (and consistency) are achieved and maintained in the face (or because) of multiplicity.

Indeed, both boundaries and ecologies (declined as plural) lead to multiplicity as a category unifying the character of IIs and their scalability, a category shared by many of this book's chapters.

The concept of ecology and that of multiplicity are inherently intertwined, as exemplified by Monteiro *et al.*'s working definition of IIs (Monteiro *et al.* 2013, 576): "openness (…) interconnections (…) multiplicity of purposes, agendas, strategies (…) ecosystem of systems" are key characteristics of IIs. All of them emphasize a crucial movement from singularity/stabilization of technoscientific innovation and sociotechnical assemblies to their constitutive openness and multiplicity of aims, uses and consequences (in one word: their diversity) across a myriad of sites, locales and articulations.

The Feminist turn, especially in the version intersecting and informing STS and through the work of scholars like Haraway and Barad, contributed very much to this movement or shift, pushing the boundaries of the disciplinary frontiers, up

> "to decenter sites of innovation from singular persons, places and things to multiple acts of everyday activity, including the actions through which only certain actors and associated achievements come into public view."[10]

[10] Suchman 2009, 1, at http://www.informatik.uni-bremen.de/soteg/ gict2009/ proceedings/GICT2009_Suchman.pdf

De-naturalizing everyday life and its obviousness through phenomenological and ethnomethodological standpoints in the light of Feminist epistemology, embraces here the STS enterprise of de-constructing the grand narrative of the isolated genius innovator, then going even further:

> "the study of how new technologies emerge shifts from a focus on invention, understood as a singular event, to an interest in ongoing practices of assembly, demonstration, and performance. The shift from an analysis in terms of form and function to a performative account, moreover, carries with it an orientation to the multiplicity of technoscience objects or, to put it differently, to the achieved nature of objective singularities" (Suchman, Trigg and Blomberg 2002, 164).

To inquire into this "achieved nature of objective singularities" needs a further shift, pursued and realized only when focusing on the *constitutive multiplicity* of technoscience and IIs. Such a multiplicity requires a re-vision of innovation processes, design and organizations, but also of other crucial sites of enactment, and amongst them, body and space, which means embodiment and topology. So what is usually conceived of as a singular and unitary system – the body - shifts and multiplies through multi-sited enactments of an ordinary illness (Mol 2002). Fragmentation and dispersion of what is named as "the body" can be avoided just because of so many BOs and IIs which not only mediate but also enable the multiplicity of reality-in-practice.

Such a multiplicity extends to space, once again taking over boundaries which do not fit emerging articulations. The space of networks – Latour's space of 'immutable mobiles' (Latour 1986) – is a space of immutability which circulates and goes on and around. Law and Mol (2001) go beyond this, showing how immutable mobiles are only a part of technoscience, plausible and useful to consider as long as

> "The focus was on control: on the work needed to hold a configuration stable; on the effort required to create a wider network fit for the transmission of immutable mobiles. But there is another problem: it is, quite simply, that often enough ideas, facts, information, even technologies, turn out to spread in a manner that is much more fluid. It is precisely a lack of rigidity that most helps movement" (Law and Mol 2001, 619).

Law and Mol's fluid and fire topologies are another example of how taken for granted boundaries can shift because of emerging ecologies and multiple articulations.

In the end, multiplicity is guaranteed and preserved through and because of IIs and BOs. Since fixed boundaries and isolated (social) worlds do not survive for a long time, translation and cooperation are continuously required.

Multiple ecologies and ecologies of multiplicity are always at work in more or less invisible ways, constituting a light but decisive texture in our lives. Sometimes, this texture becomes indistinct, some others we can look through and inside of, because of scattered holes which put in the foreground what is usually "in-between".

As editors of this book, we hope to have provided some modest key to looking at those holes and stitching together again the texture of IIs and BOs.

5. A path to read this book

Few enterprises in academic and scientific writing provide opportunities to blossom, strengthen and consolidate cooperation and knowledge shared among scholars. Edited and co-authored books are surely a very good example of this kind of enterprise, notwithstanding recent trends in ranking and evaluation de-valorize the effort they require from their editors. Translation and cooperation involved in and mobilized by an edited book are incredibly dense and – more often than not – very energy draining. In all of the above regards, this book is not an exception.

Editors and contributors of this collection were inspired not only by the ongoing research work on infrastructures, and particularly on IIs, but mainly by the intersectional and boundary character of them, as well as of organizations, objects, and other forms.

During their research, all the authors have met the powerful concepts of BO and II, and this encounter is at the very origin of this book, its character, and hopefully its richness.

Despite the wide range of case studies displayed in the chapters, the contributors often developed common themes or reflections about a recurrent problem: how to analyze a sociotechnical intersection, and its forms. Therefore, we decided to introduce some of these common themes, rather than to briefly introduce any chapter.

All the contributors had to analyze phenomena where some pervasive enabling resource (Bowker *et al.* 2010) such as IIs, BOs, or some intermediary element, provided a thread to follow. These resources can saturate relationships, or just connect social worlds or individuals. The thread has often different characteristics and enacts different research focuses. In Stefano Crabu's interests, the focus is represented by organizational properties of infrastructure, whereas in Klein and Schellhammer's research, it is the capability of an II to include a constellation of different practices. II's multiple and often contradictory character was pointed out and thematized not very differently in various contributions. In Pellegrino's view, fragility is an inescapable character of IIs, similar to Denis and Pontille's idea of their elusiveness. Francesco Miele interprets the same lack of interpretive unity of "objects" infrastructuring different worlds as a "resource for stability" of those alignments, because they increase objects' flexibility. In Pellegrino's study on disasters, in particular, vulnerability, understood as the lack of capability in adapting to circumstances, is coupled to contingency, understood as an event able to change, intervene or interfere in a process. A contingency as an accident, or a disaster, opens an *ex post* interpretive activity, not necessarily controversial, upon the characters of the hybrid ensembles put together by IIs, or BOs.

These ensembles are conceptualized as convergences by Alessandro Mongili and Giuseppe Lugano, both interested in design. In Mongili's view, convergence is a field where different actors develop some activity in order to attain interoperability between different devices, or alignment of heterogeneous entities. In Lugano's interest, convergence is coupled with divergence as multifaceted terms very important for interpreting design activity. Lugano understands combining, aggregating and connecting as convergent tasks, whereas clustering, separating and filtering as divergent in the design practices. Not differently, Mongili pinpointed the packing and unpacking activities as pillars of the originators' work in their social networks.

Other twofold aspects in infrastructuring are the divide between users and developers and its construction. As Denis and Pontille assert, the boundary between designers, "contributors", "originators", and users is uncertain. Users are a fragmented group, not always adhering to their representation, and participate in infrastructuring - Simona Isabella asserted. Their representations vary from a mocking anthropology (Isabella, Mongili) to an image of the "flesh and blood" users as "parasites" regarding designers and design activity (Denis and Pontille). In fact, as Giacomo Poderi points out, designers often emphasize the need not to "fake" a design. They work "within a limited sphere of knowing and acting" (Suchman 2002), which Mongili called "design-defined enclosure". From such an enclosure built up using formal categories to construct "users", both humans and non humans are marginalized, as in the case of machines producing renderings in Denis and Pontille's research. In a sort of double boundary work, developers too are constructed as a group; and this despite the fact that users are involved in user-led processes while designing: this configures those practices as continuous design (Poderi). Following Denis and Pontille, inclusion in categorical work is the way through which a boundary is created between "parasite" and "relevant" users. Federico Neresini and Assunta Viteritti preferred to highlight the role of articulation work and continuous learning in creating such a twofold "binding" link between designers and users, and they are deeply aware of the recursive link between practices and performances in infrastructuring. Such a link characterizes also Lazzer and Giardullo's chapter, which looks at on line reading and writing practices in the context of the Web as II.

Infrastructuring, as we have previously noted, is an activity directed towards solving tensions between different dimensions. Klein and Schellhammer analyze in their study the tension existing between infrastructure design and infrastructure emergence in a heavily institutionalized system such as the German public health system. In this case, they point out the role of regulation push and of collective action in managing those tensions. Moreover, this book's contributors had to cope with other kind of tensions regarding the situated infrastructuring activities they studied. In mobilities system infrastructuring, both Mitrea and Denis and Pontille had to deal with the intertwining of virtual (information) and physical (urban or territorial) infrastructures, and to follow their not always easily understandable recursive relationships. Mitrea conceptualizes such relationships through a "dispositive model of flexible mobility for travelling human and technologies". Denis and Pontille instead drew our attention to the classification and categorization work as the most important collaborative activity involving all the actors, despite their formal role (such as users/designers) in infrastructuring.

The various authors of this book had to deal especially with the "door to infrastructuring" (Pellegrino) represented by BOs. Their use of this concept has been sometimes canonical, sometimes creative, in the sense that they assigned the character of boundary object to diverse entities. Francesco Miele reminds us that, following Leigh Star (2010), a boundary object is not necessarily an object, a material piece of something, but "something with which and towards which people find themselves acting". Miele applied this concept to an entire organization understood as a stabilizing element in an action-net. Drawing from Czarniawska (2004) Miele interprets the action-net move as creating an organization at its inner, a spin-off, and looks at this organization as a boundary object. Mauro Turrini, in his research on diagnoses in the case of Advanced Maternal Age, found the risk thresholds as "numerical" BOs. Similarly

Michela Cozza, in her research on the Italian Scientific and Technological Parks network, which she interpreted as an intermediary organization, identifies some boundary infrastructures and some BOs used in order to face tensions between a local and a general level of scale.

Working on medical research and developing two different laboratory studies, Crabu as well as Neresini and Viteritti cope with the kit as an infrastructuring object, acting as a component of everyday laboratory life. In Crabu's analysis particular attention is devoted to the articulation work of this infra-structuring object within a local setting, throughout the production of protocols as a local literary inscription which inserts the kit in local laboratory practices. Similarly Isabella found, in another context - i.e. an inbound staff devoted to the management of users through a CRM system – similar local document and inscription able to inscribe a boundary infrastructure into the local ecology.

In many contributions, finally (Crabu, Denis and Pontille, Mongili) the boundary role of IIs (and BOs) has often been seen as a resource in order to facilitate interoperability, convergence and alignment directed to everyone and not necessarily to a specific social world. The social world, or community of practice, is a setting which appears not always as the situated handhold of infrastructuring. This absence invites, maybe, consideration of the link between learning, membership and infrastructuring as not necessarily plunged only into a social world dimension, but creating also new forms of sociality saturated by cooperative activity and by infrastructuring itself.

References

Bateson, Gregory. 1972, 1987. *Steps to an Ecology of Mind.* Northvale NJ and London: Jason Aronson.

Bietz, Matthew, Eric Baumer and Charlotte Lee. 2010. "Synergizing in Cyberinfrastructure Development." In *Computer Supported Cooperative Work*, 19(2): 245-281.

Bijker, Wiebe E. 1995. *Of Bycicles, Bakelites and Bulbs*. Cambridge, MA: The MIT Press.

Borgman, Christine L. 2000. *From Gutenberg to the Global Information Infrastructure. Access to Information in the Networked World.* Cambridge, MA: The MIT Press.

Bowker, Geoffrey C. 1994. *Science on the Run. Information Management and Industrial Geophysics at Schlumberger, 1920-1940.* Cambridge, MA: The MIT Press.

—. 2005. *Memory Practices in the Sciences*. Cambridge, MA: The MIT Press.

—. 2010. "Reflections from Geoffrey Bowker." In *Science, Technology, & Human Values* , 35(5), 579-580.

Bowker, Geoffrey C. and Susan L. Star. 1999. *Sorting Things Out: Classification and Its Consequences*. Cambridge, MA: The MIT Press.

Bowker, Geoffrey C., Karen Baker, Florence Millerand and David Ribes. 2010. "Toward Information Infrastructure Studies: Ways of Knowing in a Networked Environment." In *International Handbook of Internet Research*, edited by J. Hunsinger *et al.*, 97-117, Dordrecht: Springer Science and Business Media.

Bowker, Geoffrey C., Stefan Timmermans and Susan L. Star. 1995. "Infrastructure and Organizational Transformation: Classifying Nurses' Work." In *Information Technology and Changes in Organizational Work*, edited by Wanda J. Orlikowski *et al.*, 344-370, London: Chapman and Hall.

Cambrosio, Alberto, Peter Keating, Thomas Sehlich and George Weisz. 2006. "Regulatory Objectivity on the Generation and Management of Evidence in Medicine." In *Social Science and Medicine*, 63(1): 189-199.

Cella, Gianprimo. 2006. *Tracciare confini. Realtà e metafore della distinzione*. Bologna: il Mulino.

Clarke, Adele E. 2010. "In Memoriam: Susan Leigh Star (1954-2010)." In *Science, Technology, & Human Values* , 35(5): 581-600.

Czarniawska, Barbara. 2004. "On Time, Space and Action-Net." In *Organization*, 11: 777-795.

Douglas, Mary. 1986. *How Institutions Think*. New York: Syracuse University Press.

Edwards, Paul N., Geoffrey C. Bowker, Steven J. Jackson and Robin Williams. 2009. "Introduction: An Agenda for Infrastructure Studies." In *Journal of the Association for Information Systems*, 10: 364-374.

Edwards, Paul N., Matthew S. Mayernik, Archer L. Batcheller, Geoffrey C. Bowker and Christine L. Borgman. 2011. "Science friction: Data, metadata, and collaboration." In *Social Studies of Science*, 41(5): 667-690.

Fujimura, Joan H. 1992. "Crafting Science: Standardized Packages, Boundary Objects, and Translation." In *Science as Culture and Practice*, edited by Andrew Pickering, 168-211, Chicago: The University of Chicago Press.

Giaccardi, Chiara. 2005. *La comunicazione interculturale*. Bologna: il Mulino.

Gieryn, Thomas F. 1983. "Boundary-Work and the Demarcation of Science from Non-Science: Strains and Interests in Professional Ideologies of Scientists." In *American Sociological Review,* 48(6): 781-795.

Gruber, Thomas R. 1993. "A Translation Approach to Portable Ontology Specifications." In *Knowledge Acquisition*, 5(2): 199-220.

Haythornthwaite, Caroline. 2006. "Learning and Knowledge Networks in Interdisciplinary Collaborations." In *Journal of the American Society for Information Science and Technology*, 57(8): 1079-1092.

Jewett, Tom and Rob Kling. 1991. "The Dynamics of Computerization in a Social Science Research Team: A Case Study of Infrastructure, Strategies, and Skills." In *Social Science Computer Review*, 9(2): 246-275.

Karasti, Helena, Karen S. Baker and Eija Halkola. 2006. "Enriching the Notion of Data Curation in E-Science: Data Managing and Information Infrastructuring in the Long term Ecological Research (LTER) Network." In *Computer Supported Cooperative Work*, 15(2): 321-358.

Karasti, Helena, Karen S. Baker and Florence Millerand. 2010. "Infrastructure Time: Long-term Matters in Collaborative Development." In *Computer Supported Cooperative Work*, 19: 377-415.

Kling, Robert and Walt Scacchi. 1982. "The Web of Computing: Computer Technology as Social Organization." In *Advances in Computers*, 21: 1-90.

Lampland, Martha and Susan L. Star (edited by). 2009. *Standards and Their Stories. How Quantifying, Classifying, and Formalizing Practices Shape Everyday Life*. Ithaca and London: Cornell University Press.

Latour, Bruno. 1986. "Visualization and Cognition: Thinking with Eyes and Hands." In *Knowledge and Society. Studies in the Sociology of Culture Past and Present*, vol.6, edited by Henrika Kuklick and Elizabeth Long, 1-40, London: Jai Press.

—. 1987. *Science in Action.* Cambridge, MA: Harvard University Press.

—. 1992. *Aramis ou l'amour des techniques.* La Découverte: Paris. (1996, *Aramis or the love of technology*. Cambridge, MA: Harvard University Press).

—. 1993. *We Have Never Been Modern* (transl. by Catherine Porter). Cambridge, MA: Harvard University Press.

Lave, Jean and Etienne Wenger. 1991. *Situated Learning: Legitimate Peripheral Participation.* Cambridge, MA: Cambridge University Press.

Lave, Jean. 1988. *Cognition in Practice*. Cambridge, MA: Cambridge University Press.

Law, John and Annemarie Mol. 2001. "Situating technoscience: an inquiry into spatialities." In *Environment and Planning D: Society and Space,* 19: 609-621.

Mackenzie, Adrian. 2003. "These things called Systems: Collective Imaginings and Infrastructural Software." In *Social Studies of Science*, 33(3): 365-387

Millerand, Florence and Karen Baker. 2010. "Who are the users? Who are the developers? Webs of users and developers in the development process of a technical standard." In *Information Systems Journal*, 20: 137–161.

Millerand, Florence and Geoffrey C. Bowker. 2009. "Metadata Standards. Trajectories and Enactment in the Life of an Ontology." In *Standards and Their Stories. How Quantifying, Classifying, and Formalizing Practices Shape Everyday Life*, edited by Martha Lampland and Susan L. Star, 149-165, Ithaca and London: Cornell University Press.

Mol, Annemarie. 2002. *The Body Multiple*. Durham: Duke University Press.

Mongili, Alessandro. 2007. *Tecnologia e società*. Roma: Carocci.

—. 2008. "Oggetti e allineamenti: qual è l'oggetto della sociologia della tecnoscienza?" In *Rassegna Italiana di Sociologia*, 49(3): 403-426.

Monteiro, Eric, Neil Pollock, Ole Hanseth and Robin Williams. 2013. "From Artefacts to Infrastructures." In *Computer Supported Cooperative Work*, 22(3): 575-607.

Nardi, Bonnie A. and Vicky L. O'Day. 1999. *Information Ecologies. Using Technologies with Heart*. Cambridge, MA: The MIT Press.

Orr, Julian. 1996. *Talking about Machines*. Ithaca: Cornell University Press.

Oudshoorn, Nelly and Trevor Pinch (edited by). 2003. *How Users Matter*. Cambridge, MA: The MIT Press.

Ribes, David and Geoffrey C. Bowker. 2009. "Between meaning and machine: learning to represent the knowledge of communities." In *Information and Organization*, 19(4): 199-217.

Ribes, David and Thomas A. Finholt. 2009. "The Long Now of Technology Infrastructure: Articulating Tensions in Development." In *Journal of the Association for Information Systems*, 10: 375-398.

Star, Susan L. 1989. "The Structure of Ill-Structured Solutions: Boundary Objects and Heterogeneous Distributed Problem Solving." In *Distributed Artificial Intelligence*, Vol. II, edited by Leslie G. Gasser and Michael N. Huhn, 37-54, London: Pitman.

—. (edited by). 1995. *Ecologies of Knowledge. Work, Politics and Technology.* Albany: SUNY Press.

—. 1999. "The Ethnography of Infrastructure." In *American Behavioral Scientist*, 43(3): 377-391.

—. 2010. "This is Not a Boundary Object: Reflections on the Origin of a Concept." In *Science, Technology, and Human Values*, 35(5): 601-617.

Star, Susan L. and Geoffrey C. Bowker. 2006. "How to infrastructure." In *Handbook of New Media*, edited by Leah A. Lievrouw and Sonia M. Livingstone, 230-245, London: Sage.

Star, Susan L. and James R. Griesemer. 1989. "Institutional Ecology, "Translations", and Coherence:Amateurs and Professionals in Berkeley's Museum of Vertebrate Zoology, 1907–1939." In *Social Studies of Science*, 19: 387–420.

Star, Susan L. and Martha Lampland. 2009. "Reckoning with Standards." In *Standards and Their Stories. How Quantifying, Classifying, and Formalizing Practices Shape Everyday Life*, edited by Martha Lampland and Susan L. Star, 3-24, Ithaca and London: Cornell University Press.

Star, Susan L. and Karen Ruhleder. 1996. "Toward an Ecology of Infrastructure: Design and Access for Large Information Spaces." In *Information Systems Research*, 7(1): 111-134.

Star, Susan L. and Anselm Strauss. 1999. "Layers of Silence, Arenas of Voice: The Ecology of Visible and Invisible Work." In *Computer Supported Cooperative Work*, 8: 9–30.

Star, Susan L., Geoffrey C. Bowker and Laura Neumann. 2003. "Transparency At Different Levels of Scale: Convergence between Information Artifacts and Communities of Practice." In *Digital Library Use*, edited by Ann P. Bishop, Nancy A. Van House, Barbara Pfeil Buttenfield, 241-269, Cambridge, MA: The MIT Press.

Suchman, Lucy. 1987. *Plans and Situated Actions. The Problem of Human-Machine Communication.* Cambridge, MA: Cambridge University Press.

—. 2002. "Located Accountabilities in Technology Production." In *Scandinavian Journal of Information Systems*, 14(2): 91-106.

—. 2009. Agencies in Technology Design: Feminist Reconfigurations, available at http://www.informatik.uni-bremen.de/soteg/gict2009/proceedings/GICT2009_Suchman.pdf

Suchman, Lucy, Randall Trigg and Jeanette Blomberg. 2002. "Working artefacts: ethnomethods of the prototype." In *British Journal of Sociology*, 53(2): 163–179.

Trompette, Pascale and Dominique Vinck. 2009. "Revisiting the Notion of Boundary Object." In *Revue d'anthropologie des connaissances*, 3(1): 3-25.

Turner, William, Geoffrey C. Bowker, Les Gasser and Manuel Zacklad. 2006. "Information Infrastructures for Distributed Collective Practices." In *Computer Supported Cooperative Work*, 15(1): 93-110.

Vinck, Dominique. 2009. "De l'objet intermédiaire à l'objet-frontière." In *Revue d'anthropologie des connaissances*, 3(1): 51-72.

Waterton, Claire, Rebecca Ellis and Brian Wynne. 2013. *Barcoding Nature. Shifting Cultures of Taxonomy in an Age of Biodiversity Loss.* London: Routledge.

Wenger, Etienne. 1998. *Communities of Practice*. Cambridge, MA: Cambridge University Press.

Acknowledgements

The introduction as well as the book project was conceived and composed together by the two co-editors. However, Alessandro Mongili wrote section 2, 3 and 5 whereas Giuseppina Pellegrino wrote section 1 and 4.

Eventually we want to thank all of the contributors and the reviewers of the book. Our gratitude goes in particular to Colm Walsh for his native proof reading of English; Claudio Coletta for the book editing; CSP for the support and trust in our editorial project. Last but not least, we thank Geof Bowker for his foreword to this work.

I. Designing and Articulating Information Infrastructures

CHAPTER ONE

DESIGNERS AS USERS: BLURRING POSITIONS AND THEORIES IN CREATIVE PRACTICES

ALESSANDRO MONGILI

Introduction

In Technology Studies, an *in toto* divide between users and designers is found. Computing, design and software development practices are interesting fields where this divide can be observed. The work of developers and designers is cooperative. In many ways, they do not act differently from users, especially concerning devices and their test analysis, as well as when supporting materials on the Web, and attempting to force the extant devices to serve new functions.

Describing design practices in computing drags this kind of role-type divide on. Designers and users have, by definition, two separate roles. However, users are also involved in design and/or development (Lindsay 2003), just as designers also act very often as users. This aspect is relevant, if we consider technoscience a texture of practices, where roles and devices are very frail, despite their solid presence in discourse (de Laet and Mol 2000). Carrying out research in infrastructuring, I studied many developers or designers (*progettisti*)[1], using different ethnographic approaches, and I found their roles to be not so clearly separated as in their respective narratives.

I was inspired in this path by a remark of Leigh Star on the "double bind" characterizing the relationship between users and designers (Star 2010, 610). I faced this issue working with an Italian telecommunication enterprise, *Tiscali*, from 2010 to 2012, following the development of the

[1] The participants usually speak in Italian, or in an Italianized English, and more rarely, in a minority language diffused in the fieldwork area. I will quote the original utterances and I will translate them into English.

social network *Nemus*[2]. Access was negotiated with the CEO, but it had to be renegotiated at each stage of fieldwork advancement. During the first stage (2010), a company associate gave me a tour through the spaces, and introduced me to the different activities. As the company is usually closed to outsiders, this was very useful. She introduced me to managers and executives, and I faced an institutional-oriented narrative organizing practices into a project-ruled and finalistic frame. The workplace and the spaces were displayed using a mapping-like frame. During this exploration, I negotiated the terms of the second and the third stages for my access (2011-2012), which I carried out in parallel. Firstly, I shadowed a designer working full-time with a Nemus application for video-surveillance. He was a newcomer to the company but he brought with him lengthy experience in design and development of applications for telecommunication and Internet in other companies. He was also concerned with integration in the Tiscali work environment. This shadowing was very rich and collaborative. Secondly, I observed meetings, *demos* and the everyday life surrounding the designer's activity, and I explored his collaborators in their common work towards developing an application and connecting it to the Nemus ensemble. I carried out a series of interviews with these collaborators. This third stage was quite difficult, requiring new and different negotiations in order to access the same company spaces and to obtain interviews. Finally, the activity which I observed the most was testing, even if a boundary between testing and other activities was not clearly defined.

My research then followed a diversion from the Nemus social network. This social network allows the making of phone calls and many other *Voice over Internet Protocol* (VoIP)[3] capabilities. A little group worked on the development of one of these capabilities, in order to use the technologies comprised in Nemus ensemble to develop a remote control device for domestic video surveillance. In participants' terms, they called it *clientino* ("little client"). Their attempt consisted of connecting a camera, in remote mode, to the ensemble of the Nemus social network, exploring the capabilities of the *Session Initiation Protocol* (SIP)[4] suite to work over VoIP. Even though the social network project kept on its course, the attempt to develop a video surveillance application was eventually abandoned.

---

[2] Whereas Tiscali is the real name of the company, Nemus is a fictitious one.

[3] VoIP is understood to be all the technologies for the delivery of vocal communication and multimedia resources over IP networks.

[4] SIP is a communication text-based protocol used to manage voice and video calls over IP networks.

The group working on this attempt was composed mainly of one Tiscali *progettista*, working full-time on this project, and by another five developers and designers, working on it as a shared task. Many other people assisted in this attempt on an irregular basis, both from Tiscali and from other partner companies, especially on testing activities. Their work included mostly tests, essays and demonstrations (especially *demos* and *trials*). In order to test the new application functioning, *progettisti* used cell phones, computers, and softphones (cell phone digital simulators). Softphones are virtual devices, available on the Internet. They include video functions and they can be employed exactly as a smartphone to support any use. Using such devices, they tried to simulate video calls. They forced operating systems and protocol suites to achieve interoperability between different elements. The aim of testing was mainly to assess the measure of support provided from the SIP protocol suite to the exchange of video and audio data between two softphones.

Technicians used to monitor the tests' evolution employing two different pieces of software. *Asterisk* was used to assess the flow of data exchanged during the attempt to make a video call, while another software, *Wireshark*, gave more detailed information about the operating commands and the parts of the protocol used during the attempt to exchange a video call. Asterisk is primarily an application for managing private branch exchange (PBX) services, but in this case it was used to measure tests. Wireshark is a tool which provides a comprehensive description of protocols, software and networks involved in the video call.

After the test, an interpretation of the essay was based mainly on the reports instantly produced by these two pieces of software. Technicians aimed to analyze which protocols and standards had been activated during the test, but also the dimension and the relevance of the single component of the protocol suite. In fact, a huge protocol suite such as SIP is composed of many single elements brought together, and functioning as translation units.

The second stage of this activity was the insertion of a camera into the ensemble of different devices and exchange protocols named *Nemus*. The camera is naked and the technician intervenes by hand, in order to force its possibilities (Dodier 1993). The selection of this camera was based on different negotiations with different suppliers in China, Canada and other countries. Price, suppliers' reliability and language complications influenced the final choice. Once the choice had been made, the technicians repeatedly tested the data exchange between the camera and a softphone. Every test was monitored using Asterisk and Wireshark, and the instant reports which they produced were analyzed as a foothold for the interpretation of the test, specifically in order to establish its failure or success.

1. Convergence and originators' work

I define the combination of a web of relations between human and non-human actors – specifically, the alignment of software and protocols which compose *Nemus* and the camera - as a field where convergence occurs. Convergence is a concept used by Leigh Star and Geof Bowker to define "the double process by which information artifacts and social worlds are fitted to each other and come together" (Star, Bowker and Neumann 2003; *see also* Bowker and Star 1999, 46-49). It is a concept free from any stabilization bias, and which suits very well the analysis of the double move accomplished by the camera and by the Nemus ensemble to fit with each other but also to act together in order to send images and sounds using the SIP protocol.

For practical purposes some elements, often already functioning as elements of other socio-technical ensembles, are extracted from their previous alignment, or convergence, and pushed to have new functions in this field of convergence. During the testing phase, technicians try to prompt these elements to take new tasks on, coordinated with its new ensemble. This functional translation is visible because testers use to force software, objects, or test monitoring software to take on new tasks or to re-adapt SIP protocols at some point in the OSI/ISO model.[5]

This convergence activity is developed in a specific environment, marked by the exclusion of users, which I call a *design-defined enclosure*, in which technicians can accept the lack of interoperability between different elements of the newborn ensemble. Participants are mainly involved in trying its new tasks and in the analysis of the protocols' obduracy. In fact they embody the standards for the transmissions through phone and modem, so analysis of the tested flows is crucial.

This forcing and torquing activity is developed on the designer's desktop but also in a larger laboratory when the cameras are involved ("pieces of iron" or "irons"—*ferri*—in participants' terms).

When analyzing the work of technicians involved in design and development of a project, one is confronted with different regimes of their

---

[5] The Open Systems Interconnection (OSI), or ISO/OSI Model, is a Web standard providing a logical architecture describing and standardizing its functions. Its characteristic is to distribute functions following seven layers, coming from the physical one, the data link, the network, and the transport layers to the session, the presentation, and the application layers.

organization. The first is consistent and ecompasses a holistic frame - the project, all the activities, the schedules and all the elements considered as part of a socio-technical ensemble, making sense of them in a finalistic way. This regime is mainly representational and discursive, and displays the project "as a specific kind of imagined alignment" (A. Mackenzie 2003, 369). The Nemus project becomes here an integrated system that allows, using the same application, chatting, phone calls (inside and outside the system), exchanging video and audio content, common discussions, and setting up of accounts. Thanks to the failure of the video surveillance application of this project, we can face the other organizing regime, starting from the designer's desktop practices. From the standpoint of this convergence field, we confront the "imagined alignment" through only a part of it, a camera forced to reach the whole system developing a specific task. It is a practice-oriented regime.

Such a regime acquires a contrasting visibility during the representations of the video-surveillance diversion, which is displayed, accounted for, and performed mainly in front of managers and executives. The *demo* is the main form of representation of the attempt to develop an application for video surveillance as a whole, connected to a more vast ensemble, Nemus. It is conducted by the technicians in front of the management. They have to anticipate the management narrative frame, because its result can influence the further direction of their work and of the same application. However, it maintains a character of "view event" and of "representational artifact" (Omoronyia *et al.* 2009, 526; Rosental 2007; Tory *et al.* 2008, 312). Demos take place projecting a video or some slides where the capabilities of the applications, the attempts to force some parts of it into the general project, and some phases of the testing activities are displayed, accompanied by explanations which usually try to translate the desktop practices in terms of the management need for temporal deadlines.

The second standpoint can be called an originating regime, because it aims towards the realization of an early-stage idea, and looks for a *mise-en-contexte* (Latour 1992, 115), but also produces an ensemble which is new, despite being composed of already functioning elements, with different tasks, in other socio-technical ensembles. In this regime, the project as a whole, its temporalities, the importance of prices, and the users are absent. Realization here means only to make convergent elements interoperable in a new ensemble, or in a part of it. This originating regime needs a remark about how technicians are considered. If the holistic regime produces a design-defined enclosure, where lay users are excluded, and where technicians are *progettisti* (a term which means designers, but also developers, *sviluppatori*), the originating regime cannot be explained without discussing those roles, because the same idea of project is powerless and marginal here. Technicians work with some elements ("little bricks", *mattoncini*) in order to produce prototypes, demos, and scripts describing users' roles and tasks in a technological set. Their activity drives them toward testing, but also coding. Adrian Mackenzie (2006) proposed to call them *originators* in order to define this kind of heterogeneous engineering directed toward products which are malleable, fragile and not available for current users outside the design-defined enclosure. I agree, even if the term is unknown to participants.

These regimes define a field which excludes ordinary users, but I consider the everyday practice of the originators with the "little bricks" and other resources as a type of use, and that this ambiguous position challenges the same image of design in technology.

Moving across these different regimes, we can observe their strength as a tool for organizing practices, narratives, times, places, prices and roles. Entering a holistic regime, we find purposeful practices, and a master narrative, in which the *clientino* is only an adaptation of the Nemus ensemble. Explanations shift all the articulation phase:

> "It is a project finalized to realize a client for a PC, to make calls and video calls, among others, and to share the identity through Wi-Fi. … And this matter marries very well with video surveillance. In fact, this is what is called 'enabled'; I get on the computer and, typically, I watch my place. So it will be a very good matching, if everything moves forward." (A designer).

This narrative orders all the elements as separated. Software and devices are in a relationship with humans and do not act together. In computing, design is traditionally framed into a "one person, one terminal" image, which connects the software agency to individuals. Following the fieldwork, this parallel view of the agency is substituted by a common movement, distributed in times, spaces, humans and non humans, quite heterogeneous among them, and in any phase of a device or software originating process. For these reasons this professional image of design has already been judged as unsatisfactory (Star, Bowker and Neumann 2003). It includes only a series of cognitive operations generating a certain agency or an extension of an extant agency to a new field. Coding and testing are reduced to software development as a disembodied reality (Bucciarelli 1994; A. Mackenzie 2006). This image muddies the fact that software cannot exist without connecting itself to other elements, in order to work, and so in order to exist. For this reason, software does not exist in isolation; it is intimately fragile, and it is inserted in a field of heterogeneous tensions (A. Mackenzie 2006, 8, 12).

This frail condition appears following the originators' practices. Shifting to the originating regime, the whole finalized project order, the linear temporality, and the master narrative with its need for rational motivations and purposeful intentionality lose its importance. Originators "don't have a strategy; they get their battle plans, contradictory ones, from other actors" (Latour 1992, 162). The uselessness of a description of design practices rooted in a holistic regime narrative derives from the fact that it does not order the actual computing design practices (Suchman 1987, 4).

Nevertheless, to loosen the knots of common sense interpretations of the design work is only a first step. Ethnographically, following originators means to scale up from the set observed. Invisible infrastructures, hidden elements, and remote resources saturate the observed set. The question "What is the proper unit of analysis for design: group, organization, community, network?" (Star, Bowker and Neumann 2003) is correctly posed because scaling-up is the originators' core business.

Kling and Scacchi (1982) proposed to refer to a *web of computing* as the place where the computing design activity is reachable, underscoring the distributed character of these practices. Leigh Star (1999, 379) considering the design set, emphasized that "relatively unstudied infrastructures [...] permeate all its functions", and so shifted the point of interest from the distribution of these practices to their texture. She coherently developed the view that infrastructures are "without absolute boundary or a priori definition" and "relational." (Star and Ruhleder, 1996). Originating a device in this respect is also an activity constantly saturated by infrastructuring, despite the confusion it can cause the ethnographer at first glance, since her/his set is apparently composed only of a person and a device. "Infrastructuring" seems then the more adequate concept to describe this process-like character (Bietz *et al.* 2010; Edwards *et al.* 2007; Star and Bowker 2002).

Originators act in a world that they cannot know entirely, that they do not master. In their practice, knowing is continous. Lave (1988) underscored that any practice, including design and originating devices, is characterized by a continuous learning. This activity of knowing has a social character and it is shared into a web of relations, and it is neither reducible to a solely individual activity producing an accumulation of stable knowledge, nor to other concepts as the cultural capital. In fact,

> "Educational contexts are not separated from work contexts… the individual to whom an activity has to be taught is not the unit of analysis

> anymore, but also technologies, tools, coordination with other people and overall relations among these elements" (Viteritti 2012, 43, *my transl.*).

This shift emphasizes the social character of originating practices, but also the relevance of the originators' user-like tasks. At the same time, the new devices appear as not completely new things, but as an ensemble of elements which need to be learnt because they already exist. In Star and Ruhleder' words (1996, 113), they are "sunk into, inside of, other structures, social arrangements, and technologies."

Originators have to know the characteristics of the elements which would converge in a new device, or in a new ensemble. These characteristics are proper to any element, but also distributed in relations, computing and infrastructures: they are relational. Their task is to develop interoperability among them and to advance a convergence, just as a possibility *in statu nascendi*. I consider that this work is different from alignment, as Bietz *et al.* (2010, 253) assert, because the design-defined enclosure where they act shrinks the dimension of this activity to some actors, and does not encompass the articulation work.

During my research, in order to differentiate the user-like practices from a proper *quid* of originators' work, I set apart the use of different tools –taken for granted *vs* controversial -, the successive forcing or torquing, and also its rest. I chose only three tools in order of their relevance and frequency of their activity: the SIP protocol and both testing assessment software packages, Asterisk and Wireshark. The intimacy with these tools in everyday activity differentiated originators from other people.

In any case, I could not miss the opportunity to follow and to take note of every moment of their ordinary activities, attainable during my fieldwork. Doing that, I could explore the backstage and look at infrastructures in the making (Goffman 1956; Bowker *et al.* 2010).

However, in order to understand the infrastructures role, I asked for explanations and took notes of the movements of the designer I shadowed. Our interactions reached the point of him taking my computer to perform some tests. Our typical interactions were like the following one:

> "Do you see this line/row that it emphasizes? It is me… Actually, Wireshark is saying to me, 'Keep calm.'" (Designer).

My methodological practice in this fieldwork was negotiated and fitted to the analysis patterns of cooperation and design - relevant for the emergence of an ongoing infrastructure or a part of it (Bietz *et al.* 2010; Ribes and Finholt 2008).

2. Originating a device

Most of the originators' practices are constituted by nonverbal schemes, conventions, but also by the use of devices or routines, information infrastructures containing forms of classification and standards within them, and procedures. It is difficult to deduce them from narratives and conversations. But these forms of unexpressed and tacit knowledge have a deep conditioning influence because they are shared among actors and they classify practices, discourses and priorities (Mol 2003, 15). Many studies emphasized how this tacit or personal knowledge is characterized by informal ways of transmission and learning, and by its implicit dimension (Polanyi 1958; D. MacKenzie 1996).

Following originators' work, this embodied knowledge appeared during the testing, when they analyzed at a glance the reports released by Asterisk and Wireshark, which described the dimension and the qualities of the exhanged data during a videocall. Originators, differently from other participants, instantly assessed whether a test worked, did not work, or may have worked. Just from a glance at a software screenshot, they interpret a video call and can decide whether a failed video call was a good test, or vice-versa.

But this tacit knowledge did not mean that the tools used to monitor tests were well known, or taken for granted. The tool is not specific for the testing developed just in this set, and so it is forced in order to fit this contingent situation. To be forced, the tools have to be studied. To learn the monitoring software used in their tests, originators spend a lot of time on the Internet, downloading handbooks, participating in or reading forum threads, chatting with colleagues in different messenger applications, and watching tutorials on YouTube. During these activities, embedded and arm's-length relationships are brought together into a whole, and "self-serve" technologies are prominent, compared to e-mail, phone, and fax (Schultze and Orlikowski 2004).

It is impossible to separate learning of the monitoring software from its use. It is a hybrid practice directed to advance a convergence toward interoperability, constituting an ensemble of a recurrent activity (Schatzki *et al.* 2001). To use a tool and to learn about its uses are continuously intertwined activities and both show how originators are also users in many constitutive parts of their practices (A. Mackenzie 2003, 381) which do not need a clear-cut boundary to be understood (de Laet and Mol 2000, 227).

The authoritative relationship between an author and her/his work seems eventually to be almost frail and the collective work overwhelming

(Edwards 1994, 239). Authorship is less important than the project as a discursive frame. Following a designer at his desk, one encounters many collaborations with other people or other resources. Testing a device concerns the device itself and the Nemus ensemble, as well as a whole range of resources that make possible its interpretation insofar as it is included in a technical ecology which the originator is able to browse. (S)he does not clearly know all its elements then moves and acts as a user. The professional dimension of originators filters initial ideas and company interests and makes the realization of a device possible. If the initial project is important in order to assess intentionality, the successive activity can be described as articulation work. It consists of testing and producing interpretations of every test: using and forcing specific software which produce reports upon which interpretations are based. This activity tests the capabilities of the SIP protocol to support the video calls.

Testing is iterative, and can affect the project as a whole or some parts of it. It faces the obduracy of things, and the obduracy of relations. Displaying or assessing the interoperability between things, it can help to transform a convergence in an alignment and can create a foothold for a more vast assembly-making activity outside the design-defined enclosure. Testing is also incremental because it advances in some direction, which is not always a linear one, and often diverges from the project frame (Segal 2008). The process can advance toward the interoperability between elements, the software execution, requirements gathering, or some relational reconfiguration.

The interpretive activity is continuous in testing, and it is an important part of the same tests. It is based upon a common set of knowledge and a professional culture which define originators as computing people, necessary to know in order to be accepted as partners, collaborators, colleagues, but also customers and suppliers. In our case, the exchanges between the camera and the rest of the Nemus socio-technical ensemble were mediated by the SIP protocol standards. But the SIP too is "sunk" in different levels and layers of the OSI model, specifically the fourth layer (transportation) and the third layer (network), corresponding to a good knowledge of routing and Wi-Fi. So, a good knowledge of such aspects is essential for any participant. But it would be too easy to describe it as a corpus of knowledge:

> "[Referring to the SIP protocol handbooks and texts] there are really countless recommendations, obviously, because it is a genesis [*sic*]; these kinds of protocols were born and develop, rise, add functionalities, modify logics. It is a continuous evolution and it will be thereby ever. To say that one knows all SIP is impossible: myself, I don't know it entirely. You have

to begin from the main buildings and after… a lot of field experience." (A project officer, and designer).

Nevertheless, in the originators' activity continuous learning is important as well and occupies a large amount of their work. They continuously look up tutorials, documents and files available on the Internet and, more rarely, exhange emails, phone calls, or messages with colleagues inbound or outbound:

He's going on fring/support [mobile application] and opening a .pdf handbook. At the same time, he's browsing through the cell and, with another cell, is speaking with a colleague, allowing himself to be led by him. (From my fieldnotes).

"When x-Lite [a softphone] died, it produced a file where it explained why, but right now I can't find it… The research function doesn't function: I didn't find it, so I say, 'To hell with you…' If it doesn't find the file, a catastrophe can be threatened… It didn't write the file, this damned one!" (Designer).

So, continuous learning adds a new dimension to the established knowledge, and mobilizes some elements of this basic set of knowledge according to the participants' needs. It is a form of learning aimed at forcing the SIP protocol and the software into a new configuration. "These things have to be broken up, to be understood," said an originator, showing an attitude similar to an end user forcing a device to function. Learning also means breaking up, forcing, or modifying the specified file pages or code, not only in order to make interoperability possible, but also to understand "things".

Even if the originators I studied described their work merely as "testing", it seems more complex, and not only from the learning dimension. The device which they try to originate is conceptually unitary, but it can exist only if connected and crossed by networks, information infrastructures, software, as well as material and social elements. All these elements were part of previous alignments, where they played other roles and executed other tasks. Therefore, it is not surprising that their work is not limited to mere tests. Very often, originators work on hardware, or they code or work on code execution, or even work to "unpack" (*spacchettare*) elements from previous alignments in order to adapt them in the new one they are working on. Passages from mere testing to these adaptive and forcing activities is seamless, and was described, in participants' language, as "packing and unpacking the little bricks" (*impacchettare e spacchettare i mattoncini*). These elements extrapolated from previous ensembles have to take a role in a new ensemble. For this purpose they studied and in this case forced to execute a part of the common agency attributed to the new ensemble. The knowledge of the new elements is mainly an analysis of its behavior in this agency, and it is the reason that why originators have to study the protocols' features, the data standards, but also the hardware and software specifications.

All these "unpacked" elements enter the originators' activity, being situated and convergent to their work, but they were also sunk into networks and infrastructures and had to be replaced in other ones. Any convergent element has to become interoperable, and so modified in some parts. In our case study, both the SIP protocol and the monitoring software resist reconfiguration, showing their obduracy. Anyway, originators tried to force them, playing with different versions, and to implement them on different operating systems in order to make digital inscriptions easier to interpretate.

The intangibility of some software and even of operating systems is not ontological, and they can be modified at other levels of scale. Albeit originators are used to describing their work as mainly "testing", and affirm to leave the coding to outbound partners, they often code, or even re-write parts of some operating system (OS). To recompile parts of Linux or Android OS seems to be more and more diffused in order to force them to have a task in the future alignment. The participants wondered about it, but they made it:

> Designer D: "I can't believe you're re-compiling Android?"
> External Collaborator EC smiles in response.
> D and EC laugh together because it is "crazy stuff" (*una cosa fuori di testa*).
> EC works on the Android Open Source Project website.
> D observes the work of EC on the Android website, admiring it.
> (From my fieldnotes).

The observed activities actually remain unclear without analyzing the remote collaborations and resources. The projection of observed practices toward a remote and infrastructural horizon entails a reflection on the status of the originators. In most of their activity, they actually appear not much different from users of remote devices: like them, to develop their ordinary activities, the originators use some elements of the technoecologies of the worlds to which they belong. They just intervene, as the users do, to connect some element that was not connected before and eliminate material obstacles to the interoperability of some other convergent "little bricks".

3. Projects and Days

The originators' practice is only one point in the collective activity of infrastructuring (Star and Ruhleder 1996; Star and Bowker 2002; Bietz *et al.* 2010), where some entitled power users challenge extant ontologies and re-assemble things and people *in absentia* of end users. This condition of infrastructural work is their "daily work." Specific components are visible to participants when some parts are brought to their attention, but this attention, and the correspondent ideal of a design-defined enclosure, is ideological (Bowker *et al.* 2010; Star and Ruhleder 2006; Star 1999; Edwards *et al.* 2007). Some characteristics of computing reinforce the collective aspect of these practices. Software modules actually hide the details of their implementation. This is the principle of "information hiding," prescribing to expose to other developers only the implementation details that are not likely to change through data encapsulation, separation of interface specifications and their implementation, and polymorphism (de Souza and Redmiles 2009, 446). In fact, they take for granted many infrastructures crossing the new one they are originating.

Organization conditions are also influential for this job. Most importantly, except for those of a "dedicated" designer, all the activities concerning this project happen in multitasking conditions. Surely, the dedicated work in designing, which I have observed while shadowing a new designer, is rare. Every project is followed by different originators in temporally different tasks. These differentiated tasks mirror the fragmentation of the components of the device and reinforce the idea of the convergence of different entities in a process (not necessarily correspondent to interoperability or finalization) as a concept stronger than network or alignment in describing sociotechnical processes (Star, Bowker and Neumann 2003). Partial collaborations, when someone collaborates once, are "blind", i.e. they do not consider the project as a whole but only their portions. The position of the dedicated designer is exalted. When partial collaborators come from external enterprises, they are usually in charge of some complete tasks, then the collaboration seems more balanced and the dedicated designer less powerful. When the originators have to produce a demo for the management, all the technical activity is suddenly obliged to face commercial and schedule requirements, and the position of the dedicated designer takes the biggest role, considering his/her mediation position between Tiscali and external collaborators.

In this situation, originators seem to have a shifting membership, "as different skills and abilities were needed at different stages of each project" (Harding 2008, 79). The dedicated designer dramatically changed his performance at his desk while collaborating with external professionals or facing his superiors. Membership is not an undifferentiated quality because it corresponds to different knowledge, conventions and routines. Originators have to be implicated users in chat lines, managers toward executives, technicians involved in computing theory, and learned users with external collaborators or colleagues.

These fragmented tasks are integrated in time in some aspects of originators' practice and not necessarily over the whole project. The dedicated designer is disconnected enough from the overall pace of the organization, where it is more normal to work on different projects in periods that are not clearly isolated. If the project's purpose is also on the shoulders of a dedicated designer (in our case, he was also the project manager), his coordination is more evident in the accomplishment of the single tasks than at the general level. At this level, he is a crucial cross point, and a conspicuous part of his activity is to exchange messages, emails, and phone calls or share some resources on the Web. The work pace of a dedicated technician, in a world of multitasking, is in any case different compared to that of people present only for some tasks.

The open spaces of Tiscali seem nevertheless to accommodate internal collaborations, even occasional ones.

> He activates a virtual machine on his computer. He asks for help from Saturnino, at a close desk occupied by people working with business clients and billing. Initially, Saturnino says that he "just played" with Asterisk in the past. The problem seems to be that the SIP command is difficult to access.… Saturnino is able to guide him in Asterisk to bring the SIP command back. (From my fieldnotes).

Open spaces allow one to check other people's presence and activity to collaborate. On the other hand, being the workspace correspondent to a hierarchical position and task, it reinforces a certain narrative order that sorts tasks and people out. This narrative suggests the designer's space as a rational (and rationalist in its architecture) *hortus conclusus* where a design-defined enclosure can blossom. However, collaboration between originators continually uses chats and messengers. Akin to email, they allow mediated interaction, chronotopically out of step, without compresence, but differently from email, they allow one to check a colleague's online/offline position. Email, mailing lists, forums, and phone calls are less frequently used modes, or they are at least used in an accessory way. The Tiscali open spaces are obviously only *between* places (A. Mackenzie 2003, 373, 377), enfolding infrastructures, communications, and all kinds of socio-technical convergences. They are displayed and built as crucial passage points, but this role is not permanent; it depends on the activity, and it permanently disguises invisible elements. In contrast, originators' bodies really mediate engineering objects, spaces, and infrastructures (Suchman 2000, 12), forcing and linking things and infrastructures with their incorporated abilities and their hands. However, often, the entire body posture plays an important role in mediation: its lowering toward the desk, its movement looking for casual collaboration in the open space, its compression in an uncomfortable position. The originators, as the scientists, obtain a learned body, a mixture of responsiveness and action and an important element for their creativity (Viteritti 2012, 68 *et passim*).

Disembodiment is important in the narratives of their activity, especially in managers' discourse. They return what really occurs in originators' practice to a trajectory that finds meaning in some of its results. They are calibrated with the displayed purpose of a process rather than with the same process. This master narrative "does not problematize diversity. This voice speaks unconsciously from the presumed center of things" (Star 1999, 384). Management is really committed to the accomplishment of tasks in a strict temporal order, while technicians are more entangled in a distributed practice where they have to resolve problems in the field (Orr 1996, 149 *et passim*). These kinds of problems for technicians are often theoretical, so they are committed to the use of open source materials and with the maintenance of a border among application software packages. It is relevant for originators to limit their jobs to specifications and avoid coding, as well as to challenge the OS and the protocol suites. For originators, the epistemological status of computing (*informatica*), as a discipline, is an object of continuous criticism. It is considered "out of control." A perceived absence of fixity in its elements is considered negatively. It seems that a *mare magnum* within people is lost, and it is difficult to curb. In its implementation, it is understood as a system of exchanges more than an ensemble of objects, regulated by standards and protocols for "packing" data following forms that make it weighty to "unpack" them for a use in new contexts.

Following their routines and accidents, the process as a whole fades away:

> "Suddenly, it becomes unfeasible and the previous one feasible,… Now this too is once again feasible, so I cannot understand… Maybe everything is feasible or everything is unfeasible… Sincerely… the fact is that "they" say: 'No, this solution, absolutely we can't talk about it,' and afterward, suddenly, it is back in business. But, well, let's see how it will end up." (Designer).

Toward all this heterogeneity in practices and discourses, the organization of work as a project is a resource to give them unity in representations and self-description and to establish a hierarchy (A. Mackenzie 2003, 375), but not really to understand it except in a historical view because a project can offer a good unit of diachronic analysis. The main limit is that this kind of project-like image reduces the whole story to a designed-organization one (Wenger 1998, 241). The whole "constellation of practices," the networks of relations and computing built or used for task-resolving purposes, the datasets that converge in these fields, with their need to be coordinated through a common interpretation of concepts included in metadata standards, are forgotten (Karasti *et al.* 2010, 386). This fracture between a practice-dense activity for technicians and a linear purposeful orientation in executive approach appeared when an external collaborator was scrutinized by the Tiscali executives:

> Guido from Treviso Software shows a series of slides underscoring the small company use of open-source resources and its way of designing. Tiscali executives question him on the project they did, stressing success or times. They consider openly useless all this "stuff" displayed in Guido's slides and propose to the small Treviso Software a contract with Tiscali in exchange for strict respect of a project timetable. They don't care very much about the camera connection to the Nemus infrastructure. (Fieldnotes of a meeting between Tiscali executives, the dedicated designer, and an external collaborator).

We observe a parallel presence of different organizational principles in this process when looking at technicians and executives. John Law (1994, 1) proposes ordering as an organizing principle different from order: "Perhaps there is ordering, but there is certainly no order… orders are never complete". The ordering that dominates in the case of originators and technicians is focused on task accomplishment and on the strength of provisional alignments and of networking. Among executives, a holistic regime and a discursive order (the "project") with a strong linear, organizational (entrepreneurial), temporal, and hierarchical view, ruled. Executives have an interest in establishing their kind of ordering as dominant, as an obligatory passage point for everybody. Then, "once the process has established [it], the job then becomes to defend it against other translations threatening to displace it" (Star and Griesemer 1989, 391). Obviously, ordering should not be confused with extant lines of action.

Tiscali executives were really concerned about schedules and temporal anxiety. Karasti *et al.* (2010, 386, 398-99) underscore the importance of short cycle times in Web-oriented software development and the difference with a less compulsive time scale in infrastructure development. This difference produces tension between temporal scales because originators are more involved in infrastructuring than in ordering following a purposeful logic. Their job is entangled in networks of infrastructuring that make possible the accomplishment of single tasks. It is not readable on a project basis, but interacts with ongoing "projects". In the absence of a mediation between two activities that are differently oriented, a conflict is possible, and the imposition of the executive ordering can explain the death of many "projects".

4. Practices of use

The design-defined enclosure, this *hortus conclusus* from which users are excluded, faces a classification of users as *end users*, *lay end users*, and *users implicated as actors in design* following their own inclusion in design practice (Casper and Clarke 1998; Saetnan *et al.* 2000; Clarke 1998; Oudshoorn and Pinch 2003). This form of discourse probably induced the muddying of the fact that originators have to *use* devices, software, web resources, technical components, scripts, prices, and so on in order to advance their tasks and make convergence interoperable. They often force these elements, dismount them or discuss them. They act as power users (even if they are not always such), and they are able to adapt the elements, to take them apart and to readapt them in another ensemble. They often interpret them, or reduce their complexity. If they want to fit convergent elements in a new ensemble they have to mix them up and to re-distribute them (Latour 1992, 37) and, for that purpose, they have to act as users. Obviously, they are particular and skilled users, in the sense that they force most of the elements or resources they use, except the SIP protocol. Apart from this case, when they play the role of end users, they are more often similar to lay end users or to implicated actors in the design (Casper and Clarke 1998).

But it would be impossible for them to express it in their talks. Originators, managers, and executives describe their activity and their roles referring mainly to a linear model of innovation. This model acts as a powerful narrative and it frames the problems occurring to any innovation process referring to the limits of the projects or to the users' backwardness (Latour 1989, 231; Godin 2006; Neresini 2000, 184). In general, this model devalues the articulation work, emphasizes the importance of entitled technicians and justifies many exclusions of lay experts and users from the innovation process and from design (Scacchi 2004). Articulation is viewed as a mere process of diffusion, as a mere passage from an ordered and rational domain to the chaos and backwardness of the users' world (Mongili 2008, 405). End users are seen mainly as a danger. In Italy, computing people use a specific word, *utonto* (from *utente*, user, and *tonto*, dumb) to stigmatize end users and to mock them. *Utonti* are represented as tale characters: in an unrealistic way they use devices in a very stable way, and they do not change their technological habits as long as they are not obliged to do so. Change occurs mostly when a device (or a function of a previously used device) is replaced by a new one, or following a cost/benefit calculation, framed by a strict purposeful rationality. Anthropologically, *utonti* are set in a world very different and very contrastive compared to the designers' and developers' one. In any innovation process their subjectivity is excluded: they are only the final target (see Isabella, *in this volume*). At the same time, such an image reinforces the idea that the originators' work is outstanding, and can encourage avoidance of the changes that the designers have to introduce in the design (Star and Bowker 2002, 101). The linear model of innovation acts as a professional ideology. It is very powerful in reducing users to tale characters and to justify their exclusion from the design-defined enclosure, but it is frail in explaining the originators' everyday work and their activities.

In their self-representation, originators classify themselves as very different people compared to the *utonti* because of their skills and attitudes toward innovation and technology. Therefore, observing practices, their user-like activities were important. Making different convergent elements interoperable is a set of activities which implies the use of infrastructures, devices, and also continuous learning and mobilization of a web of relations.

This divide has already been challenged in order to reduce or reconsider the difference traditionally established between users and designers (Suchman 2002; Sefyrin 2012, 711), and sometimes in general terms (Oudshoorn and Pinch 2003).

Without using devices, knowledge, and social relations, the origin of new artefacts and infrastructures does not seem possible, at least in originators' practices rooted in enterprises that are conventionally arranged for producing "technology". This partition has also been challenged in a very clear way:

> "In attempting to understand technology as continually socially and physically constructed, it is useful to discriminate analytically between human action which affects technology and that which is affected by technology. I suggest that we recognize human interaction with technology as having two iterative modes: the design mode and the use mode. I

> emphasize that this distinction is an analytical convenience only, and that in reality these modes of interaction are tightly coupled." (Orlikowski 1992, 408).

In our case, a design-defined enclosure and a professional organization reinforce, at a conventional level, the idea that participants' activity would exclude the "use mode" among them. Despite that, passing from a convergence to an alignment *in statu nascendi*, and observing the activities of the people working on it, these "modes of interaction" appear as inextricably intertwined:

> "I'm trying to launch that 'little' software (*programmino*) that it gave me to check the camera remotely, just to put in the correct parameters. The problem is that this *programmino* doesn't work on my Linux because it was recompiled; i.e. an executable was created starting from a source code of another kind of Linux, so its source code is lost and recompiled afresh, which converts it into an executable for our system. When you finish a software program, there is the problem that a lot of pieces are lost, so you convert it from a source to an executable. It actually leans on things that it already knows are present in the operating system… so you use not only the source but also all the pieces you expect to find there. If these pieces are absent, you have to install them, so finally it is a total mess (*casino totale*)." (Designer).

So, the originators' practices show a high level of similarity with practices attributed to users implicated in design, or to power users. They are only framed differently in the professional classification and following the conventional level. A difference can be found, considering their practices as a whole, at a more general level. In fact, analyzing their work from the point of view of the advancement which they produce in the innovation process, they aim to translate a convergence field in a series of interoperable links between things and people in order to produce a future alignment.

5. Conclusion

From the analysis of this case study it emerges that originators can be considered users too in the majority of their activities, despite their traditional framing as designers and developers. In STS, much room has been reserved for the study of users, for their configuration in the design practices, for their domestication of devices and in general for power relationships in the socio-technological ensembles, and for articulation work in general (see Oudshoorn and Pinch 2003; Isabella *in this volume*).

But the role-like division between users and designers has been preserved. This conceptual frame does not work very well when analyzing the originators' practices, because they must occupy a user position in a very large amount of their everyday activities (Millerand and Baker 2010). As I have noticed very often, they use many instruments without which designing, testing, coding, or producing a demo would be unthinkable. I observed that only the protocol and the standards escape the forcing and the attempts to "unpack", but only at the level of scale which I studied.

If they, as originators and innovators, produce some important frame or script concerning projected end users (Lindsay 2003), nevertheless they are also configured as users of infrastructures, software, cameras, and so on. Their professional skills do not impede them being users, even if the absence of users defines their own workplace and their oustanding and, in some sense, sacralized social role as the innovation heroes in contemporary societies.

During their work, they often shift from a rare design-mode to a dominant user-mode. At this point, it is necessary to define what is the *quid* of this design mode, if most of the originators' activities are occupied by tasks very similar to those which any user does. The first aspect is the professional classification, and its consequences in interpreting their roles and in obscuring the role occupied in their work by the user-mode. This aspect is very important in master narratives, in self-presentations, and in taken-for-granted images of their job shared in public space. Nevertheless, if one wants to analyze their job and its specific character, the parts of their tasks which differentiate them from users, and not only on the formal or professional level have to be taken in consideration. This part is to demount elements coming from previous ensembles, circulating in a convergence field or newly introduced by the same originators, and make them interoperable in a new ensemble. This activity is legitimated by tests and demos, and in some sense over-shadowed by them, but it occupies a meaningful part of their jobs.

Despite the master narratives, which assign a dominant role to projects, initial ideas, or corporate purpose, I think that the originators' practices can be understood as an original work aimed at creating some interoperability between things and infrastructures. The dominant narratives, which assign defined roles to designers and developers toward users, became doubtful here. A concept like *originator* becomes instead very useful when describing the user-mode space present in the work of people classified as designer or developer. It encompasses the design-mode activities but does not over-shadow the important part of their job, which can be summarized as articulation work and user-mode. Originators work between technical and social elements which converge between them. They just advance this convergence and build the foothold for a subsequent work of articulation and alignment, outside and beyond their enclosure.

References

Bietz, Matthew, Eric Baumer and Charlotte Lee. 2010. "Synergizing in Cyberinfrastructure Development." In *Computer Supported Cooperative Work*, 19: 245-281.

Bowker, Geoffrey C., Karen Baker, Florence Millerand and David Ribes. 2010. "Toward Information Infrastructure Studies: Ways of Knowing in a Networked Environment." In *International Handbook of Internet Research*, edited by J. Hunsinger *et al.*, 97-117, Dordrecht: Springer Science and Business Media.

Bucciarelli, Louis. 1994. *Designing Engineers*. Cambridge, MA: The MIT Press.

Casper, Monica and Adele Clarke. 1998. "Making the Pap Smear into the 'Right Tool' for the Job: Cervical Cancer Screening in the USA, circa 1940-95." In *Social Studies of Science*, 28(2): 255-290.

Clarke, Adele. 1998. *Disciplining Reproduction*. Berkeley: University of California Press.

de Laet, Marianne and Annemarie Mol. 2000. "The Zimbabwe Bush Pump: Mechanics of a Fluid Technology." In *Social Studies of Science*, 30(2): 225-263.

de Souza, Cleidson and David Redmiles. 2009. "On The Roles of APIs in the Coordination of Collaborative Software Development." In *Computer Supported Cooperative Work*, 18: 445–475.

Dodier, Nicolas. 1993. "Les arènes des habiletés techniques." In *Raisons Pratiques,* Special issue "*Les objets dans l'action. De la maison au laboratoire*": 115-139.

Edwards, Paul. 1994. "Hyper Text and Hypertension: Post-Structuralist Critical Theory, Social Studies of Science and Software." In *Social Studies of Science* 24(1): 229-278.

Edwards, Paul, Steven Jackson, Geoffrey C. Bowker, and Cory Knobel. 2007. *Understanding Infrastructure*. National Science Foundation Report.

Godin, Benoît. 2006. "The Linear Model of Innovation. The Historical Construction of an Analytical Framework." In *Science, Technology, & Human Values*, 31: 639-667.

Goffman, Erving. 1959. *The Presentation of Self in Everyday Life.* London: Allen Lane.

Harding, Sandra. 2008. *Sciences from Below*. Durham: Duke University Press.

Karasti, Helena, Karen Baker and Florence Millerand. 2010. "Infrastructure Time: Long-term Matters in Collaborative Development." In *Computer Supported Cooperative Work*, 19: 377-415.

Kling, Robert and Walt Scacchi. 1982. "The Web of Computing: Computer Technology as Social Organization." In *Advances in Computers*, 21: 1-90.

Latour, Bruno. 1987. *Science in Action*. Cambridge, MA: Harvard University Press.

—. 1992. *Aramis ou l'amour des techniques*, La Découverte: Paris. (1996, *Aramis or the love of technology*. Cambridge, MA: Harvard University Press.).

Lave, Jean. 1988. *Cognition in Practice*. Cambridge, MA: Cambridge University Press.

Law, John. 1994. *Organizing Modernity*. Oxford: Blackwell.

Lindsay, Christina. 2003. "From the Shadows: Users as Designers, Producers, Marketers, Distributors, and Technical Support." In *How Users Matter*, edited by Nelly Oudshoorn and Trevor Pinch, 29-50, Cambridge, MA: The MIT Press.

Mackenzie, Adrian. 2003. "These Things Called Systems: Collective Imaginings and Infrastructural Software." In *Social Studies of Science* 33(3): 365-387.

Mackenzie, Adrian. 2006. *Cutting code*. New York: Peter Lang Publishing.

MacKenzie, Donald. 1996. *Knowing Machines*. Cambridge, MA: The MIT Press.

Millerand, Florence and Karen Baker. 2010. "Who are the users? Who are the developers? Webs of users and developers in the development process of a technical standard." In *Information Systems Journal*, 20: 137–161.

Mol, Annemarie. 2002. *The Body Multiple*. Durham: Duke University Press.

Mongili, Alessandro. 2008. "Oggetti e allineamenti: qual è l'oggetto della sociologia della tecnoscienza?" In *Rassegna Italiana di Sociologia*, 49(3): 403-426.

Neresini, Federico. 2000. "Scienza/Pratica scientifica." In *Parole chiave*, edited by Alberto Melucci, Roma: Carocci.

Omoronyia, Inah, John Ferguson, Marc Roper and Murray Wood. 2009. "Using Developer Activity Data to Enhance Awareness during Collaborative Software Development." In *Computer Supported Cooperative Work*, 18: 509–558.
Orlikowski, Wanda. 1992. "The Duality of Technology: Rethinking the Concept of Technology in Organizations." In *Organization Science*, 3(3): 398-427.
Orr, Julian. 1996. *Talking about Machines*. Ithaca: Cornell University Press.
Oudshoorn, Nelly and Trevor Pinch (edited by). 2003. *How Users Matter*. Cambridge, MA: The MIT Press.
Polanyi, Michael. 1958. *Personal Knowledge*. Chicago: The University of Chicago Press.
Ribes, David and Thomas Finholt. 2008. "Representing Community: Knowing Users in the Face of Changing Constituencies." In *Proceedings of the CSCW'08*: 107-116.
Rosental, Claude. 2007. *Les capitalistes de la science*. Paris: CNRS Editions.
Saetnan, Anna, Nelly Oudshoorn e Marta Kirejczyk (edited by). 2000. *Bodies of Technology*. Columbus: The Ohio State University Press.
Scacchi, Walt. 2004. "Socio-Technical Design.", in *The Encyclopedia of Human-Computer Interaction*, edited by William Sims Bainbridge, 656-659, Great Barrington: Berkshire Publishing Group.
Schatzki, Theodore, Karin Knorr Cetina and Eike Von Savigny (edited by). 2001. *The Practice Turn in Contemporary Theory*. London: Routledge.
Schultze, Ulrike and Wanda Orlikowski. 2004. "A Practice Perspective on Technology-Mediated Network Relations: The Use of Internet-Based Self-Serve Technologies." In *Information Systems Research*, 15(1): 87–106.
Sefyrin, Johanna. 2012. "From Profession to Practices in IT Design." In *Science, Technology, & Human Values* , 37: 708-728.
Segal, Judith. 2008. "Models of scientific software development." In *SECSE 08, First International Workshop on Software Engineering in Computational Science and Engineering*.
Star, Susan L. 1999. "The Ethnography of Infrastructure." In *American Behavioral Scientist*, 43(3): 377-391.
—. 2010. "This is Not a Boundary Object: Reflections on the Origin of a Concept." In *Science, Technology, and Human Values*, 35(5): 601-617.

Star, Susan L. and Geoffrey C. Bowker. 2006. "How to infrastructure", in *Handbook of New Media*, edited by Leah A. Lievrouw and Sonia M. Livingstone, 230-245, London: Sage.

Star, Susan L. and Karen Ruhleder. 1996. "Toward an Ecology of Infrastructure: Design and Access for Large Information Spaces." In *Information Systems Research*, 7(1), 111-134.

Star, Susan L., Geoffrey C. Bowker and Laura Neumann. 2003. "Transparency At Different Levels of Scale: Convergence between Information Artifacts and Communities of Practice." In *Digital Library Use*, edited by Ann P. Bishop, Nancy A. Van House, Barbara Pfeil Buttenfield, 241-269, Cambridge, MA: The MIT Press.

Suchman, Lucy. 1987. *Plans and Situated Actions. The Problem of Human-Machine Communication*. Cambridge, MA: Cambridge University Press.

—. 2000. "Organizing Alignment: A Case of Bridge-building." In *Organization*, 7(2): 311-327.

—. 2002. "Located Accountabilities in Technology Production." In *Scandinavian Journal of Information Systems*, 14: 91-105.

Tory Melanie, Sheryl Staub-French, Barry A. Po and Fuqu Wu. 2008. "Physical and Digital Artifact-Mediated Coordination in Building Design." In *Computer Supported Cooperative Work*, 17: 311–351.

Viteritti, Assunta. 2012. *Scienza in formazione*. Milano: Guerini Scientifica.

Wenger, Etienne. 1998. *Communities of Practice*. Cambridge, MA: Cambridge University Press.

CHAPTER TWO

CONTINGENCY IN INFRASTRUCTURES: VULNERABILITY, DUCTILITY, RESILIENCE

GIUSEPPINA PELLEGRINO

Introduction: Infrastructure as *epoché*

What is the relationship between contingency and vulnerability in the evolution of infrastructures? How do they enact flexibility and stability of Boundary Objects (BOs) and how do the latter shape infrastructures?

Departing from these research questions, this chapter aims to investigate the relationship between contingency and vulnerability in the evolution of infrastructures through three stories in which very different types of infrastructures, technologies, and social settings are at work. The different relationships between infrastructures and BOs emerging from these stories are used to argue the role of contingency and vulnerability in shaping action and ontology, occurrence/situatedness and resilience/resistance of infrastructures. The notorious case of the accident of the Italian cruise ship "Costa Concordia", the design of sensor technologies for transport and health detection, classifications and protocols for treatment of Non-Hodgkin lymphomas are presented as different examples of how contingency and vulnerability can shape the evolution of an infrastructure and how BOs are enacted through a specific mixture of flexibility and consistency which makes them both ductile and resilient.

The theoretical standpoint of this contribution is phenomenological, assuming the relationship between routine and innovation, as depicted in the Schutzian analysis of everyday life. Everyday life consists of an attitude, which is based on "putting the doubt into brackets" (*epoché*) . Such an attitude allows the taking of ordinary action for granted and obviously working out (Schütz 1945). Extending such an approach to sociotechnical devices and information infrastructures (see Suchman 1987) the role of doubt in ordinary working of complex sociotechnical environments is played by breakdowns, malfunctions, misuses, errors and unforeseeable events which put the infrastructure at risk. It is then that what is usually obvious, invisible and taken for granted becomes evident, visible and questionable.

In fact sociotechnical artefacts, as well as information infrastructures, come out of collaborative processes where hybrid, socio-material networks allow multiple social worlds to communicate to each other, both re-affirming and overcoming their local, situated character (Star and Griesemer 1989). Vivid presence and resistance "offered" by objects in everyday life, however, as well as the attainability of experience could not be there without a hidden "enabler". Actually, any set of *pervasive enabling resources* can be seen as an infrastructure (Bowker *et al.* 2010), as long as it is embedded, transparent, embodied in standards, linked with practice, large scaled and visible upon breakdown (Star and Ruhleder 1996).

Infrastructure is something "between" people, tools or rules, which makes it difficult to study it (Star 1999):

> "[Infrastructure] becomes visible upon breakdown. The normally invisible quality of working infrastructure becomes visible when it breaks (…)" (Star and Bowker 2006, 231).

In this sense, infrastructure is the object of the *epoché* of the natural attitude, described by Schütz as a far diverse exercise of the radical Cartesian doubt. Instead of suspending the belief in the outer world, what is suspended is just the belief that the world might be *otherwise*. This different way of doubting, or different suspension of belief, distinguishes the phenomenological *epoché* from the *epoché* of the natural attitude (Schütz 1945, 550-51).

Like the sedimented base of knowledge which typifies our future anticipations (Schütz 1945; Schütz and Luckmann 1973), infrastructure is built on an installed base, and there are many hidden interdependencies which connect it to pre-existing infrastructures. This characteristic can also be referred to as "interoperability" and "saturation" (Bowker and Star 1999; Star and Bowker 2006). For example, there is more of an integration, than a substitution, between old and new media, technologies and infrastructures. Evidence is linked to the media history (Marvin 1990 *amongst others*), which shows how each time a new medium appears on the scene, it is shaped through dreams and fears of substitution, but more often it integrates with older media already socially appropriated.

The focus of this contribution is on processes which shape the *epoché* of the natural attitude with reference to infrastructures. Two concepts are introduced in order to shed light on these processes in infrastructures, namely contingency and vulnerability, which are often intertwined or at least linked with each other. Furthermore, they make clearer the role played by BOs as a "door", or "privileged point of access" to the evolution and resilience of infrastructures (see section 3).

The chapter is structured as follows: first, it presents three stories of infrastructures in which contingency is differently declined and articulated: as a possible event or circumstance, but difficult or unlikely to occur (the Costa Concordia accident); as a circumstance to avoid and "expel" from the infrastructure in the making (ubiquitous computing design); as a constitutive characteristic in the infrastructure of oncological research and care (Non-Hodgkin Lymphoma).

Then, the relationship between contingency and vulnerability is articulated, referring to infrastructure as ecology, a metaphor which enriches the analysis of infrastructure as a complex arrangement and alignment of heterogeneous components, none of which has either epistemological or operational primacy. However, a specific and crucial role in this ecology is played by BOs, whose combination of flexibility and consistency is focused on with reference to contingency and vulnerability and in terms of resilience and resistance of infrastructures.

1. Stories of contingency in infrastructures: an analytical comparison

Materials collected and presented hereafter are as heterogeneous and diverse as infrastructures are (Star and Ruhleder 1996). Of the three stories narrated, only the second one presents "canonical" elements which qualify it as an empirical research, being based on observations and interviews carried out over several years. The first stage started at Klagenfurt University in 2008, and lasted two months, with researchers in the Embedded System field; a second stage was conducted with research groups and a spin-off company in the field of ubiquitous computing and mobile applications at Darmstadt Technical University (Germany) for three months in 2012.

Instead the first story is based on second hand data only, mainly mainstream media as well as web reports on the Costa Concordia disaster, and subsequent events. These accounts are framed through the lens of infrastructure and BOs.

Eventually, the third and last story comes from the author's experience and self-observation as a patient affected by Non-Hodgkin Lymphoma. Data used in the account are both referred to literature review on analysis of standards and practice in oncological pathologies, and to personal communications and observations which occurred in the hospital during the different stages of the illness (diagnosis, hospitalization, treatment and re-evaluation).

Different but complementary dimensions of contingency as well as of infrastructure emerge from this heterogeneous set of accounts. In general, contingency implies an emergency or exigency that is regarded as possible but uncertain of occurrence (http://iword.com/idictionary/juncture). The etymology of the word goes back to late Latin *contingentia* (circumstance), referring from then onwards to "a future event or circumstance which is possible but cannot be predicted with certainty", as well as "the absence of necessity; the fact of being so without having to be so (philosophical sense)" (http://www.oxforddictionaries.com/definition/english/contingenc y?q=contingency).

Diverse manifestations of unnecessary, unpredictable and possibly occurring circumstances are presented in the following stories, which are based on different types of infrastructure.

In the case of the Costa Concordia, the infrastructure at stake is the material and technological side of the cruise ship as a means of transport and part of the broader emergent industry of cruise tourism. Indeed, cruise ships are popularly seen as "floating luxury houses" more than as carriers: "the accommodation and related resort facilities comprise 75% of the ship with the remainder devoted to its operations" (Dowling 2006, 3). However, not only the operations environment, but also most of the facilities are made possible by complex sociotechnical infrastructures, whose saturation is increasing (see Lindauer 2002 on cruise ship theaters).

The case of ubiquitous computing is far diverse, as this is a typical example of sociotechnical infrastructure in-the-making, an infrastructure conceived since the very beginning as distributed, invisible, unobtrusive, embedded; all characteristics which, by definition, belong to infrastructure as a concept. The role played by classifications and classifiers is analyzed as bearing the capability for the new infrastructure to be really pervasive, ubiquitous and able to detect the behavior which has to be distributed across different environments and hybrid ecologies (Crabtree and Rodden 2008).

The world depicted in the case of Non-Hodgkin Lymphoma is the most "classical" in terms of infrastructure and BOs: around the definition, investigation and treatment of this heterogeneous set of pathologies of the immune system, different social worlds act and negotiate. Their work depends on classifications and standardized protocols; BOs which allow the infrastructure of onco-hematological care to develop and adapt to local contexts and even idiosyncratic cases.

The comparison carried out among the three cases is an analytical one, where emphasis is more on differences than similarities.

The heterogeneity of contexts considered (transport, ICTs and medicine), while providing an account of "superficially" diverse infrastructures, allows us to see their "deeper" similarity as based on the intervention of BOs, small social worlds highly specialized (skilled and drilled communities are in charge of the infrastructure maintenance), ecological texture of reach or scope of infrastructure and the kind of enablement they provide.

The comparison is conducted on few variables as illustrated in table 2-1.

	Costa Concordia	**Ubiquitous Computing**	**Non-Hodgkin Lymphoma**
Stage	Disrupted	In the making	In expansion
Field	Transport	Communication	Medicine/ Healthcare
Contingency	Breakdown	Embedded	Embedded
BOs	External/global standards (e.g. emergency procedures)	Internal/local standards + global (coincident boundaries)	Global standards locally applied (protocols) (coincident boundaries)
Vulnerability	High	Not known yet but low in theory	Low in theory
Type of enablement	Movement and glocal im-mobility	Continuous connectivity (ubiquity)	Consistent treatment of vulnerabilities (local universality)
Resources/ Communities involved	Captain and crew staff; passengers	Engineers and designers	Doctors, nurses and clinical researchers; patients

Table 2-1: comparative axes of the three case studies.

1.1 The Costa Concordia disaster: breakdown, contingency, fragility

On the night of 13th January 2012 the cruise ship Costa Concordia, owned by the Costa Crociere and part of the international group "Carnival", impacted on superficial reefs very close to the tiny island of

Giglio, just off the Tuscan coast line. What seemed an apparently "strange" contingency and an unlikely circumstance, rapidly turned out to be one of the most serious accidents in the recent history of big cruise navigation, finally resulting in 32 fatalities (plus a professional diver who was trapped while inspecting the ship relict in February 2014). The partially sinking ship sat in front of the tiny island for one year and a half, after which a spectacular and advanced technological "re-adjustment" of the ship happened through a very sophisticated control room[1].

Thanks to social web technologies and mobile Internet used by passengers, on the spot records of the event were available from the very first moments of the wreck. In the immediacy of the impact, along with the sinister sound of a siren, two things happened: somebody asked "But where have we crashed?", along with the Captain's spokesperson attempting to reassure passengers of a little electric inconvenience going on, then classified as a "black out"[2]. That means, an attempt to make sense of where the ship was moving on, and the enactment of a "comfortable", "plausible" classification of the event. Since electric energy shut down, this was the first framing of the event. Very likely, such a framing of the critical moment was aimed at disguising the impact with the reefs: during the trial, the cruise director assistant testified that

> "I was told to tell passengers it was a black out (…) I announced what they asked me to say, that 'we have an electric problem with the generators."[3]

Of the two classical elements of the conundrum "human *vs* technical error" so typical of deterministic reasoning concerning science and technology (see Winner 1977) the Costa Concordia represents very well the first one. Responsibilities of the Captain immediately appeared very tough: he was not leading the ship, he was not in the pilot house (17 meters of "pure" technology aimed at monitoring all of the ship, see Minella 2012), he ordered the "bowing", the salute to the island as a "touristic play"-or showboating-to entertain the passengers aboard. Last but not least, during the rescue operations, he left the ship with his second official, as testified by a notorious phone call with the Port Authority[4].

---

[1] http://video.repubblica.it/dossier/costa-concordia-il-recupero/costa-concordia-dentro-la-control-room/140101/138636
[2] http://www.corriere.it/cronache/speciali/2013/costa-concordia-un-anno-dopo/
[3] http://tg24.sky.it/tg24/cronaca/2013/11/25/costa_concordia_processo_grosseto_testimoni_isola_del_giglio.html
[4] http://www.youtube.com/watch?v=dNsqDTdykrc

However, the “distracted and out-of-place Captain” is a little bit of a far broader puzzle. Not by chance, to defend himself he blamed for the technology, stating neither the nautical charts nor the ship’s instruments signaled the “Scole” reefs. So, what sense of direction and orientation in a such a complex environment can be trusted as human skill or ability to maintain a straight route? Is this separable or separated from the hypersophisticated infrastructure of a gigantic cruise ship pilot house? And what about the technical training of the other officials acting on behalf of (or in substitution for) the Captain?

On the other hand, it is certain that the first version of the accident as a “black out” was communicated not only to passengers but also to the Port Authority; that the alert was given with a conspicuous delay, likely decisive in terms of the final number of victims and consequences.

According to one of the most impressive statements reported by the media, that of passenger Benji Smith [recounted on Saturday] making his own rope ladder to save himself and his wife,

> “It was the Marx brothers, watching these guys trying to figure out how to work the boat," (…) I felt like the disaster itself was manageable, but I felt like the crew was going to kill us.”[5]

What turns, then, a “manageable disaster” into a tragedy?

Amongst other things, BOs like classifications and standards (Star and Griesemer 1989; Bowker and Star 1999) are of crucial importance to the immediate organizing of an emergency situation. In the case of a cruise ship, BOs guarantee (im)mobility going smooth. The flexibility and consistency of classification and standards is then crucial to the activation of emergency procedures. It becomes of the greatest importance how they are interpreted and who does (not) share them.

In the case of the Costa Concordia wreck-a manageable, but not managed disaster-the immutability and stability of the network goes disrupted, due to multiple gaps: a lack of sharing, a fragility of authority as well as the missed enactment of processes of learning and training. A clamorous, on line, recorded conflict takes place between the Captain and the Port Authority, but inconsistencies and ambiguities pervade the whole role of the ship owner, the Costa Crociere company, with reference to if and how the chain of command was activated. The Captain, in an intercepted phone conversation, said that “one of the managers was asking him to showboat”. Furthermore, no formal “may day” was sent to the Port

---

5 http://www.cnn.com/2012/01/15/world/europe/italy-cruise scene/index.html?iid=article_sidebar

Authority and one hour passed between the clash and its formal communication[6].

This short account of an infrastructure disruption, therefore, points straight to the absence of working BOs and of a common language: under these circumstances, no intercommunication is possible (some passengers testified that the crew staff did not speak Italian) and infrastructure (informational and physical, from the "may day" to the watertight doors which seemed not adequately resistant) falls to pieces, not only in a figurative, but in a literal and concrete sense. In this case, an unlikely and relatively trivial event unveils the high vulnerability and fragility of a big piece of infrastructure, due to the failure of both articulation and categorical work (Bowker and Star 1999; *see also* section 3 of this chapter) in the face of contingency.

1.2 Ubicomp design: classifying motion, marginalizing contingency

Contemporary sociotechnical infrastructures are characterized by sometimes contradictory, co-existing trends summarized through key-words or adjectives, like "convergent", "ubiquitous" "embedded", "distributed", "smart", and so on. Most of them go towards the pursued or desired evidence of technologies able to saturate and surround bodies as well as daily environments, in a more or less prosthetic way. Again, with regard to infrastructures, the importance of saturation becomes clearer just when it stops working out, whenever any kind of breakdown, interruption, misuse or unexpected use occurs. It can happen, for example, with the mobile phone as ubiquitous technology virtually accessible everywhere/anytime, whose saturation increases expectations of continuous availability/attainability of participants to the communicational process. Furthermore, the concept of saturation describes well the way our bodies and environments are intertwined into inextricable chains of socio-technical relationships, like in the "everyware" texture of ubiquitous computing, imagined as a technology capable of colonizing surfaces and settings of everyday life (Greenfield 2006).

Ubiquitous computing traces back to Weiser's seminal vision of an unobtrusive, embedded technology (Weiser 1991) but like many other words born to envisage future technologies and information infrastructures, it

---

6 http://firenze.repubblica.it/cronaca/2012/01/15/news/corsa_contro_il_tempo_rumori_al_ponte_3_nella_notte_salvati_due_intrappolati_in_cabina-28144252/?ref=search

has become a "buzzword", an "umbrella" term with many ramifications and affiliated concepts, e.g. pervasive computing, distributed computing, embedded systems, cloud computing, Internet-of-things, participatory sensing technologies, and so on: a set of close relative technologies, which can be inquired and analyzed looking at design as a privileged site of future-in-the-making (Pellegrino 2011).

> "The emergence and growing shift towards ubiquitous computing has seen digital technologies become increasingly embedded in the physical world that we inhabit. This has been manifest in a range of environments that combine mixed reality environments with ubiquitous computing infrastructures, applications and devices to support collaboration across the digital-physical divide" (Crabtree and Rodden 2008, 1).

From several interviews and visits to University research laboratories where some of these technologies are studied and researched, definitions (sometimes contradictory ones), limits and potentials of this infrastructure in process emerged.

> "Ubiquitous computing is not a technology but a paradigm (…) It means you get rid of the dependence between digital artifacts and physical tools" (R., researcher and PhD candidate).

The result is a mixed reality, which allows the designer or researcher "(...) to give users the means to organize their artifacts" (R.). The same researcher explained the meaning of ubiquity as follows:

> "you know you have a pool of personal resources and information digitally stored; you know conceptually the place where you want they to be, but they are no more device-dependent".

Limits and constraints to the achievement of this complete independence (which implies a tight interdependence of other components of the ubiquitous infrastructure) are also linked to politics and policies, as clarified by another researcher:

> "If I want ground for new applications like videostreaming , efficiency of existing wireless communication must be increased (…) It's about politics, those who get the bandwidth, those who own it, those who can pay more (…) We are allowed to use a very small bit of the spectrum for wireless communication experiments. And our aim is to make the best of it" (W., research group leader).

Beside these "macro" problems, I was interested in understanding the minimum components of this kind of infrastructure, which are represented by sensor-based technologies, or "embedded systems" technologies, part of the ubiquitous computing world, that means

> "automatic detection of data by machines with sensors of different kinds. Information science is about collecting information from small sensor devices" (K., professor and research team director).

These sensors can be implemented and used for recording different types of movement or behavior, e.g.

> "what type of movement bi-polar disorder affected people do during their everyday life. So the psychiatrist can infer from recorded data to make a diagnosis" (K.).

Therefore, information collected through sensors is a boundary object of the type "coincident boundaries", namely a type of object having the same boundaries but different contents to those who interpret and use it (Star and Griesemer 1989, 410-411).

Another example of application for sensor-based technology is a project called "Characterizing sleeping trends from postures", in which sensors are used to extract information out of sleeping motion, whose ultimate aim is

> "(…) to have a sensor to wear 24 hrs. to detect different information also during day-time activity, not just during sleep" (M., PhD student).

The most interesting part of the project is the way it builds a model,

> "looking at occurrences and building a classifier (…) People were interested, we recorded people without disorders for a month saying them they slept and how. Weekend patterns are different from weekday ones" (M.).

Classification work is the crucial part of all this data detection, and building up a classifier is the indispensable step towards the construction of reliable sensor-based infrastructures:

> "A classifier is a program capable of saying this is sleep, this is not sleep, and what happens in both the cases. A classifier is about detecting movement and its specific features. For each action, a specific characteristic (pattern, feature) [is identified and recorded], so this set of

> features is close to this activity, and so on. There is also an approximation" (E., PhD student).

The crucial function of a classifier, then, is to extract information from the data recorded by sensors and recognize their patterns. As in all classification processes, "the inherent problem for any classifier system is any unexpected event (e.g., one night without sleeping when tracking sleeping postures)" (E.).

An unexpected event is exactly the core issue of contingency (see next section) addressed by this contribution. "Training" the classifier implies, then,

> "to look for repeating patterns during the dancing, identifying a characteristic motion, used as a feature to detect it [in the case of flamenco, moving the fan] (…) trying to find a feature for whatever activity" (E.).

Therefore, building up an infrastructure of this kind means marginalizing contingency as much as possible, aiming to "extract as much possible information from the sensors" (E.) The latter is the ultimate aim of E.'s project on "Rhythm model of long-term activity data", whose objective is overcoming multiple trade-offs: quantity of data *vs* space; data compression/abstraction *vs* consistency of classification; data downloading *vs* what signals reveal.

> "Data to detect trains [can be] peaks corresponding to wheels, number of wagons, or number of vibrations. Depending on how my parameters and algorithms are set, different things will be detected (…) For dancing motion the feature is not the number of peaks but the shape of patterns (…) Vibrations are features on the fly, detectable directly on the sensors, which is possible for gesture detection but not for other long term activities (…) like monitoring of bi-polar disorders" (E.).

All these differences and variations, depending on the type of movement or event studied as well as on the sensor technology used, are managed through the key-component of the ubiquitous infrastructure: the classifier. This is a classical boundary object, as it allows a dialogue among designers, users, and buyers of the system, and also a classification tool which presents the challenge to maximize information detected while minimizing and governing the unexpected (contingency), in order to prevent potential fragilities of the in-built infrastructure.

1.3 Protocols for Non-Hodgkin Lymphoma treatment and patient care: from discretized standards to discretional action

> The young female doctor who has just finished the aspired needle procedure looks at me saying: "You shouldn't be worried. Even if it were a lymphoma, it's not like it was in the past. There is a trend towards not considering it as a neoplastic illness, rather re-classifying it as a disease or a syndrome. I know patients coming to this hospital are frightened by the word... but it's curable" (Author's self-observation).

> "(...) and the radiologist looked at me so happily: Madam, it's only a lymphoma!"... and she hugged me tightly" (Patient).

(Re-)classification is an integral part of medical practice as well as of medical science and research. As Bowker and Star (1999) showed, classification is the core of any information infrastructure, even more when objects classified are symptoms, diagnoses and illnesses, as in the case of pathologies listed by the ICD (International Classification of Diseases), the global standard for epidemiologists, doctors, clinicians. As mutually exclusive and consistent, complete and exhaustive, classifications are the result of a continuous work aimed at building up categories and related boundaries to enable interaction among different social worlds.

Classifications and standards are the result of categorization as both an everyday activity and an infrastructure in/across professional communities. Classifications are what make information infrastructure work well.

> "Formal classification systems are, in part, an attempt to regularize the movement of information from one context to another; to provide a means of access to information in time and space. The ICD, for example, moves information across the globe, over decades, and across multiple conflicting medical belief and practice systems" (Bowker and Star 1999, 290).

An analogous function is played by therapeutic protocols in oncological pathologies: the development of a common standard based on specific "pharmacological mixes" is the therapeutic regimen for treating different types of cancer. In this respect medical protocols, according to Timmermans and Berg's definition, are a technoscientific script which can be followed and re-arranged during design and use (Timmermans and Berg 1997).

The case here analyzed shortly, that of Non-Hodgkin lymphoma (NHL from now onwards), is exemplary of how in medical science and practice multiple classifications systems, not always consistent with each other, are in play; but it also shows how implementation of new therapies in the treatment can "informally" shift the classification and consideration of the illness and its severity.

NHL was born as an "oppositional", negative category, to include all of what was non-Hodgkin (from the scientist's name who discovered and described the illness in 1832). Such a residuality was articulated as a new set of classifications, which internalized an exclusion, giving voice to what used to be "otherness" and "orphanage" in existing classifications (see Bowker and Star 1999 and introduction *to this volume*).

Both scientific literature and more informative, patient-oriented accounts of NHL always emphasize the "heterogeneity" and "constitutive diversity" of this hematological pathology, along with the complexity and controversy of its classification (see for example Hernandez *et al.* 1997; Armitage and Weisenburger 1998; Armitage 2005; Bioncology 2010).

Feller and Diebold (2004) revise the history of NHL classification analyzing three main classifications systems: the Kiel classification; the REAL classification proposal; the WHO classification which is currently the most used and established, given the high consensus and degree of relative universality it reached amongst researchers and clinicians. Differences between Europe and the US in the elaboration of classes and categories of NHL, along with diverse parameters to set up the system (morphology and/or immunochemistry; cytogenetics and/or size of cells; grade and prognosis, and so on) were overcome over the decades shifting from separated systems (Rappaport in 1966 then Lukes and Collins in 1974 in the US *vs* Kiel 1974 in Europe) to provisional agreement in the '80s (the so called "International Working Formulation", or IWF) and culminated in the Revised European American Lymphoma (REAL) classification which founded the basis for the current authoritative standard, the World Health Organization – WHO classification.

> "Thus today, the definition of a clinicopathological lymphoma entity is given not only by morphology but also immunophenotypic and genetic aberrations, as well as clinical data and lymphocyte development. Nevertheless, it is interesting that most of the newly described entities were primarily recognized" (Feller and Diebold 2004, 9-10).

Why did so many classifications overlap and evolve in less than half a century? First of all, the need for classification obeys to a clinical need and urgency, as

> "The purpose of further classifying non-Hodgkin's lymphomas into specific categories is to describe the individual behavior and develop appropriate treatment strategies for each type of lymphoma"[7].

Being a diffuse large B-cell lymphoma (DLBCL) patient, I rarely encountered a similar pathology while doing therapies and talking to patients and their accompanying persons. Indeed, the type of therapy was the most evident and recognizable subject for conversation. Talking about therapies happens more frequently than talking about the type of lymphoma, which is generally referred to as "non-Hodgkin" or "Hodgkin", sometimes with a more specific qualification, like "follicular" or "indolent" rather than "aggressive" (another line of classification). However, the conversation often stopped at the very first distinction of the classification, given the prevalence of elderly patients and their attitude of not inquiring the specificities of their own illness: "I don't want to know anything, I just tell the doctors: please take care of me!" (Patient).

When I was hospitalized, I was told "an ID-like picture" (Nursing coordinator) of my illness had to be compiled and this was the staging process, which took a few crucial days and exams. The staging process is another indispensable boundary object in clinicians' work, and not less complicated than classifications in its evolution (Armitage 2005). While working on this chapter, I also discovered how recent crucial progress in the treatment of DLBCL is (Hallas *et al.* 2005; Hernandez-Ilizaliturri *et al.* 2013) and was confirmed with what biologists, doctors and ex patients told me during the treatment, namely that "the monoclonal antibody completely changed the prospects of care, healing and survival for this type of lymphoma" (Clinical biologist).

> "Chemotherapy is typically delivered in regimens, or groups of chemotherapy drugs, (…) used because different chemo drugs attack cancer cells in different ways and at different stages of the cell cycle (…) R-CHOP has been prescribed as chemotherapy treatment for a number of non-Hodgkin's lymphomas (…) The R-CHOP acronym is composed of the following drugs (Rituximab Cyclophosphamide Hydoxydaunorubicin (doxorubicin) Oncovin (vincristine) Prednisone). Rituximab is a new type of drug known as a monoclonal antibody, meaning it's 'trained' to do a very specific job within the body—something like a 'magic bullet'. Rituximab's narrow job is to seek out B-cell lymphocytes by finding a certain protein on the surface of the cell, and kill them" (http://www.lymphomainfo.net/blog/general-lymphoma-blogs/chemotherapy-drugs-101-the-r-chop-regimen).

---

7 http://www.texasoncology.com/types-of-cancer/non-hodgkins-lymphoma/lymphoma-classification/

Despite the use of standardized regimens,

> "each patient is a case of his own… it is not possible to make comparisons and these would not be fruitful for the patient" (Psychologist).

Heterogeneity of the pathology and the variability of patients' condition, their age, gender, lifestyle, concurrent pathologies make each therapeutic plan a unique combination of factors, so that even if protocols are the same for any specific pathology, their design and implementation is not at all the same across sites. I even experienced differences from one doctor to another while doing the same protocol with reference to single drugs, and observed high latitude during the treatment. My doctor continuously took crucial decisions concerning routine blood exams, drug combinations, dates of treatment and re-evaluation of the illness.

As a consequence, despite the standardization of therapeutic regimens and treatment, the infrastructure of care for the single patient and the single site of care is never exactly overlapping or comparable: multiple contingencies require high flexibility but also coherence and consistency of the protocol, showing that

> "universality always rests on real-time work and emerges from localized processes of negotiations and pre-existing institutional, infrastructural, and material relations" (Timmermans and Berg 1997, 275).

As in the case of a bridge building (Suchman 2000), NHL treatment infrastructure is made up of information standards and material networks, as well as of persuasive story telling frames.

So beside classifications and protocols, infusions and therapeutic plans, as a NHL patient I have encountered what I like to define as "the power of healing stories". This reminded me of Orr's war stories, circulating among photocopier repair technicians at Xerox corporation (Orr 1996).

"Healing stories" represent a common repertoire – a "repository" in Star and Griesemer' terms, among relatives and acquaintances of ex patients and current patients. What impressed me when I was diagnosed with NHL was the fact that the majority of people I talked to-even before being hospitalized-knew a case of NHL ending happy through treatment, because of "therapies having made enormous advances over the last years".

These stories are enacted and shared as long as elicitation of the problems happens by chance or intention through a medical consultation or small talk. Furthermore, they confirm somewhat of the global diffusion of malignant lymphoma (the fifth most widespread cancer in the US,

Jhanwar and Straus 2006) despite the most popular oncological pathologies covered by the mass media and scientific divulgation usually not including hematological cancer.

Ultimately, "healing stories" are not BOs (Star 2010), and also their definition as *intermediary object* (Vinck 2009) is doubtful. However, stories have a flexible part (the individual patient's history and condition) accompanied by a stable core (the "healing" happy ending). It is this stable core that makes them recognizable and functional to be exchanged and used between old and new patients, and their close relatives and friends.

2. Contingency and vulnerability of infrastructures as ecologies

In all of the stories presented above, there is the intervention of a contingency: a set of conditions, contextual circumstances and occurrences which act as a "trigger" to unveil the relational texture of infrastructure and its vulnerable dimensions or relational components, them being material artifacts (watertight doors), organizational procedures (emergency protocols), a research problem aimed at minimizing the impact of contingency in future infrastructures (the sensor information classifier), specific groups or social worlds (individual patients as subjects made vulnerable by the illness)

What is, therefore, the relationship between contingency and vulnerability? If contingency is an event capable of changing, intervening or interfering in a process, vulnerability can be seen as the lack of capability in adapting to circumstances.

Vulnerability is not an "absolute" or "standardized" measure: it is linked with social context and national or local technological cultures (Bijker 2006, 2010) and it is also bound according to different disciplinary traditions[8] so that the environmental vulnerability is not exactly the same as the social one.

> "Indeed, for both natural and technological disasters the protection and response measures are often similar. From this perspective, a natural disaster, in a pure sense does not exist; rather there is the interaction of changes in physical systems with existent social conditions. The disaster itself occurs within society and not within nature" (Weichselgartner 2001, 86).

---

[8] See Weichselgartner 2001 for a comprehensive review of the concept in the field of environmental disasters.

To the aim of the present arguments, the core of the concept of vulnerability is double-sided: on the one hand, it is related to the "exposure" of a system or infrastructure to unforeseen events and damages (as in the Costa Concordia case); on the other hand, it is the degree of weakness or the sensitivity of individuals and groups to losses due to social or natural hazards (as in the case of NHL patients), referring to

> "the characteristics of a person or group and their situation that influence their capacity to anticipate, cope with, resist and recover from the impact of a natural hazard (an extreme natural event or process). It involves a combination of factors that determine the degree to which someone's life, livelihood, property and other assets are put at risk by a discrete and identifiable event (or series or 'cascade' of such events) in nature and in society" (Wisner *et al.* 2003, 11).

In particular, the following definition of vulnerability seems very adequate to illustrate the case of infrastructures:

> "Vulnerability is the degree to which a system acts adversely to the occurrence of a hazardous event. The degree and quality of the adverse reaction are conditioned by a system's resilience (a measure of the system's capacity to absorb and recover from the event)" (Timmerman 1981, in Weichselgartner 2001, 88).

This definition includes a crucial dimension in infrastructure's ontology, that means its resilience and resistance to adverse contingencies and the capacity to maintain itself while adapting to change according to circumstances (*see also* section 3 *in this chapter*).

The gap of adaptation and subsequent inadequate response, therefore, goes hand in hand with exposure and fragility at facing unforeseen or unpredictable events (e.g. the Costa Concordia accident or other cases of systemic disruption in complex sociotechnical arrays).

On the other hand, fragility of infrastructures goes back to the metaphor of ecological texture (Bowker and Star 1999; Star and Bowker 2006) and to those characteristics of an ecology devoted to maintain a dynamic equilibrium. When analyzing information as ecology, five elements characterize the use of ecology as a metaphor according to Nardi and O'Day (1999): system, diversity, co-evolution, keystone species, locality. Using this metaphor in a loosely-bounded perspective (Pellegrino 2008) three of the five elements above are selected as crucial with reference to infrastructure: *diversity*, *keystone species* and *locality*.

Diversity is the stratification and co-existence of different relations and components on the installed base of the infrastructure. This implies that contradictions can emerge, as well as opportunities for enhancement of the infrastructure itself. It is the case of different classifications and protocols for treatment of NHL, but also the continuous adjustments and approximation to build up the sensor technology classifier.

Keystone species are crucial to the survival and maintenance of the equilibrium of the ecology: without them, the existence of the ecosystem is at risk and subject to very tough transformations. Such a role in infrastructures is played by BOs, especially classifications and standards. Without them, infrastructures cannot be either installed or developed and used. The equilibrium of the (im)mobile ecology of a cruise ship depends on interdependent keystones (e.g. the captain's role, the crew skills, the watertight doors, the pilot house infrastructure), none of which is in itself sufficient to keep the Costa Concordia in dynamic equilibrium. The classifier and its relations is the keystone in the sensor technology infrastructure; the protocol and its networks in the case of NHL treatment.

Locality of ecology goes back to the temporal dimension in infrastructures, which is also important to situate them in a historical perspective and to understand their rich, potentially disruptive, political meanings and functions (see Bowker and Star 1999; Timmermans and Berg 2003). The showboat practice and the missed enactment of appropriate security guidelines constitute the local dimension of the Costa Concordia accident. So it is the uniqueness of the infrastructure of care in NHL, its "local universality" (Timmermans and Berg 1997), and the same locality that characterizes the research carried out to build up the sensor-based classifier.

To sum up, contingency can emphasize vulnerability of groups and exposure of infrastructure to break down or transformation. This process is led by the role of BOs in governing or contrasting contingency. It depends on the combination of flexibility and consistency which is the core of BOs, which in the following section I define as their "ductility". Such a term identifies the flexibility through which contingency can be governed and not simply ignored (also in its potentially fruitful consequences), keeping BOs consistent and autonomous, therefore recognizable by different groups.

3. The ductile door to infrastructuring: BOs between flexibility and consistency

Since Star and Griesemer's seminal article on BOs (1989), the term has entered the vocabulary of Science and Technology Studies (STS), proofing itself as flexible and adaptable as the concept it depicts.

In their analysis of the Museum of Vertebrate Zoology at Berkeley, a historic case of information science in a multi-disciplinary professional community, Star and Griesemer analyze "the problem of common representation in diverse intersecting social worlds" (Star and Griesemer 1989, 388). By definition, BOs are

> "both plastic enough to adapt to local needs (…) yet robust enough to maintain a common identity across sites. They are weakly structured in common use and strongly structured in individual-site use" (Star and Griesemer 1989, 393).

Even if standardized forms are the most common and widespread BOs, there are also other types (repositories, ideal types, coincident boundaries; Star and Griesemer 1989, 410-411). Functions provided by BOs are, therefore, translation, coherence, cooperation and control across different sites and inside organizations. The concept has become somehow popular in the field of information management, where functions of negotiation and control with reference to organizational identity have been emphasized (see Gal *et al.* 2008; *see also* Miele *in this volume*).

However, BOs can also provide the crucial function of a "door" to the work of infrastructuring: not by chance, sets of different BOs shared by diverse communities of practice and common to multiple sites, memberships and discourses are named as "boundary infrastructures": "objects that cross larger levels of scale than boundary objects" (Bowker and Star, 297; *see also* Neresini and Viteritti *in this volume*). This inner (or upper) level of the in-built, sunk infrastructural environment allows to the governing of divergences and tensions emerging from diverse interpretation by different communities of practice or social worlds, without losing coherence to any of them.

Of the three cases illustrated above, classifications and protocols for NHL are a classical boundary infrastructure, which guarantee for consistent interpretations by individuals and collectives across multiple sites.

Instead both emergency procedures for cruise ships and the classifier for sensor technology are examples of BOs part of larger boundary infrastructures (e.g. ubiquitous computing).

To work well, BOs have to be built and used as ductile: ductility is a property of materials which do not lose their core characteristics but are able to adapt and take different shapes according to the environment or scope they are used for.

Similarly, resilience is the capability not to be destroyed by stressful events and occurrences, and to repair damages through a self-conservative capacity: in physics, resilience is the ability of a material to absorb energy under the pressure of an elastic deformation and release that energy upon unloading.

Ductility and resilience are maintained through the balance between multiplicity and naturalization: categorical work (Bowker and Star 1999) is the core of infrastructure, which makes cooperation and distribution of work possible. Categorical work is "partly about managing the mismatches between memberships and naturalization" (Bowker and Star 1999, 311); an interpretation has to be flexible *and* consistent for everybody and many at the same time, in order to work as a boundary object.

So contingency can create but also prevent from vulnerability, depending on how BOs provide appropriate and shared means for categorical work. In this respect, the absence of a shared vision and the fragmentation of command and control can end with disruption (the Costa Concordia or more notorious failures of complex sociotechnical systems, like the case of the Chernobyl accident). But contingency can also be "marginalized" and avoided – this being a goal to pursue rather than a concrete result-as in the design of ubiquitous environments based on sensor technologies. And still, contingency can be an embedded value for the infrastructure, like in the case of NHL classification and treatment. In any case, the management of anomalies constitutes a tracer:

> "Anomalies or interruptions, the cause of contingency, come when some person or object from outside the world at hands interrupts the flow of expectations" (Bowker and Star 1999, 311).

Rather than an exceptional event, therefore, contingency constitutes a frequent occurrence in a world where belongings and interpretations are multiple and variable. However, when contingency is not governed or "welcome", extreme cases of misinterpretation occur, generating vulnerability and fragility, but also peculiar marginalities, namely *orphans* and *monsters* (see Bowker and Star 1999). When any belonging or membership is denied, residual categories are black-boxed as "otherness" (rare illnesses are orphans difficult to adopt by both the clinical research and the pharmaceutical industry) whereas denial of multiple memberships and resistance to naturalization bring about monsters. These cases are far from being infrequent but this does not imply they are simply a "physiological by-product" of categorical work and infrastructuring. In fact, such a "naturalization" would prevent from seeking for alternative ways of interpreting, translating and naturalizing people and objects, transforming contingency into a "condemnation" rather than an opportunity for transformation and change of infrastructure. Also, vulnerability and fragility point to the crucial work of care and maintenance of the socio-material world (Denis and Pontille 2013), without which most infrastructure would not survive the wear and tear of time. On the other hand,

> "infrastructure work frequently entails frequent and ongoing articulation work in order to enable continued functionality" (Bowker *et al.* 2010).

This work is real time, invisible, performed in the face of the unexpected and able to accommodate unanticipated contingencies (Bowker and Star 1999, 310). This is the hidden texture which allows us to face contingency as a "natural" occurrence in the infrastructure. If BOs are too flexible or too rigid, the texture of articulation work is also compromised and damaged.

Classification work is challenged when contingency re-classifies situations and standards, so exposing the infrastructure to the unpredictability and uncertainty of coping with "new" events.

"Ordinary" and established classification work cannot suffice to coping strategies, and may be restructured because of peculiar circumstances to which classification work has to be adapted and flexibly "tailored" or "curved". New routines in the repertoire or even new repertoires can enrich and enlarge classification work, making contingency an occasion and a motive for mobilizing resources devoted to maintenance and repair of infrastructures. So disruptive contingencies (like in the Costa Concordia disaster) are the result of the invalidation of usual classification work. Similar to the case of a big transport infrastructure, ubiquitous computing design attempts to foresee the most of possibilities and potential occurrences in the stage of the classifier construction. In this way, the infrastructure tries to embed and figure out the landscape of wanted and expected changes. On the other hand, where contingency is continuously solicited by the conditions of the infrastructure (as in NHL patients' vulnerability), classification work is able to welcome changes occurring under recurrent though diverse circumstances.

Temporally, therefore, categorical work is enacted differently upon contingency. It can occur in the stage of construction of infrastructure (ubicomp design), but also whenever heterogeneity occurs (NHL treatment), or upon breakdown and malfunctioning (Costa Concordia).

4. Final remarks: ductility for resilience?

Some final although tentative remarks can be pointed out at the end of this analysis of contingency in infrastructures.

Looking at infrastructure as *epoché*, its enabling character emerges clearly along with its standing-in-the-background, sunk and in-built constitution. The phenomenological standpoint allows the tracing of trajectories of evolution in infrastructures according to the role played by BOs in enacting and interacting with the texture of categorical work under particular circumstances, namely when infrastructure is faced with some kind of contingency. In most cases contingency can be managed through a more or less ordinary articulation work, but this is not always the case. As in the peculiar accident of the cruise shipwreck, this articulation work can fail because of the lack of translation, cooperation and coherence, that means the crucial functions played by BOs. Infrastructures in the making are built upon other kinds of work, a long work of classification and distinction of relevant events which attempts to avoid and foresee contingency through exhaustive and consistent BOs (e.g. the classifier for sensor technology in ubiquitous infrastructure). Local universality and embodiment of contingency are instead typical of boundary infrastructures like medical classifications and protocols (as shown in the story of NHL treatment).

In any case, different social worlds, communities and actors have to negotiate, cooperate and confront with objects which have to be naturalized.

The most interesting things happen when this naturalization is impeded, the ecology of infrastructure is damaged, and new categorical work can occur. BOs are decisive in defining trajectories of translation, cooperation and coherence, depending on how the balance between flexibility and consistency, or what I have named as "ductility", is managed by single individuals in specific sites and over larger scale communities. Too loose interpretations, or too rigid ones contrast with this ductility and prevent infrastructures from facing with contingency, increasing vulnerability and fragility across them.

If infrastructures are ecologies, their equilibrium depends on ductility of BOs taking part in them, and this ductility is a sort of pre-requisite or precursor of their global resilience. In other words, infrastructuring is a continuous process of the re-shaping of BOs through articulation and categorical work, which can benefit from contingency and enhance the infrastructure, or end with vulnerability, fragility and disruption of it.

Since resilience is a self-conservative capacity, is it possible to build it in infrastructures by enhancing ductility of BOs, sensitizing social worlds to accommodate contingency by taking care of the most vulnerable and fragile species of the ecology? And beside the metaphor of ecology, are there other ways to depict and inquire infrastructure in its evolution and trajectories? What about the metaphor of the "immune system", a conceptual hybridation to describe the potential for self-repair, maintenance and resilience of infrastructures?

In this respect, few considerations can be drawn from the three cases discussed in this contribution. Far from a normative orientation, STS calls for a beyond-descriptive approach, where attention to policy and politics is practiced to make research results operationally relevant.

Especially from the Costa Concordia case, it can be inferred that the best way to overcome vulnerability is to orient infrastructures towards *resilient invisibility*. The opposite of vulnerability is not obduracy but smoothness of working, achieved through accuracy of design and respect of ecological texture of the infrastructure. Therefore, even one of the biggest and most impressive transport infrastructures – the cruise ship – needs the invisible work of coordination and negotiation ensured by BOs. This invisible work is more evident in the design of sensor technology infrastructures and in the case of NHL medical protocols. Then how would transport infrastructures and logistics look when inspired by "lighter" infrastructures like sensors and ubiquitous computing, or local universes like the situated treatment of globally standardized pathologies?

Amongst others, these possible directions for further research are not exempt from risks of determinism and value-based judgments concerning the design and use of infrastructures. However, these risks may be worth running as it is undisputable that most of our current and future everyday life relies on diverse and complex arrays of enabling resources. Designing for preserving their multiplicity and ductility could be of the greatest importance to shape forthcoming sociotechnical environments.

References

Armitage, James O. 2005. "Staging Non-Hodgkin Limphoma." In *CA A Cancer Journal for Clinicians*, 55: 368–76.

Armitage, James O. and Dennis D. Weisenburger. 1998. "New Approach to Classifying Non-Hodgkin's Lymphomas: Clinical Features of the Major Histologic Subtypes." In *Journal of Clinical Oncology*, 16(8): 2780-2795.

Bioncology. 2010. *Non-Hodgkin's Lymphoma. A histopathologic and prognostic evaluation*. USA: Genentech.

Bijker, Wiebe E. 2006. "The Vulnerability of Technological Culture", in *Cultures of Technology and the Quest for Innovation*, edited by Helga Nowotny, 52–69, New York: Berghahn Books.

—. 2010. "Different forms of Expertise in democratising Technological Cultures. Experiences from the current societal dialogue on

Nanotechnologies in the Netherlands." In *Tecnoscienza Italian Journal of Science and Technology Studies*, 1(2): 121-132.

Bowker, Geoffrey C. and Susan L. Star. 1999. *Sorting Things Out: Classification and Its Consequences*. Cambridge, MA: MIT Press.

Bowker, Geoffrey C., Karen Baker, Florence Millerand and David Ribes. 2010. "Toward Information Infrastructure Studies: Ways of Knowing in a Networked Environment." In *International Handbook of Internet Research*, edited by Jeremy Hunsinger, Lisbeth Klastrup and Matthew Allen, 97-117, New York: Springer.

Crabtree, Adam and Tom Rodden. 2007. "Hybrid ecologies: understanding cooperative interaction in emerging physical-digital environments." In *Pers Ubiquit Comput*, January.

Denis, Jerôme and David Pontille. 2013. "Material Ordering and the Care of Things." In *CSI Working Papers Series*, 34.

Dowling, Ross K. 2006. *Cruise Ship Tourism*. Wallingford: CABI.

Feller, Alfred C. and Jacques Diebold. 2004. *Histopathology of Nodal and Extranodal Non-Hodgkin's Lymphomas*. Berlin-Heidelberg: Springer-Verlag.

Gal, Uri, Kalle Lyytinen and Youngjin Yoo. 2008. "The dynamics of IT boundary objects, information infrastructures, and organisational identities: the introduction of 3D modelling technologies into the architecture, engineering, and construction industry." In *European Journal of Information Systems*, 17(3): 290-304.

Greenfield, Adam, 2006. *Everyware: The dawning age of ubiquitous computing*. Berkeley: New Riders.

Halaas, Jeffrey L., Craig H. Moskowitz, Steven Horwitz, Carol Portlock, Ariela Noy, David Straus, Owen A. O'Connor, Joachim Yahalom and Andrew D. Zelenetz. 2005. "R-CHOP-14 in patients with diffuse large B-cell lymphoma: feasibility and preliminary efficacy." In *Leuk Lymphoma*, 46(4): 541-47.

Hernandez, José, Jo Elle Cottey Krueger and Eli Glatstein. 1997. "Classification of Non-Hodgkin's Lymphoma: A Proposal." In *The Oncologist*, 2: 235-44.

Hernandez-Ilizaliturri, Francisco J. 2013. "Diffuse Large B-Cell Lymphoma (Non-Hodgkin Lymphoma) Treatment Protocols." http://emedicine.medscape.com/article/2005945-overview. Accessed February 2014.

Lindauer, Michael, 2002. "When Down Isn't Down." In *Entertainment Design*, June: 26-33.

Minella, Massimo. 2012. "Quel gigante dal cervello hi-tech così fragile se si avvicina alla costa."

http://ricerca.repubblica.it/repubblica/archivio/repubblica/2012/01/15/quel-gigante-dal-cervello-hi-tech-cosi-fragile.html?ref=search. Accessed April 15, 2013.

Nardi, Bonnie A. and Vicky L. O'Day. 1999. *Information Ecologies. Using Technologies with Heart.* Cambridge, MA: The MIT Press.

Orr, Julian. 1996. *Talking about Machines*. Ithaca: Cornell University Press.

Pellegrino, Giuseppina. 2008. "Convergence and Saturation: Ecologies of Artefacts in Mobile and Ubiquitous Interaction." In *Integration and Ubiquity. Towards a philosophy of telecommunications convergence*, edited by Kristof Nyíri, 75-82, Vienna: Passagen Verlag.

—. 2011. "Timescapes and Sociotechnical Infrastructures: Current and Future Ubiquity of Communication." In *Sociologia@DRES, Quaderni di ricerca*, 2: 4-13.

Schütz, Alfred. 1945. "On multiple realities." In *Phenomenology and Social Research*, 5(4): 533–76.

Schütz, Alfred and Thomas Luckmann. 1973. *The Structures of the Life-World*, vol. I. Evanston: Northwestern University Press.

Star, Susan L. 1999. "The Ethnography of Infrastructure." In *American Behavioral Scientist*, 43(3): 377-391.

Star, Susan L. and James R. Griesemer. 1989. "Institutional Ecology, "Translations", and Coherence: Amateurs and Professionals in Berkeley's Museum of Vertebrate Zoology, 1907–1939." In *Social Studies of Science*, 19: 387–420.

Star, Susan L. and Karen Ruhleder. 1996. "Steps toward an ecology of infrastructure: Design and access for large information spaces." In *Information Systems Research* 7: 111–34.

Star, Susan L. and Geoffrey C. Bowker. 2006. "How to infrastructure." In *Handbook of New Media*, edited by Leah A. Lievrouw and Sonia M. Livingstone, 230-245, London: Sage.

Suchman, Lucy. 1987. *Plans and Situated Action: The Problem of Human–machine Communication.* Cambridge, MA: Cambridge University Press.

—. 2000. "Organizing Alignment: A Case of Bridge–Building." In *Organization*, 7: 311–29.

Timmerman, Peter. 1981. *Vulnerability, Resilience and the Collapse of Society: A Review of Models and Possible Climatic Applications.* Toronto: Institute for Environmental Studies, University of Toronto.

Timmermans, Stefan and Marc Berg. 1997. "Standardization in Action: Achieving Local Universality through Medical Protocols." In *Social Studies of Science*, 26: 769–99.

—. 2003. *The Gold Standard: The Challenge of Evidence–Based Medicine and Standardization in Health Care*. Philadelphia: Temple University Press.

Vinck, Dominique. 2009, "De l'objet intermédiaire à l'objet-frontière." In *Revue d'anthropologie des connaissances*, 3(1): 51-72.

Weichselgartner, Juergen. 2001. "Disaster mitigation: the concept of vulnerability revisited." In *Disaster Prevention and Management*, 10(2): 85-95.

Weiser, Mark. 1991. "The Computer for the 21st Century." In *Scientific American*, September: 94-104.

Winner, Langdon. 1977. *Autonomous Technology*. Cambridge, MA: The MIT Press.

Wisner, Ben, Piers Blaikie, Terry Cannon and Ian Davis. 2003. *At Risk. Natural hazards, people's vulnerability and disasters*. Second Edition. London: Routledge.

Acknowledgements

The first stage of research on ubiquitous computing was carried out in Austria while visiting the IAS-STS in Graz (2008), thank to the funding of the Regional Council of Calabria under the European Programme POR "Voucher programmi di alta formazione rivolti a ricercatori calabresi".

The second stage of the same research took place in Germany during the visiting fellowship funded by the "Topologie der Technik" Program at the Technical University in Darmstadt in 2012.

Alessandro Mongili, friend and co-editor of this book, always encouraged and supported me to contribute to this common project.

Many thanks to my colleagues and friends Assunta Viteritti and Mauro Turrini, for their precious comments on different stages of this chapter.

My deepest gratitude goes to doctors, nurses and operators of the Hematology Day Hospital in Rome, who are still accompanying me. I dared to reflect on their skilful competence and compassion. Without them this chapter would not exist as such.

CHAPTER THREE

INFRASTRUCTURE INNOVATION IN HEALTH CARE: THE EXAMPLE OF AUTOMATIC DRUG DISPENSING IN GERMANY

STEFAN KLEIN AND STEFAN SCHELLHAMMER

Introduction

Comprehensive medication management for polypharmacy patients has been recognized as a key area of health care in need of improvement and innovations: it affects a growing number of patients, has huge financial implications and ties into broader issues such as patient health and medication safety, medication records, coordination across different medical specialists and cooperation between medical doctors and pharmacists. It fits the characteristics of an infrastructure as a constellation of practices.

Given the prevalence of national regulation, we have been studying the public discourse on about improving medication management in Germany over the course of the past 10 years. Specifically, this chapter investigates the development of an initiative to introduce industrial automatic dose dispensing (ADD), i.e. the large scale production of patient specific blister packs for solid oral medicines, as an infrastructure innovation in the German healthcare system. We trace the arguments of both opponents and proponents of this innovation and theorize the outcome so far.

ADD has been introduced in a number of countries, namely Northern Europe, the Netherlands, the US, Canada, Japan and Australia (Bell *et al.* 2013). While some of the effects of ADD are contested, overall it has been positively assessed. Since regulatory preconditions for the industrial creation of ADD in Germany have been put in place, trials were run by two industrial ADD operators. Additionally, a number of smaller blister centers have been set up by pharmacies at a regional level. Yet, in contrast to other countries, Germany still lacks a regulatory and reimbursement framework to facilitate routine use of ADD at a national level. The goal of this chapter is to provide explanations for the impediments of infrastructure innovation in the field of medication management and to illustrate ways of theorizing that can be transferred to other areas as well.

Actors in the health care sector have sided either with the opponents or proponents of ADD as a solution to medication management. Due to this situation, we find that attempts to introduce ADD as a solution to existing problems with medication management are unlikely to succeed on a national level anytime soon.

Our analysis of the discourse that has unfolded over recent years reveals a number of challenges characteristic of infrastructure innovation: (1) The ADD initiative illustrates the tension between design and emergence in infrastructure innovation. (2) Digital disruption (or disruptive innovation) plays a major role in explaining the reluctance to infrastructure innovation. (3) Infrastructure innovation as institutional innovation relies heavily on collective action. As such the discourse is characterized by the forming of coalitions striving to frame the debate and gain ownership of meanings. Yet, the complexity and dynamics of infrastructure evolution put tight constraints on our ability to forecast the impacts of particular designs. Our explanation thus rests on two pillars: the contested issues in the public debate illustrate different approaches of sense making and framing among opponents and proponents of the solution. At the level of institutional innovation we can interpret the debate as part of a political process to facilitate or inhibit regulatory changes.

The chapter is structured as follows: The next section explains the rationale for innovating medication management in Germany. The second section introduces the theoretical lenses and explains the methodological approach. Section three provides an overview of building blocks of the key ADD proposal and a timeline of events from 2003 to 2013. The fourth section reconstructs the discourse on medication management during the named period. The discussion section brings theoretical lenses and findings together.

1. Need for innovation in medication management

Medication safety and compliance are major issues in the management of medication for patients on a complex medication regime (polypharmacy patients). Polypharmacy patients are patients who regularly have to take 4 or more distinct types of medication and they are typically suffering from diseases such as coronary heart disease, congestive heart failure, hypertension and diabetes mellitus. Medication safety addresses specifically adverse drug reactions and critical interactions among medicines. Compliance or adherence[1] focuses the patients' behavior in particular in long-term medication therapies.

In order to foster compliance, dose administration aids (DAA) have been designed such as the 7 * 4 pill box, with separate compartments for the taking of medication 4 times per day. However, from the patients' or caretakers' point of view, multi-dose drug dispensing (MDD), i.e. filling pill boxes, is a tedious and therefore error-prone process (Lauterbach *et al.* 2007). Adherence is in particular a problem for chronically ill elderly patients which constitute the largest group of polypharmacy patients.

Given the sheer number of medication and over-the-counter drugs (OTC) there is a high risk of critical interactions. Adverse drug reactions and critical interactions among medicines can often be identified and resolved before the actual administration of the drug takes place. However, accurate identification of risks relies on complete information about the current and past medication regime of the patient. At the moment, different specialists may prescribe medication with little or no coordination between them.

The response to this set of problems across different health care systems varies. Yet, there is a broad consensus about the key components of a solution:

1. A comprehensive patient medication record to document a patient's medication.
2. A control for critical interactions based on the medication record.
3. Monitoring of the medication effects over time: it has sometimes been suggested that doctors and pharmacists do this collaboratively.
4. Dose administration aids to support patients and their helpers to follow the medication regime (dosage and timing).

While local practices have been developed in response to obvious needs, none of these issues have been systematically addressed in Germany so far. First, there is currently no mechanism in place to share medication records among health care professionals on a routine basis. Second, each health care professional is obliged to control for critical interactions. Given the sheer number of medication and the fragmentation of medication records, errors are likely to occur. Third, there is no clear

[1] Adherence is the broader concept, which encompasses acceptance (redeeming the prescription), persistence (continuing the medication therapy) and compliance (following the prescriber's instructions) (Düsing 2006, 11). Throughout this paper we will use adherence and compliance synonymously.

division of labor between pharmacists and doctors regarding the monitoring of medication effects over time. Both professions rely on the vigilance of patients and their helpers. Fourth, most dose administration aids require individuals to single out pills from larger packages. While this manual process is already error-prone in itself, it represents an even greater challenge for non-professionals like patients and their helpers.

2. Theoretical and methodological framework

In order to conceptualize the suggested improvements of medication management in Germany we use a combination of infrastructure and institutional innovation lenses. Both address changes at the level of heterogeneous actor constellations. The notion of infrastructure provides conceptual tools to deal with complex constellations of heterogeneous actors and technology. Specifically, we conceptualize infrastructures as constellations of practices. As health care is a highly regulated industry, institutional innovation provides conceptual tools to study processes of institutional and regulatory transformation. It looks at the development of networks or coalitions in order to bring about change when individual actors do not have sufficient resources, power, or legitimacy to do so by themselves (Hargrave and van de Ven 2006).

2.1 Infrastructure as a constellation of practices

The health care sector has witnessed a plethora of initiatives aiming to improve quality and efficiency of care. Most of these initiatives employ ICT as a means to interconnect health care providers. These changes do not only aim to automate existing practices of data exchange, rather they allow for and indeed require new forms of collaboration among health care professionals.

While the discourse about information or digital infrastructures emphasizes the specifics of digital goods and markets, a major part of infrastructure innovation takes place in the field of informated infrastructures, i.e. traditional infrastructures, such as health care, are transformed in conjunction with and through an increasing use of ICT. In other words we observe processes of the reciprocal shaping of traditional and information infrastructures. The information infrastructure affords visibility of physical flows, documentation of financial flows and administrative information, and control and analysis across all the information collected and stored. Machine readable information such as barcodes become embedded in the physical infrastructure in order to facilitate monitoring and control. From a design point of view, both infrastructures need to be integrated and aligned. From a usage point of view they are typically perceived as one in the specific area of application (health care).

From a historical point of view, infrastructures are emerging phenomena. They typically result from the scaling of services to a higher level. Energy was initially generated locally; but over time regional, national and even international infrastructures have been built, which allow for more efficient provision of energy. Similarly the proposal to establish industrial ADD implies the scaling of local or regional practices to a national level. In this way infrastructures have been characterized as hallmarks of modern societies (Edwards 2003).

Infrastructures are not created on a green field but evolve under the impression of a socio-technical installed base (Aanestad and Jensen 2011). Installed base is not to be mistaken for artifacts and technologies. Star and Ruhleder (1996) conceive infrastructure as a fundamentally relational concept in order to stress that infrastructure emerges in relation to organizational practices. The interconnection and mutual alignment of multiple, heterogeneous actors suggests conceiving infrastructures as constellations of practices (see Wenger 1998). It is an assembly of distinct, yet interrelated practices that together give shape to the medication management infrastructure in Germany. The particular shape is an expression of an ongoing (re-)definition of boundaries between communities.

In line with Monteiro *et al.* (2013) and Star and Ruhleder (1996) we conceive medication management as infrastructure exhibiting the following general characteristics:

- **Shared**: medication management relies on a complex configuration of technological artifacts and organizational arrangements: production facilities, electronic and paper-based documents and their exchange, dedicated software as well as organizational rules and routines;
- **Openness and heterogeneity**: Medication management is not only a multi stakeholder arrangement but is open to a wide range and number of users: a constellation of patients and their helpers, physicians, pharmacists, hospitals, nursing homes, nursing companies, health insurance companies;
- **Heterogeneous and evolving installed base**: Medication management brings together a range of different practices of health care professionals and other actors in the sector. It draws on standardized components (prescriptions) and local adaptations

(dose administration aids). This web of relationships between different actors and artifacts brings forth a constantly evolving constellation of practices for medication management. Medication management is nested in the larger infrastructure of the health care sector. It is one aspect of the role and division of labor between the actors.

The dynamics of infrastructure development have been characterized as a combination of emergence and design; on one side, infrastructures are built on an "installed base" and are subject to conscious design and planning; on the other side, infrastructure development is inherently emergent as infrastructures become appropriated and used by a large variety of types and groups of users and for a broad range of uses, often beyond the imagination of the designers. Wenger refers to the two sides as "designed organization" and "lived organization" (Wenger 1998, 241). He conceives "design" as a systematic, reflexive and planned endeavor to achieve control over the creation of artifacts or social processes (*Id.*, 228). Practice on the other hand does not lend itself to design. Instead, practices emerge in response to design. One can envision and design the roles of health care professionals but their identities and their practice emerge outside the control of the design effort.

> "Unexpected adaptations of the design [in practice, ...] do not necessarily indicate a lack of specification. In fact, they may very well indicate a healthy response […]." (*Id.*, 233)

From a practice perspective we can look at infrastructure as background which not only enables the seamless performance of routines but also functions as a reference point for sense making. Infrastructure comes to the foreground if it breaks down. It can be brought to the foreground during periods of development, innovation and transformation. We will look at infrastructure development as the development of a constellation of practices or communities of practice (Wenger 1998). In our case, infrastructure innovation, design and emergence show up in the design of material, normative and ideational structures, in the transformation of established practices, including practices designed to connect and align other practices, and in the (re-)alignment of practices (Reimers, Johnston and Klein 2012).

The digitalization or informating of infrastructures in healthcare affords new practices at the individual level and up to a global scale. It is not a mere extension of existing practices into the digital realm. Riemer and Johnston (2013) have extended the concept of disruptive innovation by Christensen–they call it digital disruption-and linked it to the notion of infrastructure. As digitalization and informatization in health care advance, new business models and new ways to deliver care can emerge. In this way new infrastructures, such as mobile communication, enable new ways of engaging with the world and allow us to view ourselves and our environment differently. Infrastructure innovation represents a disruption of a relatively stable arrangement (constellation) of practices, implying a process of re-arranging, re-positioning of existing actors alongside potential new entrants. For that very reason, forecasting the effects of infrastructure innovation is quite limited as it changes over time what we take for granted.

2.2 Institutional innovation

Hargrave and van de Ven (2006) have developed a collective action model – based on a synthesis of technology innovation literature and social movements literature – to explain institutional innovation as a result of dialectical processes in which the issues at hand are framed differently and coalitions are formed in order to mobilize support in a contested political arena. The model

> "views institutional change as a dialectical process in which partisan actors espousing conflicting views confront each other and engage in political behaviors to create and change institutions." (Hargrave and van de Ven 2006, 864).

It takes a historical perspective in order to explain the temporal sequence of events and looks for generative mechanism.

Framing contests highlight attempts by opposing actors to control the issue, to provide explanations and reasoning in line with their own goals and interests.

The **construction of networks** or coalitions reflects the needs of individual actors to extend their resource base and political influence. Scale and scope of the innovations are such, that individual actors cannot enact them and need to find support.

Enactment of institutional arrangements refers to the need to shape regulation, standards or other parts of the institutional infrastructure in order to be able to enact and stabilize the intended change.

Finally the **collective action process** describes the overall sequence of events and initiatives, the ongoing dynamics and dialectic.

In our view, Hargrave and van de Ven's model is useful to reconstruct and highlight the dialectical dynamics of infrastructure innovation.

2.3 Method

This chapter is based on secondary research. It looks at published studies, Web sites, blogs, discussion fora, policy briefings and media coverage results about ADD in Germany over a period of more than ten years as a basis for reconstructing the debate. While we are aware that a significant part of the debate has happened behind closed doors, we are particularly intrigued in the publicly visible part of the controversy. Moreover, we have included a number of international studies in order to contextualize the debate.

Both authors have done research in the field of health care for years and have thus developed some familiarity with the industry.

3. ADD as a proposal for infrastructure innovation

Based on a brief introduction of the building blocks of ADD, we will sketch a timeline of events triggered by 7x4 Pharma's proposal and pilot for Automatic Dose Dispensing.

3.1 Conceptual building blocks of Automatic Dose Dispensing

The concept of automatic dose dispensing (ADD) changes medication management in a number of ways. ADD builds on established practices of arranging medicines in pill boxes. The widely used 7x4 pill box has four separate compartments for pills. Each compartment may contain several pills which are to be taken at the same time during a day (morning, noon, afternoon, evening). From a patient's perspective ADD replaces the 7x4 pill box with sachets or blister packs each of which contain the pills of the pill box compartments. These sachets are produced and sealed on an industrial level. Thus, on the most basic level, ADD removes the manual provision of pill boxes and augments the blister packs with information about patient, medication and schedule for administering the medication. In order to do so ADD needs to be embedded in the larger context of medication management involving a multiplicity of actors. In order to operate on an industrial level, ADD assumes:

- electronic information exchange between ADD operator and physician or pharmacy (some kind of *electronic prescription*);
- a *medication list* (depending on the model of implementation);
- a patient specific, comprehensive *medication plan*.

The creation and exchange of these documents implies a transformation of existing practices and involves – apart from the ADD operator – patients, physicians, pharmacies, possible pharmaceutical wholesalers, caretakers or nursing homes and health insurances as illustrated in fig. 3-1.

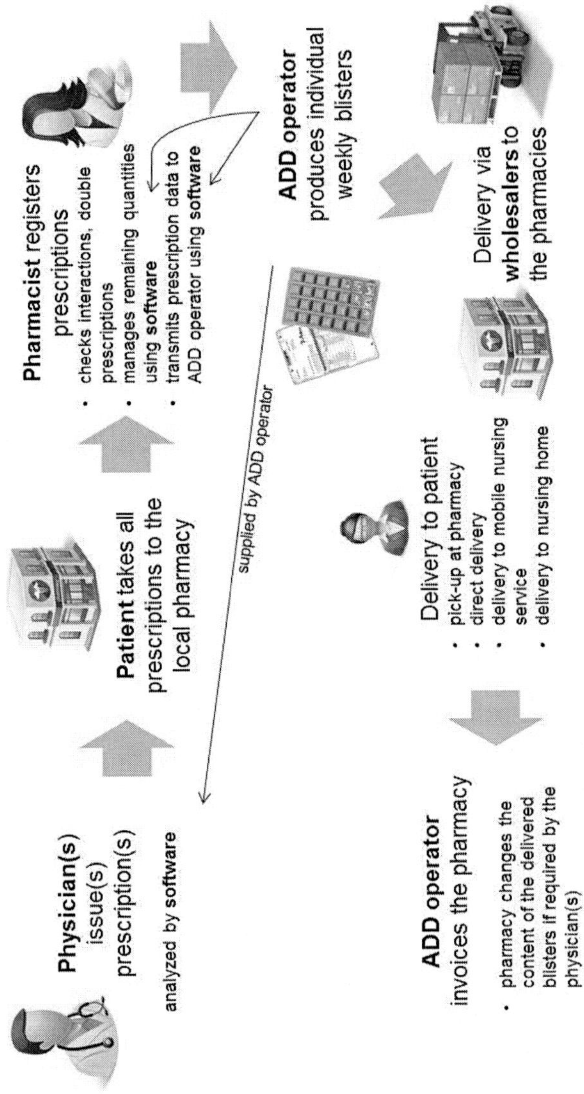

Fig. 3-1: ADD process flow (based on Kohl 2010)

3.2 The history of ADD in Germany – from concept to initiatives

In 2003, Kohl Medical AG, a pharmaceutical wholesaler, founded 7x4 Pharma in order to produce weekly blister packs for polypharmacy patients at a large scale. In 2005, the regulatory preconditions for the industrial creation of ADDs were created. Subsequently, two industrial ADD operators – 7x4 Pharma and AvidiaMed – have set up production sites and run trials. In parallel a number of blister centers have been set-up by pharmacies at a regional level, which produce blisters for a small number of participating pharmacies. Moreover, a number of pharmacies offer the (manual) production of DAAs as an additional, usually complimentary service.

Between 2004 and 2007 a number of studies – commissioned by Kohl Medical AG, 7x4 Pharma's parent company – were published, which examined different facets of ADD and provided the rationale for industrial ADD at a national level (Lauterbach *et al.* 2006, 2007; Lauterbach, Lüngen and Gerber 2004; Pharmafacts and Glaeske 2007; Lauterbach, Gerber and Lüngen 2004).

In 2006, Wille and Wolf (2006) published a study – commissioned by the association of research active pharmaceutical companies (VFA) – on the costs and benefits of secondary blisters, which concluded that ADD is neither cost efficient nor effective.

In 2009, the 7x4 Pharma facility, designed for the production of up to 100.000 weekly blister packs per day, went live (http://www.apotheke-adhoc.de/nachrichten/nachricht-detail/kohl-gibt-startschuss-fuer-7x4-box/). Subsequently pilot studies – based on industrial as well as regional ADD – were conducted in order to assess the effects of ADD in life settings.

At the beginning of 2011, the results of two pilot studies, one based on industrial ADD (Leker and Kehrel 2011), the other on blisters produced by pharmacies (Neubauer 2011; Neubauer and Wick 2011) in cooperation with health insurance providers, were published. The studies provide evidence that ADD contributed to improvements of both medication safety and compliance and cost saving of up to €31 per patient per week (Neubauer and Wick 2011).

In the Spring of 2011, KBV and ABDA published a proposal for improved medication supply in Germany, which addressed the same issues of compliance of polypharmacy patients and medication supply. While the proposal can be seen as complementary to ADD, it refrains from even mentioning the issue of drug administration.

In August 2011, the association of statutory physicians and the association of pharmacies for the state of Brandenburg (Landesapotheker- und Landesärztekammer Brandenburg 2011) issued a position paper, which assessed and rejected ADD. The association of patient individual blister companies (BPAV 2011) issued a critical and angry rebuttal. The association of pharmacies supplying nursing homes (BVKA (Schumbach 2013a)) has also articulated criticism against ABDA's blockade of ADD:

In November 2011 7x4 Pharma was sold. The fact that the ABDA-KBV concept paper (ABDA and KBV 2011a) has been published in 2011 and that 7x4 Pharma was sold during the same year may be more coincidence.

In mid 2013, the second industrial provider, AvidiaMed (AvidiaMed 2013), closed its ADD operation. A pilot study based on the ABDA-KBV proposal, originally scheduled to start at the end of 2013 has been delayed (Ziegler 2013).

Figure 3-2 provides an overview of the different organizations and associations that took part in the discourse: patients, pharmacies and physicians form the center of the medication management infrastructure. Only polypharmacy patients as a subgroup are in the focus of ADD initiatives. The pharmacies and physicians are represented in the discourse via their associations (ABDA, KBV – federal level, LAKBB, KVBB – state level). Pharmacies supplying nursing homes and clinics bundle their interests in a separate association on the federal level (BVKA). AvidiaMed and 7x4 Pharma are the most prominent companies offering ADD on an industrial level. Several regional operators exist as well. The federal association of patient-individual blisters (BPAV) and the Blisterunion are their representative bodies. Kohl Medical is the founder of 7x4 Pharma and operates as a pharmaceutical wholesaler and parallel importer in the German market. Manufacturers take part indirectly in the discourse via the representative body of research active pharmaceutical companies (VFA).

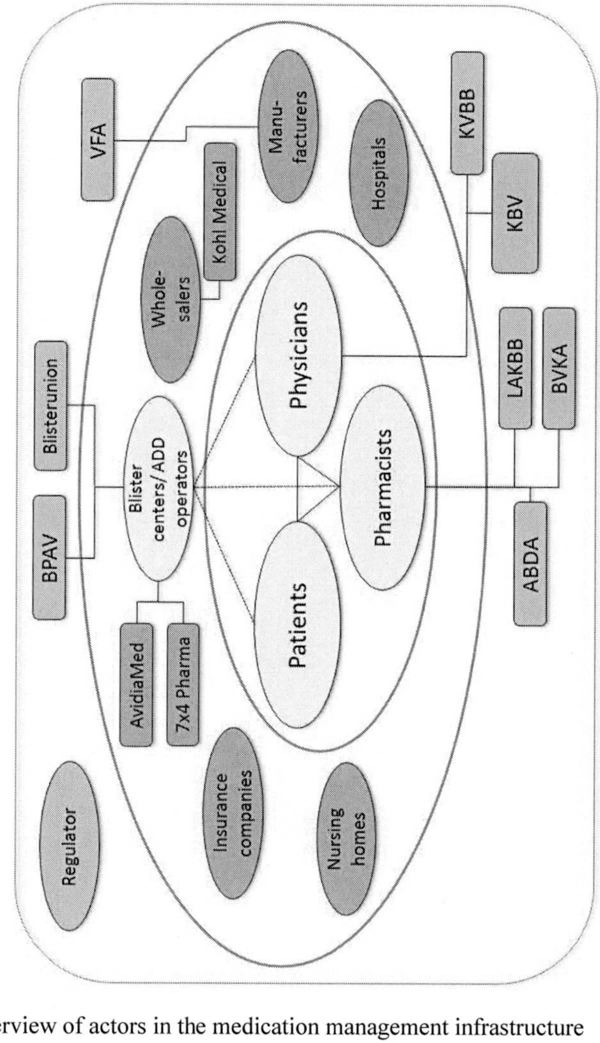

Fig. 3-2: overview of actors in the medication management infrastructure

4. Discourse on medication management in the German context

This section introduces the key arguments put forward by proponents and opponents of ADD respectively. The reports and studies mentioned previously serve as sources and means for justifying the arguments. For the purposes of this chapter we grouped the different claims according to the three problem fields mentioned earlier: compliance, safety, and efficiency. We conclude the section by describing the status quo of the debate.

4.1 Proponents of ADD – increased patient safety and economic efficiency

The key claims, which the proponents of industrial ADD make, are higher efficiency, higher levels of safety and higher levels of compliance (for a detailed summary of the arguments in favor of patient-specific blisters see also BPAV 2012):

The **compliance** (or adherence) claim is not specific to automatically generated DAAs, but applies to other, functionally similar DAAs as well. One could argue though that automatically generated DAAs draw on the established practices of 7*4 pillboxes and facilitate somewhat higher compliance as more patient specific and medication specific information can be printed onto the DAAs (see http://www.deutscheblisterunion.de/schlauchblister.htm). Moreover, once an ADD system has been technically and administratively introduced into a national health care system, a larger ratio of polypharmacy patients will have access to DAAs. Compliance addresses both the quality of health care and avoiding costs related to non-compliance.

The **safety** claim rests on the design and monitoring of the ADD systems in line with established manufacturing practices (GMP), including barcoding to assist tracking and tracing. In comparison, manual creation of DAAs be it by the patients, their helpers or in pharmacies, is more error prone. Moreover the claim also refers to related changes such as the introduction of a comprehensive patient medication plan as the DAAs are supposed to cover all solid oral medicines. The medication plan is not only a coordination device between physicians (prescribers), pharmacies, ADD operators, and perhaps patients, it also facilitates control over critical interactions between the prescribed medication and the patient specific blisters

The **efficiency** claim is based on the industrial production of DAAs in comparison to the creation of DAAs in pharmacies, nursing homes or by the patients and their caretakers. Additional arguments in favor of increased efficiency are a lower rate of discarded medication if medication packs are not fully consumed, e.g. due to a changed medication regime, and – depending on the design of the solution – lower prices of the medication due to preference to generic medication and larger packages.

The essence of the argument is the following claim; we can provide patient specific blisters efficiently and safely at a large scale. However, the claim is based on a number of preconditions: (1) It requires collaboration with patients, physicians and pharmacies, which implies several changes in their roles, linkages and practices. (2) It requires regulatory adjustments, so that the choice of patient specific blisters becomes a defined option, including rules for reimbursement.

Multiple stakeholders (patients, physicians, pharmacies, wholesaler/ logistics partner, blister unit) take part in the process flow. Each one of them needs to adhere to standards and processes for information collection, processing and exchange. In addition, existing practices of prescription and dispensing of medicines need to be extended. In order to effectively address the problem of compliance and safety, the proposal relies on a (complete) medication record (or plan). Efficiency and economic viability of the proposal require a limited assortment of 400-500 distinct medicines that can be prescribed and used for ADD. The economic viability of operating such a solution rests upon the ability to offer the service at a large (national) scale, which assumes a regulatory framework and standardization.

The proposal strives to gain credibility by arguing from an installed base of experience, production and distribution facilities. Studies, documenting pilot runs, have been commissioned to add credibility and legitimacy to these claims.

4.3 Opponents of ADD – a threat to quality healthcare

The opponents argue that each problem area – compliance, safety, and efficiency – will be negatively affected by introducing ADD.

In terms of **compliance** (or adherence) the study commissioned by the association of research active pharmaceutical companies (VFA) (Wille and Wolf 2006) finds no clear evidence that individualized blisters will increase adherence. A joint position paper by the associations of physicians and pharmacists of the State of Brandenburg (Landesapotheker- und Landesärztekammer Brandenburg 2011) argues that ADD violates principles of good medication therapy and dispensing. The limited assortment and bias towards solid oral medication might not suit elderly patients, who prefer fluid medicines. Furthermore, they claim that nursing homes would still need to break blisters in order to administer them as well as other forms of medication (fluids, salves).

With regard to **safety** the position paper sees additional risks associated with ADD. The additional process step of creating blisters by itself increases the risk as it increases the complexity of administering medication. Furthermore, it would delay the availability of medicines by up to two weeks depending on the schedule of the ADD operator. Moreover, changes in prescriptions would take weeks, while manual changes are not only cumbersome but risky. By forwarding patient specific data to the ADD operator, patients' privacy will be constrained – unless they explicitly consent. Moreover, concerns have been raised regarding the flow of information and manufacturers' liability issues. The study by Wille and Wolf (2006) finds it difficult to assess qualitative effects as a result of modified processes, roles and responsibilities between prescribing physicians, pharmacies, the ADD operator, patients and their caretakers.

Lastly, both, the study as well as the position paper cannot support the proponents' claim for higher **efficiency**. On the contrary, they highlight the threat of additional costs for the health care system. Wille and Wolf (2006) do not see a customer segment that is large enough. Calculated cost savings in the range of 0,28 to 0,59 EUR with a likely value of 0,35 EUR per blister are compared to the total assumed costs of production and distribution of at least 3 EUR. Furthermore, they point to the threat that Kohl Medical could achieve a monopoly position. In conjunction with the limited assortment of about 400 distinct medicines pharmacies would be forced (contractually and economically) to use 7x4 Pharma's services due to the price advantage it is able to offer (economies of scale). In addition to this rebuttal, the position paper disagrees on the economic value of repackaging because pharmaceutical manufacturers would already offer therapy adequate package sizes. In case pharmacies would be reimbursed for their efforts in an ADD scenario, customer specific blisters would be unlikely to reduce the overall costs. Instead, additional packaging waste would be caused.

In conclusion, the study by Wille and Wolf (2006) finds no economic benefits of industrial ADD, nor proof of the hypothesized increase in compliance. The limited assortment is viewed critically. In addition they raise the industry political concern that 7x4 Pharma might become a monopolist The position paper (Landesapotheker- und Landesärztekammer Brandenburg 2011) builds on and extends this critique.

4.4 Status quo and critical assessments of the claims

For the past 2-3 years, the Federation of German Associations of Pharmacists (ABDA) seems to have been avoiding even the discussion about ADD and more broadly reimbursement for DAA as if the case had been settled and was no longer an issue, despite ongoing attempts by the associations of blister producing pharmacies (BPAV 2011; BPAV 2012) and AvidiaMed (Breu and Grözinger 2012) to revitalize the public debate. Individual politicians support ABDA's position that blister creation is like a "dead horse" (Schumbach 2013a).

Instead, the representative bodies of physicians and pharmacists developed a joint proposal for reforming medication management in 2011. While it targets the broader issue of medication management, it avoids any reference to the former ADD initiative or DAAs, even though BPAV has claimed early on that the creation of blisters and monitoring of patients need to be combined (Schumbach 2013a). The joint concept paper on medication management by the Federation of German Associations of Pharmacists (ABDA) and the National Association of Statutory Health Insurance Physicians (KBV) (ABDA and KBV 2011a; ABDA and KBV 2011b) rests on three pillars: (1) a medication catalog, i.e. the specification of medicines for particular diseases in primary care, and explicitly no "positive list" (an indication of admissible medicines), (2) the prescription of active ingredients instead of branded medicine, and (3) patient specific medication management, i.e. creating a comprehensive medication plan, reducing unintended effects, e.g. due to critical interactions, and fostering compliance. Medication management is focused on chronically ill polypharmacy patients (taking at least five medicines, without OTC medicines), for whom a physician and a pharmacist will take responsibility for one year. They will consult the patient, create a comprehensive medication plan and will try to facilitate a high level of compliance. The candidates for medication management will be identified by a physician, a pharmacist or the health insurance company. Joint reimbursement is suggested at a level of €360 per patient per year. The concept paper addresses non-compliance (up to 50% for long-term medication therapies) and polypharmacy (26% of patients with statutory insurance coverage), including a high level of (additional) self-medication based on OTC medicines (22% of OTC medicines are for patients over 65, figures based on ABDA and KBV 2011b). A pilot implementation has been planned for Saxony and Thuringia starting in the second half of 2013.

Given the arguments directly opposing the ADD initiative and the exclusion of ADD as a possible part of the 2011 concept paper, it is obvious that the National Association of Statutory Health Insurance Physicians (KBV) and the Federation of German Associations of Pharmacists (ABDA) have clearly positioned themselves against ADD.

The termination of ADD operations by the two industrial ADD providers, the delayed pilot projects based on the ABDA-KBV concept and the reluctance of the regulator to take a more active stance make it unlikely that ADD will be introduced at the national level anytime soon.

5. Discussion

It is not the intention of this chapter to assess the arguments brought forward by proponents and opponents. Rather we conceive the different approaches to frame ADD and the need for innovation in medication management exemplary for challenges in infrastructure innovation in general.

5.1 ADD initiative as disruptive innovation (digital disruption)

Medication management in its current form depends on the interlinking of several actors. Together, the distinct but related practices of pharmacists, physicians, patients and others contribute to the performance of medication management. At first glance, ADD seems like an obvious component of a remedy to a widely recognized and accepted problem in health care, which is likely to become more important in the future. Given the authorship and the scale of the proposal, it is not surprising that the particular proposal for implementing ADD has been contested, even though the notion of ADD seems to be reconcilable with the positions of the opponents. This raises the question: why has ADD has become a non-issue for German health care providers?

The notion of digital disruption is useful to explain the opposition because the ADD initiative challenges established assumptions about the roles and linkages of key actors of the health care system. The established constellation of practices is challenged and would require a re-orientation and re-alignment of actors. ADD operators would have to be included as new entrants in the constellation.

In the case of medication management, aspects like freedom of choice of medication, splitting prescriptions across pharmacies and a right of privacy to one's own medication record have been taken for granted by health care providers and patients likewise. 7x4 Pharma's ADD proposal provides an alternative vision of affordable, safe and patient centered provision of medication. This vision implies less choice but higher efficiency and promises lower total cost despite the production and distribution cost of blisters. It assumes a different view of health care, which values electronic communication and documentation in order to provide improved patient care.

The building blocks of the ADD initiative assume a redefinition of how practices of the different actors are aligned with regard to polypharmacy patients. In fact, by identifying specific characteristics (polypharmacy) and linking these to issues such as medication safety and compliance, it creates and provides criteria to identify a new category of patients: polypharmacy patients, eligible for ADD. ADD implies a restructuring of the distribution chain with the introduction of a new actor (ADD operator or blister center) and a redefinition of responsibilities and professional boundaries of existing actors. Such changes might trigger a broader momentum of innovation and change.

Infrastructure innovations like ADD call into question the very position an actor has in a constellation of practices. Thus, one way of making sense of ABDA's resistance against ADD is that they perceive it as an emergent infrastructure, i.e. once regulation is in place that foresees industrial production of weekly blisters, an evolutionary dynamic of subsequent changes and adaptations might follow, way beyond what is part of the proposed design right now. This might refer to the market structure and a near monopoly of 7x4 Pharma, which is one concern also raised by Wille and Wolf (2006), or a discussion about the role and remuneration of pharmacists in the context of medication management reform.

Overall, the ADD initiative illustrates the duality of infrastructure design and emergence. The ADD proposal is an example of infrastructure design; a design that addressed the constellation of involved practices with a range of material, normative and ideational assumptions. Yet the outcome shows the emergent character of infrastructures: the existing constellation of practices prevailed and de facto hindered the diffusion of the innovation.[2] The potentially disruptive, unpredictable and uncontrollable effects of ADD are one possible explanation of the incumbents' behavior.

5.2 ADD initiative as attempt for institutional innovation

The ADD initiative challenges established assumptions about the roles and linkages of key actors of the health care system. In this highly regulated environment, infrastructure innovation requires regulatory adjustments and thus turns into a political process. The discourse

---

[2] We are not making a judgement about the quality of the proposal nor the likelihood of its success, rather we are interested in patterns of transformation or persistence.

illustrates that the impact of infrastructure innovation is difficult to predict. Still – or even because of this – it causes concerns about shifting power, control and influence. The required regulatory changes imply financial benefits for a relatively small group of blister centers and ADD operators, while pharmacies and doctors would end up empty handed and potentially weakened in their roles.

Hargrave and van de Ven (2006, 865) depict the process of institutional innovation as a

> "dialectical process in which opposing actors in the organizational field frame issues and construct networks in an attempt to introduce new institutional arrangements."

We have been able to identify the patterns of framing contests, construction of networks, enactment of institutional arrangements as well as collective action processes.

We have depicted the historical reconstruction of the ADD initiative as a dialectical process of public discourse and profound opposition to Kohl Medical's initiative (**a framing contest**). The role of ADD to contribute to achieving these goals is empirically substantiated (through studies and pilot implementations) by Kohl Medical. The studies helped to frame the issue clearly, to position it into the context of international practices and gain legitimacy. While there is a consensus about the key goals, namely compliance, efficiency and safety, the proposal by Kohl medical has been contested by VFA, ABDA and KBV. Wille and Wolf's (2006) study that VFA has commissioned aims at undermining the legitimacy and validity of the results of Kohl Medical and the researchers they have involved. While the title sounds generic ("creating patient specific blisters"), the study looks only at the proposal made by Kohl Medical. A stylized summary of the findings could be: Kohl Medical's proposal is inefficient at best, if not even outright dangerous. Wille and Wolf's (2006) study has not been directly contested and has subsequently become the unquestioned reference study for the opponents of ADD. Particularly contested issues appear to be the freedom of medication and the related opposition against any positive list. ABDA may be concerned about the influential market position of Kohl Medical AG and benefits for pharmacies specialized in patients in nursing homes, which suggests that there is even a debate within the group of pharmacists. Surprisingly, the suggestions and the proposed model by AvidiaMed GmbH, which does not include a positive list and has individually packaged medication, has neither been broadly discussed nor assessed more positively even though it mitigates key points of critique against 7x4 Pharma's ADD pilot. The remaining argument is "unproven efficiency", which is almost used as a killer argument without providing empirical evidence or a rebuttal against the calculations based on the pilot studies which suggest significant cost saving as a result of ADD. It is therefore still unclear why ABDA and KBV are so strongly opposed, especially in light of their own proposal. We did not find evidence that the opposition is interested in engaging in a critical dialog with the ADD proponent and discussing specific design options. The ABDA and KBV initiative in 2011 is another example of reframing the problem: the objectives show a high level of consensus with the ADD proponents, however ADD and the entire issue of providing some form of patient medication aids is conspicuously absent from the proposal.

Kohl Medical had formed a coalition of health insurance providers, collaborating physicians and pharmacies with *de facto* support of a few academics who conducted studies on their behalf (**construction of networks**). This coalition succeeded in conducting a pilot project as a proof of concept study, however its composition was not suitable to build political support to amend the regulation. Kohl Medical has not gained the support of either pharmacy or physician associations.

The group of proponents of weekly blisters have created two associations (Blisterunion - http://www.deutscheblisterunion.de and Bundesverband Patientenindividueller Arzneimittelverblisterer e.V. www.blisterverband.de/), which coordinate public relations and political lobbying. The "Blister movement" is a network of mainly regional actors, which are working towards institutional reform, i.e. amendment of health care regulation to establish weekly blisters as a recognized option for patients and their doctors with clear rules for reimbursement. While 7x4 Pharma was a highly visible proponent of the Blister issue, we found no evidence that they actually liaised with the regional Blister centers, often run by pharmacies or wholesalers with a focus on nursing homes, which might have regarded 7x4 Pharma as a competitor in the blister market. This would suggest a fragmentation of the Blister movement.

The opposition is represented by ABDA and KBV or their regional members and (indirectly) supported by the association of research active pharmacy manufacturers.

The **enactment of institutional arrangement** would be an adjustment of the regulatory framework. According to Lauterbach, Gerber, and Lüngen (2004), ADD is most advanced in countries that have a nationally controlled health care system such as Sweden. In contrast, Germany has a corporatist governance structure in which the regulator collaborates with the self-governing bodies within the statutory health insurance system

(Andersen and Woyke 2003). Therefore, the opposition of ABDA and KBV effectively blocks any regulatory initiative by the legislator.

As the pilot studies have shown, ADD is not viable without a proper regulatory framework. While the existing regulation allows for bilateral agreements between ADD operators, physicians and pharmacists, scaling of the model to a national level would require standardized rules and procedures, which the stakeholders simply can apply or refer to, e.g. patient consent forms, ADD as option on prescriptions, rules for reimbursement of cost, and standardized contracts between the stakeholders. At the same time, regulation would have constraining effects on the ADD operator(s) in terms of regulating competition and imposing quality standards.

In the case of ADD, the corporatist governance structure of a consultative and consensus building process with representative bodies of the health care system, has led to intense lobbying activities and a highly politicized debate. Thus, as a result of the corporatist structure, initiatives need to mobilize broad support if they hope to avoid any stalemate and to achieve swift regulatory action.

Both groups have achieved small successes in terms of support for their pilot studies, however, the more profound issues of regulatory amendment such as level of reimbursement for blisters or medication management have not been addressed yet.

Our historical reconstruction has illustrated the dialectic **collective action process**, public debates and power games. Enacting infrastructure innovation in medication management does not only require engineering and business competence or the ability to innovate medication distribution chains but also to mobilize political support by a number of stakeholders in order to enact a political process of adjusting regulation.

6. Conclusion

This chapter reconstructs a ten year period of attempted medication infrastructure innovation in Germany. Against the backdrop of long delays in key health care infrastructure projects such as the patient health card, we have focused on a small segment of customized dose administration aids for polypharmacy patients, which have been identified internationally as a distinct and growing risk group.

By combining the lenses of infrastructure and institutional innovation we are able to reconstruct the dialectical process between blister proponents and opponents: we find different ways of framing the problem and using scientific results next to the formation of different networks and coalitions in order to mobilize resources, build political clout and legitimacy. Inevitable uncertainty about the evolution of infrastructures is reflected in framing contests in the form of academic studies which suggest predictability. It is surprising that the opponents do not even see the need to engage in a constructive dialog about the pros and cons of ADD, different design options or alternative forms of creating dose administration aids. Instead blisters have been framed as a non-issue despite a long history of ADD use in numerous countries. The alternative proposal for medication management addresses the critical issues quite clearly but omits the daily practice of administering medication by and to patients, and it implies higher cost. Moreover, its introduction has been significantly delayed and key components such as remuneration for participating physicians and pharmacists have not yet been clarified (Ziegler 2013). Given the salient role of regulation in health care, we attribute the lack – or delay – of innovation to a lack of regulatory activity justified by the governance rules of the national health care system. That this is in line with the interests of one of the coalitions (ABDA, KBV and indirectly VFA) has become quite obvious. The institutional change processes have taken place at two levels: at the level of the pilot study, 7x4 Pharma has been successful to form a coalition, however at the political level they have not succeeded.

Conceptually, we have tried to show how infrastructure innovation can be linked to the disruptive innovation discourse. The infrastructure design clearly had the potential to become a disruptive innovation not for one company but for the system of distribution of medication. Infrastructure evolution and emergence limit the scope of predicting the future and the effects of changes. Yet the logic of infrastructure itself can be seen as an invitation to reflect on our visions for the future and the need to provide governance forms that enable innovation.

References

Aanestad, Margunn and Tina Blegind Jensen. 2011. "Building Nation-Wide Information Infrastructures in Healthcare through Modular Implementation Strategies." In *The Journal of Strategic Information Systems*, 20(2) (June): 161–176.

ABDA and KBV. 2011a. "Zukunftskonzept Arzneimittelversorgung - Gemeinsames Eckpunktepapier.". Berlin. http://www.kbv.de/ais/38730.html.

—. 2011b. "Zukunftskonzept Arzneimittelversorgung."

http://www.abda.de/fileadmin/assets/Pressetermine/2011/ABDA_KBV/Zukunftskonzept_07_10_2011_KBV_ABDA_Homepage.pdf.

Andersen, Uwe and Wichard Woyke. 2003. "Gesundheitspolitik." *Handwörterbuch Des Politischen Systems Der Bundesrepublik Deutschland*. Opladen: Leske+Budrich.

AvidiaMed. 2013. "Pressemeldung: Körber fokussiert Kernaktivitäten der Sparte Pharma-Verpackungssysteme." http://www.avidiamed.de/fileadmin/content/www.avidiamed.de/130130_PM_AvidiaMed.pdf

AvidiaMed GmbH. "Der MediFalter - Patientenindividuelle Arzneimittelversorgung - MediFalter." http://www.medifalter.de/fachkreise/pav.html.

Bell, J. Simon, Kristina Johnell, Barbara C. Wimmer and Michael D. Wiese. 2013. "Multidose Drug Dispensing and Optimising Drug Use in Older People." *Age and Ageing* (July 21). http://ageing.oxfordjournals.org/content/early/2013/07/19/ageing.aft080.extract.

BPAV. 2011. "Entgegnung des BPAV auf das gemeinsame Positionspapier der Landesapotheker- und Landesärztekammer Brandenburg vom August 2011." http://www.blisterverband.de/entgegung-bbrg-5351.html.

—. 2012. "Pharmazeutische Blister – Ein Beitrag für mehr Qualität und Effizienz im Gesundheitswesen und für eine Chance des Marktes im Arzneimittelsektor - Positionspapier des BPAV." http://www.wmscdn.de/uploads/9743/9743.pdf.

Breu, Gerhard and Thomas Grözinger. 2012. "Politikbrief." http://www.medifalter.de/uploads/media/Politikbrief_Nr03_Jun2012.pdf

Christensen, Clayton. "Disruptive Innovation." http://www.claytonchristensen.com/key-concepts/.

Düsing, Rainer. 2006. "Medikamentöse Therapie mit verblisterten Arzneimitteln: Möglichkeiten und Chancen." http://www.kohlpharma.de/_data/mediapool/download_daten_assist/1171892367_070125_Duesing-Studie.pdf

Edwards, Paul N. 2003. "Infrastructure and Modernity: Force, Time, and Social Organization in the History of Sociotechnical Systems." In *Modernity and Technology*, edited by T.J. Misa *et al.*, 185–225. Cambridge, MA: The MIT Press.

Hargrave, Timothy J. and Andrew H. van de Ven. 2006. "A Collective Action Model of Institutional Innovation." In *Academy of Management Review* 31(4): 864–888.

Kohl, Edwin. 2010. “Patient Compliance and Treatment Success Individualised Blister Packs using the 7x4 Box as an Example.” http://www.costeff.eu/_shared/p_file_download.php?files_id=70.

Landesapotheker- und Landesärztekammer Brandenburg. 2011. “Positionspapier: Qualitätsgesicherte Arzneimitteltherapie und Arzneimittelversorgung von Patienten, die in unterstützenden Wohnformen (Heimen) betreut werden.” http://www.laekb.de/50ueberUns/20Beitraege/90Archiv/70Archiv2011/110509_Positionspapier110818.pdf

Lauterbach, Karl W., Andreas Gerber and Markus Lüngen. 2004. “Internationale Erfahrungen mit der Verblisterung von Arzneimitteln”. Köln. https://www.7x4pharma.com/_shared/p_file_download.php?files_id=2798

Lauterbach, Karl W., Andreas Gerber, Björn Stollenwerk and Markus Lüngen. 2006. “Verblisterung von Arzneimitteln für Bewohner von Alten- und Pflegeheimen und in der häuslichen Pflege: Beschreibung und Bewertung eines Pilotprojekts (September 2004 bis Dezember 2005) Verblisterung von Arzneimitteln”. Köln. https://www.7x4-pharma.com/_shared/p_file_download.php?files_id=2801

Lauterbach, Karl W., Markus Lüngen and Andreas Gerber. 2004. “Nutzung der Verblisterung von Arzneimitteln im Rahmen von Disease-Management-Programmen”. Köln. https://www.7x4-pharma.com/_shared/p_file_download.php?files_id=2797

Lauterbach, Karl W., Markus Lüngen, Andreas Gerber, Ines Kohaupt and Guido Büscher. 2007. “Quantifizierung der Fehlwurfrate beim Stellen fester oraler Darreichungsformen in drei Pflegeheimen”. Köln. https://www.7x4pharma.com/_shared/p_file_download.php?files_id=2799

Leker, Jens and Uwe Kehrel. 2011. “Abschlussbericht der Studie zur patientenindividuellen Arzneimittelverblisterung in Berliner Pflegeeinrichtungen - Struktur des Berliner Pilotprojektes (Präsentation).” http://www.wmscdn.de/uploads/9758/9758.pdf

Monteiro, Eric, Neil Pollock, Ole Hanseth and Robin Williams. 2013. “From Artefacts to Infrastructures.” In *Computer Supported Cooperative Work (CSCW)*, 22 (4-6): 575–607.

Neubauer, Günter and Alexander Wick. 2011. “Patientenindividuelle für Bewohner von Pflegeheimen (PIVP) Zusammenfassung der wichtigsten Ergebnisse des Modellprojekts der AOK Bayern.” http://www.wmscdn.de/uploads/9756/9756.pdf

Neubauer, Günter. 2011. “Patientenindividuelle Arzneimittel-Verblisterung Für Bewohner in Pflegeheimen – Bericht Zum Modellprojekt Der AOK Bayern.”

Pharmafacts, and Gerd Glaeske. 2007. "Analyse der Verordnungen in einer GKV-Population im Hinblick auf die Eignung für eine individuelle Verblisterung." https://www.7x4-pharma.com/_shared/p_file_download.php?files_id=2803&locale=de

Reimers, Kai, Robert B. Johnston, and Stefan Klein. 2012. "Evolution of Inter-Organizational Information Systems on Long Timescales: A Practice Theory Approach." In *Inter-Organizational Information Systems and Business Management: Theories for Researchers*, edited by Kishor Vaidya, Hershey: Business Science Reference.

Riemer, Kai and Robert B. Johnston. 2013. "What Is Digital Disruption?" In *Bbr [backed by Research] on WordPress.com*. http://byresearch.wordpress.com/2013/03/07/digital-disruption/

Schumbach, Karoline. 2013a. "Blisterapotheken: ABDA mauert." In *Apotheke Adhoc*. http://www.apotheke-adhoc.de/nachrichten/nachricht-detail/bvka-tagung-steinweg-verblistern-muss-honoriert-werden/.

—. 2013b. "Kein Honorar für Verblisterung." In *Apotheke Adhoc*. http://www.apothekeadhoc.de/nachrichten/nachrichtdetail/verblisterung-hennrich-verblistern-ist-kein-thema/?tx_ttnews%5BsViewPointer%5D=1&cHash=0315a0f1b5a6f6b6e67fcad1417b375c.

Star, Susan L. and Karen Ruhleder. 1996. "Steps Toward an Ecology of Infrastructure: Design and Access for Large Information Spaces." I*nformation Systems Research*, 7(1): 111-134.

Wenger, Etienne. 1998. *Communities of Practice: Learning, Meaning, and Identity*. Cambridge: Cambridge University Press.

Wille, Eberhard and Malte Wolf. 2006. "Neuverblisterung von Arzneimitteln." In *Gutachten Im Auftrag Des Verbandes Forschender Arzneimittelhersteller e.V. (VFA) Endbericht*. http://www.vfa.de/download/gutachen-neuverblisterung-lang.pdf

Ziegler, Juliane. 2013. "Politik: Mit neuem Namen: ABDA/KBV-Modell startet im Januar: DAZ.online." In *DAZ.online*. http://www.deutsche-apotheker-zeitung.de/politik/news/2013/10/11/abdakbv-modell-startet-im-januar/11211.html.

Acknowledgements

We would like to thank Kai Reimers and his team, Robert Johnston and Reima Suomi for engaging discussions about infrastructures and feedback to earlier versions of this chapter. Financial support by the Deutsche Forschungsgemeinschaft (grant number 887/9-1) is gratefully acknowledged.

II. Information Infrastructures as Ecological Tools

CHAPTER FOUR

THE DESIGN KNOWLEDGE MULTIPLE: DECONSTRUCTING BALANCE IN AN OPEN SOURCE VIDEO GAME INFRASTRUCTURE

GIACOMO PODERI

> *Balance is mainly a feeling in your community. People say a game is not balanced when it's 'not fair', when it's not challenging, but if it's too easy it is not balanced either, probably even when it is not fun. [...] Balance is a moving target. We have been doing this for years and we are still doing it, because whenever we change something it also has a ripple effect throughout the game. So your game is never balanced, your game is about balanced. (BfW Developer, FOSS Conference, 05/02/2012)*

Introduction

Greatly inspired by Annemarie Mol's concept of *body multiple* (Mol 2002), this chapter finds its place amid two different, yet partly overlapping, areas of interest. On one hand there is the general interest for understanding the role that infrastructures play in the production of Information Technologies (IT). On the other hand, there is the wish to investigate a specific user-led, bottom-up process: designing in Free and Open Source Software (FOSS). By relying on the tradition of Science and Technology Studies (STS), the chapter investigates the relationship between continuous design and software development distributed infrastructure in FOSS. It does so by adopting an ecological approach to infrastructures (Star 1995; Star and Ruhleder 1996) and through the ethnographically informed inquiry (Hakken 1999) of a FOSS video game. In STS, the viewpoint of infrastructures is a challenging, yet rewarding approach. Indeed, due to the pervasive and relational properties of infrastructures, this approach prompts us to rethink the typical distinction between *structure* and *agency*, to redefine the boundaries of the concerned processes and the roles of the involved human and non-human actors (Orlikowski 1991; Star 1991). At the same time, it puts researchers at the centre of a dense and differentiated layering of people, activities and things that allows them to grasp the inquired phenomenon in a richer and non-reductionist way (Suchman 2002).

The chapter is informed by the work done for a broader research addressing the relationship between users' participation and distributed design in FOSS development. The research was designed as the ethnographic study (Hakken 1999) of *The Battle for Wesnoth* (BfW henceforth): a turn-based strategy video game. The fieldwork was conducted overtly during the period between December 2010 and April 2012. Besides note-taking, participant observation (Garcia *et al.* 2009) and Internet mediated interviews (Kivits 2005) were adopted as the main techniques for data gathering. The empirical work involved long time lapses of online interactions and observations in the collective infrastructure (mainly the official Internet forum, a few IRC channels and developers' mailing-list), "in game" situations and off-line interactions, such as a couple of face-to-face interviews and the attendance of a key FOSS conference. 29 interviews[1] were conducted with 8 game developers and 21 participant users who were primarily involved in the creation of add-ons.

The contribution of this chapter is twofold. It shows how design knowledge de-constructs and (re-)enacts itself across FOSS development infrastructure and, at the same time, it highlights that, when designing is concerned, the traditional meaning of FOSS "technical infrastructure" no longer suffices for capturing the complexity of such process. The chapter is structured as follows: the next section defines the concept of infrastructure and it explains the basic foundations of an ecological approach. Section two frames Free and Open Source Software designing in relation to FOSS infrastructures and the emerging *design knowledge* that they support. The third section briefly introduces *The Battle for Wesnoth* as the investigated case. Afterwards, the empirical section shows how different areas of BfW assemblage can act as infrastructure when balance is concerned and how such concern gets differently reified. The fifth and sixth sections discuss the gains derived from an ecological approach to infrastructures.

---

[1] Since anonymity was vouched for, participants' names are replaced by aliases in this chapter.

1. An ecological approach to infrastructures

Infrastructures are systems of substrates that, by definition, are invisible and integral part of the background for other works. They can be thought of as something upon which something else "runs", such as the rail-road or highway systems upon which trains and cars (Star and Ruhleder 1996). They are invisible and taken for granted. They are there, ready-to-hand, to support other kinds of work and uses. They become visible only upon breakdowns or malfunctions. For the gardener, the water distribution system (water pipes, pumps, shafts, regulations, private and public service providers) is an invisible infrastructure supporting another kind of work – *e.g.* watering plants in the garden. For the plumber called to fix a water supply problem, part of that invisible system becomes the target object of the work – *e.g.* fixing a sink, a water pipe or locating a deeper problem. At the same time, other substrates become (the invisible and taken for granted) infrastructure for the plumber's specific work (Star 1999).

Traditionally, infrastructuring is understood as a macro phenomenon where national or supranational entities plan for and push the construction of infrastructural elements (*e.g.* new roads, speed bump sets, traffic light networks) which they will own and administer (Hughes 1987). However, this chapter is mainly concerned with infrastructures that are characterized by user-driven and self-organizing construction processes as well as collective and bottom-up management. Phenomena such as FOSS development, user-generated content (Sotamaa 2010; Scacchi 2011), online peer production (Star and Bowker 2006), end-users development (Bolmstein and Dittrich 2011) portray these kinds of infrastructures, which can be referred to as *inverse infrastructure*[2] (Egyedi and Mehos 2012).

From an analytical perspective, an infrastructure is a *relational* property and not an object taken out of its context and stripped of use (Jewett and Kling 1991). When studying a specific organizational practice or production process, we always run into the infrastructural elements which underlie and sustain them. For instance, the study of software development always involves dealing with the development tools, shared protocols, standards and communication channels that are an integral part of developers' and designers' daily work. However, the inquiry of *those* infrastructures puts us at the centre of "a dense and differentiated layering of people, activities and things, each operating within a limited sphere of

[2] Inverse infrastructures are a specific type of infrastructure and have no relationship with infrastructural inversion mentioned by Bowker and Star (1999).

knowing and acting" (Suchman 2002, 96): software development is no longer the sole process relying on or being interested in that infrastructure.

As a relational property manifesting at the crossroads of many different socio-material practices and mediating heterogeneous interests, actors and works, such an inquiry calls for a change of perspective: the most important question about infrastructure becomes "*when*" something is an infrastructure, and not "*what*" is an infrastructure (Star and Ruhleder 1996, 112). To adopt an ecological approach to an infrastructure means "refusing social/natural or social/technical dichotomies" (Star 1995, 2). This implies investigating its systemic elements and properties *when* they act as infrastructural elements and, equally important, in relation to the other components of the (eco)system. The following sections of the chapter attempts to show exactly this.

2. Infrastructure and free and open source software designing

Free and Open Source Software (FOSS) programmes develop in tight relationship with unique socio-material assemblages that are

> "vitally concerned with the material and practical maintenance and modification of the technical, legal, practical, and conceptual means of [their] own existence." (Kelty 2008, 3).

Thanks to specific software licenses, the availability of software source code, sets of shared tools and common practices, developers and skilled users come together and nurture software programmes. They use and test the software, provide bug-reports and bug-fixes, suggest and implement new features, and they do it mainly on a volunteer basis and on the principle of self-assignments of tasks (Crowston *et al.* 2007). Nowadays, such an Internet-mediated, open, collaborative, volunteer-based and user-led phenomenon sustains a complex ecosystem of software projects, communities and practices that produces and maintains a vast array of programmes which are an integral part of the contemporary IT world (Boldyreff *et al.* 2009).

From a project management perspective, FOSS assemblages rely on a technical infrastructure that supports governance[3] mechanisms by providing

---

[3] Governance can be understood as "the means of achieving the direction, control, and coordination of wholly or partially autonomous individuals and organizations on behalf of a FOSS development project to which they jointly contribute" (Markus 2007, 152).

tools for storing, accessing, sharing, managing information and artefacts, and by facilitating collaboration in software development activities. Such an infrastructure is usually divided into production (e.g. bug-trackers, version control systems) and communication tools (e.g. mailing lists, IRC channels, Internet Forum). The former are used to handle collaboratively the various pieces of code, bugs and software versions. The latter are used by different groups to coordinate their work, exchange information and stay in touch with the users' base (Rantalainen *et al.* 2011). The scope and quality of participants' involvement support the emergence of a hierarchy of roles and domains of pertinence that distribute authority over the source code and over such a technical infrastructure (Crowston *et al.* 2007). Indeed, while most of the tools and communication channels are accessible to anyone, participants' roles influence the different ways they can use and relate to them. For instance, only core developers are granted with the technical possibility to edit directly the source code, which is handled through version control systems, every one else can only *read* it.

This infrastructure is continuously at the centre of FOSS assemblages' activities and "vital concerns". Differently from closed and corporative software production FOSS development portrays a form of *continuous design* (Gasser *et al.* 2003; Henderson and Kyng 1991): there is neither a formal nor a temporal separation between the development process – specify, design, build, test, release and maintain – and actual use. Users and developers can always propose new features for evaluation, implementation and testing at any moment in the life of the software (Barcellini *et al.* 2008).

From a software design perspective, FOSS assemblages do not "fake"[4] a rational, engineered, planned design process where formalization of blueprints, decisions and project milestones bear great organizational power. Through software informalisms they rely on and enact a collective capability for continuous distributed design and organizing (Orlikowski 2002). Such informalisms are sets "of semi-structured and relatively informal knowledge-sharing tools, representations and practices" which embed *design knowledge*: information that is gathered, transformed and used for making decisions about whether and how to modify the software (Scacchi 2002).

As it is summarized in Table 3-1, such knowledge is continuously constructed, rather than formalized by experts, through open debates and

[4] "Faking" a rational design process, here, refers to the fact that even in formal, engineered and top-down managed development projects, work in practice consistently diverges from the planned rationality of the process (Parnas and Clements 1986).

confrontations among heterogeneous actors, who relate to the software in different ways and who interpret, question and evaluate it from their perspectives. The representation of this knowledge does not take the form of "official and final" artefacts, such as project blueprints, that are used in formal development processes. On the contrary, it takes the form of an informally developed, debated and continuously modified artefact that is at the centre of different social processes (Gasser *et al.* 2003). The infrastructure is the locus where such knowledge is stored, circulated, accessed, transformed and, ultimately, constructed.

Design Knowledge	*Traditional/Formal design*	*FOSS/continuous design*
Acquisition	Learned (by designers from the users, by users from the final artefact)	Constructed and negotiated while the "final" artefact is under development
Representativeness	Experts' viewpoint	Multiple viewpoints
Depiction	Formal, stable, bounded	Interpreted, contested, informal
Use	In a "linear with feedback loops" process	At the centre of network of socio-technical processes

Table 4-1: main differences between knowledge design in formal software development and in FOSS. Adapted from Gasser *et al.* (2003).

3. The case of balance in The Battle for Wesnoth

Given the above framework, the inquiry of how FOSS infrastructures sustain the rise and circulation of design knowledge becomes the empirical inquiry of how the goal of balancing video game content is pursued, sustained and constructed among the heterogeneous components of the assemblage's infrastructure, understood in its analytical and relational meaning. As described in video game theory, balance is a game design quality which refers to "an optimal equilibrium between the challenges perceived in a given situation and the skills a person brings to it" (Charles *et al.* 2005).

In the game, players build up armies by gathering resources for the recruitment of new units and to meet the objectives of the game scenarios or to challenge other players over the Internet. The strategic dimension of the game involves fighting on a favourable terrain, at a favourable time of the day and matching player's units against weaker or disadvantaged enemies. Launched in 2003 as a prototype, BfW slowly gained success in the niche area of strategy games and reached version 1.10.5 in November 2012. Currently, only a couple dozen developers are actively working on the project, but during the past years a total of 60 developers contributed to it. About 200 people are officially credited for helpful contributions and the official Internet Forum, which is the primary support channel for gamers, participant users and developers, has 24500 registered members with an average continuous presence of 35 people.

BfW main game elements include *unit factions*: collections of battle units from which players can recruit their "soldiers"; *maps*: the game-boards on which battles unfold; *campaigns*: linear stories made of several connected scenarios whose objectives players have to meet in order to win. Imbalance among these game elements can severely spoil the experience of playing a *strategy* video game and for this reason balancing the game acquired a prominent role in BfW. Players and developers always pay great attention in tuning BfW to provide a good equilibrium between the challenges perceived in a given situation and the skills a person brings to it. Simply speaking, no unit faction should be perceived by players as clearly disadvantaged against other factions. No map should clearly favour specific unit factions against others.

In BfW, balancing is a collaborative process that involves: a feedback loop among players, content creators and developers; a lot of (play-)testing activity; and a trial and error approach to it. Although with time BfW reached a balance level that is generally considered of a good quality, it is not rare that suspicions about the balance of specific elements emerge. Therefore, although balancing no longer has the same centrality it had until a couple of years ago, it continues to be an important design goal that is looked for and strived towards being maintained.

4. When something becomes an infrastructure for balancing

The endeavour of balancing game content pervades every corner of BfW infrastructure, but the following analytical parts focus on the infrastructural loci that support the process in three important moments: when it is experienced in the game, when it is discussed in the forum, and when it is implemented into the source code.

4.1 The video game – when experiencing balance

BfW provides two game modes: single-player and multi-player. In the former, players can play campaigns where they engage in a story-like journey by meeting the objectives set for each campaign scenario. In the latter, players face other human or non-human players through single matches played on specific maps and organized over the Internet. To set up a game, every player can choose from one of the six official factions[5] and, in the case of multiplayer games, from one of the 54 maps. They can also choose from the custom and user-made content that include dozens of additional factions and hundreds of maps. Such wide content availability allows for a high number of possible combinations for setting up a match and conducting a game.

Regardless of the chosen factions and maps, players expect and take for granted that their strategic skills mainly determine the outcome of battles, and not the initial settings of the match. Those game dynamics which do not conform to players' expectations are often suspected to be balance problems, but players have a hard time discerning the responsibilities for defeats between their own tactical choices and the technical design of the game:

> "[…] Something we read all the time in those kind of place...'All players are good players!' I mean, there is just no beginners! That doesn't happen. They never do mistakes, I mean, if they don't do mistakes, if they lose there is always a balance problem. All strategies should work: "I took my hero and I went straight ahead on that strands of orcs and it got killed. That's unfair"...so, which means that all defeats are due to random numbers or imbalanced game. [...] That the AI is cheating? Yeah, that [complaint] comes too a lot, that the AI is cheating and we should fix it, it's not fair too. Basically, our AI does not cheat but is very good with the numbers!" (BfW Developer, FOSS Conference, 05/02/2012)

On one hand, BfW players tend to overestimate their own skills: they prefer to find problems in the game design over faulty logic in their strategic thinking or planning. On the other hand, the game engine that determines the outcome of each battle relies on a probabilistic algorithm. Therefore, by reflecting on the traits of the units in the battlefield and other environmental characteristics, players can understand the likelihood of a

[5] These are: *Drakes*; *Knalgan Alliance*; *Loyalists*; *Northerners*; *Rebels*; *Undead.*

battle outcome, but they can never predict it in advance[6]. Both players' unique subjectivities (Talamo, Pozzi and Mellini 2009) and the game design make it difficult to identify unexpected game dynamics as balance issues out of single manifestations. Nonetheless, recurrent patterns of units' usage and game strategies sometimes emerge which highlight and consolidate such unexpected dynamics. The example of the cavalry, which emerged as an overpowered unit a while ago, is explanatory in this regard.

Due to their recruiting cost and unit traits, cavalrymen discourage their use as expendable defensive lines or isolated raiders but the two illustrations below show game situations in which they turned out to be more effective than their traits would suggest. In Fig. 4-1, two cavalry units warded off the attack from four enemy units that surrounded them. Basically they acted as a strong defensive line for the castle (at the top-right corner of the illustration). In Fig. 4-2, despite being on unfavourable terrain (a river), a cavalry unit resisted the coordinated attack led by three enemy units.

Fig. 4-1: two cavalry units just warded off the assault of four enemy units.

Fig. 4-2: a cavalry unit resisted the attack of three enemy units despite being on unfavourable terrain (river).

Singularly taken, these situations do not mean much. They could have been interpreted as "lucky" battle outcomes, or consciously risky strategic decision in a broader game strategy. However, as players started noticing that the ratio among recruitment costs, health points, attack efficiency and

---

[6] E.g. in a single match, it is possible that a unit with three available attacks and 60% chance to hit misses the enemy with all three attempts. However, in the long run and over repeated matches the ratio of hit/miss attacks will lean towards 60%.

other traits was more favourable for the cavalry units than many other battle units, they started using them abundantly in the most disparate battle situations and in many different games. As the practice kept consolidating, puzzled players started questioning cavalry's balance and raising complaints.

> "Recurring [balance] complaints are a good clue that you should do something...they're not always a good clue [...] but still, when they come back regularly that's a big clue that you should, at least, have a deep look at what's going on" (BfW Developer, FOSS Conference, 05/02/2012).

Since each imbalance claim carries the individuals' subjectivities and the contingency of singular matches, the possibility to save the played game into savegame files and to watch it again later is a crucial feature for balancing work that is supported by the video game, as the next section shows.

4.2 The Internet Forum - when discussing balance

Internet Bullettin Boards, or forums, are websites with a hierarchical structure for people to hold conversations: members' replies to discussions are displayed in chronological order and organised in threads, which fit into thematic sub-boards. In BfW, the Internet Forum is the most populated channel of the whole infrastructure and is divided into six boards: *General*, *Content feedback*, *Development (open)*, *Development (restricted)*, *Unofficial development*, and *Miscellaneous*. As of January 2013 it includes 29980 unique threads and a total of 435930 posts.

Here, potential balance issues are discussed in specific threads usually named "Balancing [Content X]". Broadly speaking, fruitful discussions leading to software changes unfold in three phases: (*i*) identify and legitimise the potential balance issue; (*ii*) identify and support an adequate fix among the many possible ones; (*iii*) implement a final solution to the initial issue. However, many of these discussions do not pass beyond the first step: they often die out of replies or, more simply, turn out to include neither a concrete balance issue nor one that can be easily addressed. The few ones which develop further and lead to balance changes go on for relatively long periods and gather replies from many different participants.

The following excerpts come from a thread which dealt with balance issues for the *cavalry* unit type. At the heart of the matter was the claim that cavalry units were too strong. Before reaching a solution, the thread went on for five months and gathered 180 replies from 42 different people. The issue got raised by a player who called for a change in the cavalry unit type. He provided a suggestion on how to improve it and the motivation for his claim.

> "UNIT CHANGES BALANCE IDEAS"
> **14/12/2010 – Player1**: After some extensive play and discussion with other high level ladder players I would like to submit these unit balance ideas for consideration in future patches.
> *Cavalry*: Reduce resistance from 40% blud and 30% blade to Horsemen levels of 30% blud and 20% blade.
> Reason: Considered to be one of the most unbalanced units in the game currently, Cavalry are too effective as a main attack unit. [...]
>
> **14/12/2010 – Player2**: I agree that some changes to cavalry would be good but I'd prefer changing something other than resistances which are fairly unique to the unit and give it some flavour. Besides decreasing blade or impact resist wouldn't do much IMO since it would still be a pretty bad idea to try take a cav out with those attacks. I'd be in favour of one of :
> 1. Reduce HP by a bit (hard for me to know exactly how much)
> 2. Reduce MP to 7 (heavier armour slows him down compared to horse?)
> 3. Reduce attacks to 5-3 [...]

During the first few days, a dozen replies arrived either in support of or against the cavalry as a problematic unit and about which aspect to modify. However, despite the many replies that followed, the issue remained unclear even though two developers stepped in to support the claim that a balance problem was present. Therefore, in order to ease the discussion they called for more concrete proof.

> **16/01/2011 – Developer1**: briefly, I'd agree that cav are great against orcs, but they have weaknesses. Anyone recruiting too many of them is going to be punished. I'd enjoy seeing replays showing them exploiting orcs, and I think it would help the debate. [...]

As Developer1 encouraged the submission of replays showing the imbalanced behaviour, the issue slowly moved forward. A few more weeks were needed for players to submit evidence, to compare and interpret them, but the certainty of a balance problem was consolidated.

> **08/04/2011 – Player3**: Here's a bunch of Loyalist wins vs Northerners that feature a prominent number of cavalry:
> http://[link to replay savegame1]
> [...]
> http://[link to replay savegame7]

<u>08/04/2011 – Developer1</u>: Those look promising, I'll take a look when I can. You're right that no one replay is likely to be a magic bullet-but it helps a lot to get a sense of what players are doing, what interesting play and/or mistakes are happening. I'm also curious what leaders are involved.

At this point, the balance issues had been identified and legitimised, but players and developers kept engaging in the evaluation of how to address it.

Player5 summarized most of the proposals previously made which spanned from reducing the Health Points (HP) to increasing recruitment costs or changing the defence type and more.

<u>13/04/2011 – Player5</u>: I think that at this stage we've reached a consensus on the fact that Cavs are overpowered (whether to a great extent or only a little bit, we can differ). I think it's time to consider more seriously the kind of changes we should make to it.

To date people have suggested:
1. Lowering its HP [health points], usually to make it conform with other scouts. This is problematic as it takes away the flavor of what is meant to be a beefier scouting type unit.
2. Lowering its movement points to 7. [...]
[...]
6. Damage. 9-3 cavs at day can be nasty, […]

<u>13/04/2011 – Developer1</u>: Thanks very much for the replays, I'm still going over them but it is helpful! I would like to get some games in personally, before commenting about what I think it should be done […]

Many more opposing positions emerged and got discussed, but ultimately the one that got adopted was to reduce cavalry's health points from 38 to 34. As Player6 reported the news to the others, it was greeted with satisfaction.

<u>10/05/2011 – Player6</u>:
In 1.9.6:
Unit changes and balancing:
* Increased the pierce attack of the Orc Archer from 5-3 to 6-3.
* **Decreased the HP of the Cavalryman from 38 to 34.**
* Decreased the HP of the Dragoon from 53 to 49.
* Decreased the HP of the Cavalier from 68 to 64.
* Changed the 'feral' trait to give 50% instead of 40% defense on villages.

10/05/2011 – Player7: Excellent, I'm glad that our feedback led to this solution!

When an issue of balance emerges in BfW, the Internet Forum becomes the substrate for the collective construction of its legitimacy and resolution. In this collective process the game replays act as spokespersons for the design knowledge matured in-game. They "speak" to legitimise the balance issue and to help the "closure of the controversy", which happens technically through the translation of this design knowledge into a modified variable in the source code, as the next section shows.

4.3 The version control system – when implementing balance

As it is typical for many FOSS assemblages, BfW version control system (VCS) includes two main versions (*branches*) of the software source code. One is the *stable* branch and includes the source code for the video game which has been thoroughly tested and stabilized before wide distribution. The other is the *development* branch and includes the source code which is under active development and maintenance. Simply speaking, the former relates to the game version which the majority of end-users adopt and install on their computers for playing, while the latter is the version which developers and skilled users work upon. After each development branch has been developed and tested enough, it gets promoted as the next stable one and a new development branch gets started. These release cycles (*i.e.* from one stable version to the next one) take slightly longer than two years, although intermediated versions within the single branches are often created and released.

Here, balance is scripted and embedded into the source code files that define the game content. Indeed, each content element has a specific configuration file defining the appearance, characteristics and behaviour of that element in the game.

The following file is placed in the "Human" folder, which also contains all the units belonging to the human race. The file content represents the source code for the cavalry unit discussed above. *Loyalist_Cavalryman.cfg* opens with a declaration of the basic characteristics of this element: it is a unit ([unit_type]), with a specific identity (id=), race and a picture which represents it (image=). Furthermore, the file defines units' traits such as hitpoints (also known as health points), movement type and speed, recruiting cost, attack type, damages for each successful hit and so forth. By changing any of these variables, developers re-define the traits of that unit, but they also alter the equilibrium among content elements as it is experienced in-game or discussed in the forum.

```
#textdomain wesnoth-units
[unit_type]
  id=Cavalryman
  name= _ "Cavalryman"
  race=human
  image="units/human-loyalists/cavalryman.png"
  profile="portraits/humans/cavalryman.png"
  ellipse="misc/ellipse"
  {MAGENTA_IS_THE_TEAM_COLOR}
  hitpoints=38
  movement_type=mounted
  movement=8
  [...]
  [attack]
    name=sword
    description=_"sword"
    icon=attacks/sword-human.png
    type=blade
    range=melee
    damage=6
    number=3
  [/attack]
  [...]
```

In conjunction with the discussion that went on in the Internet forum, Developer1 edited the copy of *Loyalist_Cavalryman.cfg* that resided in the VCS development branch and changed the hitpoints value to 34. By so doing, he decreased the amount of damage that other enemies must inflict on the cavalry unit in order to defeat it.

```
profile="portraits/humans/cavalryman.png"
ellipse="misc/ellipse"
{MAGENTA_IS_THE_TEAM_COLOR}
hitpoints=34
movement_type=mounted
movement=8
```

However, since the change was applied to a file in the development branch and since this is continuously subject to other balance changes too, when end-users adopt the next official stable version (*i.e.* after the completion of the development cycle, in approximately two years), the modified cavalry unit will reach them together with many more other content edits. The following ones are just small excerpts of the many balance changes done for BfW v.1.9.x series (the development branch at the moment of research).

CHANGELOG FOR BFW 1.9.X SERIES (DEVELOPMENT BRANCH)

1.9.2
Unit changes and balancing:
* Decreased the physical resistances of the Dwarvish Scout and Dwarvish Pathfinder to 10% and those of the Dwarvish Explorer to 20%.
* Decreased the village defense of Chocobone from 60% to 40%.
[…]
1.9.6:
Unit changes and balancing:
* Increased the pierce attack of the Orc Archer from 5-3 to 6-3.
* **Decreased the HP of the Cavalryman from 38 to 34.**
* Decreased the HP of the Dragoon from 53 to 49.
[…]
1.9.9
Unit changes and balancing:
* Lowered swamp defense of Cuttlefish and Sea Serpent from 60% to 40%
* Decreased cost of Giant Rat from 13 to 6
* Various changes to the defense, movement and resistances of the Giant Rat
[…]
1.9.10
Unit changes and balancing:
* All mounted units now have forest defines capped at 30%. This reduces their defense on forested hills from 40% to 30%

All these changes alter the equilibrium among game elements as it was previously possible to experience it and as it was previously discussed in the Forum. As Developer1 explained, such changes are never transferred to the stable branch before they have been thoroughly tested both due to an established development practice and to a technical reason:

Q: Is it true that you fix balance problems for stable series only within development series, because it would brake compatibility? Does it mean that users spotting balance issues in a stable series, will see a fix only in the new stable one?
A: Yes, we guarantee that you can play together with all minor release of the same major release. This means that if the characteristics of a unit change, both sides could disagree on the value, which would lead to the dreaded "out of sync" error. (Developer1, Interview, 16/06/2012)

When a balance issue emerges in BfW, the version control system becomes the substrate for its technical resolution. Here such a resolution takes the form of variable changes and although only developers have the authority to perform such changes, their decisions about how to concretely address them rest on the construction of balance done in the forum.

5. Design knowledge: The balance multiple

As far as BfW infrastructural elements are transparent, accessible and conveniently interconnected, when they get examined in relation to a specific activity, such as balancing, it becomes evident that this "dense and differentiated layering of people, activities and things *operate within a limited sphere of knowing and acting*" (Suchman 2002, *emphasis added*).

Players playing the game and challenging other players, experience balance in relationship to repeated encounters of puzzling game dynamics. Forum members discussing the game strengths and weaknesses co-construct balance in relationship to the evolution of on-going and open threads. Developers working on source code files alter balance by changing variables and by grounding the rationales for such changes in their knowledge of the source code base and in relationship to their interpretation of players' feedback.

If balance comes under the spotlight, the video game, the Internet Forum and the Version Control System become the infrastructure for experiencing, testing, discussing and inscribing this aspect of game design. As shown in Table 4-2, each of these infrastructural elements portray a different kind of design knowledge.

When the video game is concerned, such knowledge is acquired through repeated playing and it embeds players' viewpoints. The most typical way to represent it is by means of savegame files which, anyway, need to be interpreted. This knowledge is used to both inform game strategies and sustain claims about game design issues. When the Internet Forum is substrate for balance, the design knowledge is acquired through an open and collective co-construction process which includes the viewpoints of all the forum members who actively participate in it. Such knowledge is used to legitimise and define specific edits to the source code. Finally, where the VCS is concerned, balance design knowledge is acquired through a trial and error process in which variables are changed by developers with the assumption that they may be changed soon again. This kind of knowledge, inscribed in source code files, is mainly used to alter the equilibrium among the several content variables.

Balance Design Knowledge	Video game	Internet Forum	Version control system
Acquisition	Through repeated playing	Co-constructed through open discussions	Through trial and error approach
Representative ness	Players' heterogeneous viewpoints	Forum members' heterogeneous viewpoints	Developers' know-how and perspective
Depiction	Savegames	Forum open threads	Source code files
Use	To sustain claims about balance and to inform game strategies	To legitimise and pinpoint balance	To alter equilibrium among content variables

Table 4-2: the multiplicity of balance design knowledge in relation to the infrastructural element it builds upon: the video game, the Internet Forum and the Version Control System.

By borrowing Annemarie Mol's terminology, the spheres of knowing and acting supported by the video game, the Internet Forum and the Version Control System reify balance in very different and, ultimately, incommensurable ways: balance becomes *multiple* (Mol 2002). Similar to Mol's position on the multiple ways to reify atherosclerosis in various hospital wards, balance too is multiple in an ontological sense. It is not a matter of perspectives, it is a matter of different reifications of the same conceptual, abstract, ideal. To pursue a general design goal, FOSS assemblages rely on their technical infrastructure for performing and coordinating software development activities. However, the pursuit of this general design goal deconstructs and reconstructs itself in heterogeneous ways across the infrastructure.

6. Conclusions

Through the ethnographically informed inquiry of a FOSS video game and the employment of an ecological approach to infrastructures, this chapter investigated the relationship between continuous design and software development infrastructure. This showed how design knowledge constantly (de-)constructs and (re-)enacts itself across the assemblage's infrastructure. At the same time, it highlighted that, where designing is concerned, the traditional meaning of FOSS "technical infrastructure" no longer suffices for capturing the complexity of such processes.

On one hand, the video game itself becomes infrastructure where designing is concerned. The fact that software use is an important aspect of any software design endeavour is no surprise, especially in the case of FOSS continuous design. However, it is important to note how the vantage point of infrastructures helps in better framing it in relation to the whole designing process. Indeed, to look at software usage as an infrastructure for designing exposes the peculiarities of such infrastructural components and their influence on design knowledge. On the other hand, the traditional distinction, in functionalist terms, between "production" and "communication" tools becomes lurching, because a communication medium, such as an Internet Forum, becomes the locus for the actual co-construction and definition of design rationales. These rationales are expressed in the form of "software informalisms", but, as the chapter has shown, are rationales for software modifications nonetheless.

In addition to these theoretically-oriented considerations, the present work suggests more substantial insights for FOSS continuous design. Indeed it has shown that FOSS assemblages do work towards general design goals, but as the pursuit of these general goals becomes deconstructed, dispersed and locally performed across the infrastructure, they also translate into very different manifestations that are loaded with their own meanings and embedded into different socio-material practices each of which have their own specific logics, struggles and "trajectories". As such, the fulfilment of *one* general design goal becomes largely unachievable, yet continuously perpetrated.

Ultimately, the chapter sheds some light on the meaning of the opening quote: *balance is mainly a feeling in your community [...] So your game is never balanced, your game is about balanced.*

References

Barcellini, Flore, Françoise Détienne and Jean-Marie Burkhardt. 2008. "User and Developer Mediation in an Open Source Software Community: Boundary Spanning Through Cross Participation in Online Discussions." In *International Journal of Human–Computer Studies* 66: 558–570.

Boldyreff, Cornelia, Kevin Crowston, Bjorn Lundell and Anthony I. Wasserman. 2009. *Open Source Ecosystems: Diverse Communities Interacting*. Berlin: Springer.

Bolmsten, Johan and Yvonne Dittrich. 2011. "Infrastructuring When You Don't – End-User Development and Organizational Infrastructure." In

End-User Development, edited by Maria Francesca Costabile, Yvonne Dittrich, Gerhard Fischer, and Antonio Piccinno, 139–154, Berlin; Springer.

Bowker, Geoffrey C., and Susan L. Star. 1999. *Sorting Things Out: Classification and Its Consequences*. Cambridge, MA: The MIT Press.

Charles, D., A. Kerr, M. McNeill, M. McAlister, M. Black, J. Kcklich, A. Moore and K. Stringer. 2005. "Player-centred Game Design: Player Modelling and Adaptive Digital Games." In *Proceedings of the Digital Games Research Conference*. 285.

Crowston, Kevin, Qing Li, Kangning Wei, U. Yeliz Eseryel and James Howison. 2007. "Self-organization of Teams for Free/libre Open Source Software Development." In *Information and Software Technology* 49: 564–575.

Egyedi, Tineke M. and Donna C. Mehos. 2012. *Inverse Infrastructures: Disrupting Networks from Below*. Cheltenham: Edward Elgar Publishing.

Fogel, Karl. 2006. *Producing Open Source Software: How to Run a Successful Free Software Project*. Sebastopol, CA: O'Reilly Media.

Garcia, Angela Cora, Alecea I. Standlee, Jennifer Bechkoff and Yan Cui. 2009. "Ethnographic Approaches to the Internet and Computer-Mediated Communication." In *Journal of Contemporary Ethnography*, 38(1): 52–84.

Gasser, Les, Walt Scacchi, Bryan Penne and Gabriel Ripoche. 2003. "Understanding Continuous Design in F/OSS Projects." In *16th International Conference Software & Systems Engineering and Their Applications*. Paris, France.

Hakken, David. 1999. *Cyborgs@cyberspace?: An Ethnographer Looks to the Future*. Routledge Chapman & Hall.

Henderson, Austin and Morten Kyng. 1991. "There's No Place Like Home: Continuing Design in Use." In *Design at Work: Cooperative Design of Computer Systems*, edited by Joan M. Greenbaum and Morten Kyng, 219–240, Hillsdale: L. Erlbaum Associates Inc.

Hughes, Thomas P. 1987. "The Evolution of Large Technological Systems." In *The Social Construction of Technological Systems. New Directions in the Sociology and History of Technology*, edited by Wiebe E. Bijker, Thomas P. Hughes and Trevor Pinch. Cambridge, MA: The MIT Press.

Jewett, Tom and Rob Kling. 1991. "The Dynamics of Computerization in a Social Science Research Team: A Case Study of Infrastructure, Strategies, and Skills." In *Social Science Computer Review*, 9(2): 246–275.

Kelty, Christopher M. 2008. *Two Bits: The Cultural Significance of Free Software*. Duke University Press.

Kivits, Joëlle. 2005. "Online Interviewing and the Research Relationship." In *Virtual Methods: Issues in Social Research on the Internet*, edited by Christine Hine, 35-49, Oxford: Berg Publishers.

Markus, M. Lynne 2007. "The Governance of Free/open Source Software Projects: Monolithic, Multidimensional, or Configurational?" In *Journal of Management and Governance*, 11(2): 151–163.

Mol, Annemarie. 2002. *The Body Multiple: Ontology in Medical Practice*. Durham: Duke University Press.

Orlikowski, Wanda J. 1991. "Integrated Information Environment or Matrix of Control? The Contradictory Implications of Information Technology." In *Accounting, Management and Information Technologies*, 1(1): 9–42.

—. 2002. "Knowing in Practice: Enacting a Collective Capability in Distributed Organizing." In *Organization Science*, 13(3): 249–273.

Rantalainen, Aapo, Henrik Hedberg and Netta Iivari. 2011. "A Review of Tool Support for User-Related Communication in FLOSS Development." In *Open Source Systems: Grounding Research*, edited by Scott Hissam *et al.*, 90–105, Berlin: Springer.

Parnas, David Lorge and Paul C. Clements. 1986. "A Rational Design Process: How and Why to Fake It." In *Software Engineering, IEEE Transactions On*, 2: 251–257.

Scacchi, Walt. 2002. "Understanding the Requirements for Developing Open Source Software Systems." In *Software, IEE Proceedings-*, 149: 24–39.

Scacchi, Walt. 2011. "Modding as an Open Source Approach to Extending Computer Game Systems." In *Open Source Systems: Grounding Research*, edited by Scott Hissam *et al.*, 62–74, Berlin: Springer.

Sotamaa, Olli. 2010. "Play, Create, Share? Console Gaming, Player Production and Agency." In *The Fibreculture Journal*, 16.

Star, Susan L. 1991. "Power, Technology and the Phenomenology of Conventions: On Being Allergic to Onions." In *A Sociology of Monsters: Essays on Power, Technology, and Domination*, edited by John Law, 26–56, London: Routledge.

—. 1995. *Ecologies of Knowledge: Work and Politics in Science and Technology*. Albany: SUNY Press.

—. 1999. "The Ethnography of Infrastructure." In *American Behavioral Scientist*, 43(3): 377–391.

Star, Susan L. and Geoffrey C. Bowker. 2006. "How to infrastructure." In *Handbook of New Media*, edited by Leah A. Lievrouw and Sonia M. Livingstone, 230-45, London: Sage.

Star, Susan L. and Karen Ruhleder. 1996. "Steps Toward an Ecology of Infrastructure: Design and Access for Large Information Spaces." In *Information Systems Research*, 7(1): 111-134.

Suchman, Lucy. 2002. "Located Accountabilities in Technology Production." In *Scandinavian Journal of Information Systems*, 14(2): 91–106.

Talamo, Alessandra, Simone Pozzi and Barbara Mellini. 2009. "Uniqueness of Experience and Virtual Playworlds: Playing Is Not Just for Fun." In *Mind, Culture, and Activity*, 17(1): 23–41.

CHAPTER FIVE

THE LABORATORY KIT BETWEEN INFRASTRUCTURE AND BOUNDARY OBJECT

FEDERICO NERESINI AND ASSUNTA VITERITTI

Introduction and background

Over the years, science historians and STS scholars have focused on a number of infrastructures – such as railroads, waterworks, highways, telephony, business communication systems, bicycles, the Internet – which have been variously regarded as ecologies of artifacts (Bowker and Star 1999), distributed cognitive systems (Hutchins 1995) or information infrastructures (Edwards *et al.* 2007; Star and Ruhleder 1996; Star 1999; Jackson, Bowker and Knobel 2007; Karasti, Baker and Millerand 2010; Star and Bowker 2002; Pellegrino 2012).

Both empirical and theoretical studies have paid particular attention to the creation and development of scientific infrastructures, mostly viewing them as complex, unstable and large entities. So far, much less attention has been given to the contribution that systems and their components have been making to these processes, instead privileging an overall view where infrastructural components have largely remained in the background.

Among the infrastructural components of research laboratories, kits have become increasingly important. These objects have appeared in laboratory practices only recently and consist of assemblages of ready-to-hand substances, materials and procedures, which allow scientists to optimize relevant portions of their research work. Kits embody expert knowledge and make it usable. They are manufactured to be used in the laboratory, and continue to survive through their references in scientific articles. These tools are widely used both in molecular biology and in the biotechnological research field, as well as in other fields such as chemistry and physics. They are ready-made packages containing solutions, plasmids, chemical agents, enzymes and so on, plus small devices (pipettes, test-tubes, containers, buffers, etc.). These are ready-prepared assemblages which are often costly and allow the researcher to obtain rapidly many of the preparations adopted in the experiments, using standardized procedures. In a certain sense they resemble semi-ready meals, in that they contain the different partially-prepared ingredients as well as technical support mechanisms such as a measuring glass or a mixer. Among the various examples to be found in the biological and bio-medical fields are DNA and RNA extraction kits or those for the purification of cells or for electrophysiological analysis. They appear as easy-to-use objects, with a set of instructions to be followed, and do not seem to require particular precautions or interventions. They are everyday objects, perfectly integrated and naturalized into laboratories, to the extent of becoming almost transparent. Their use, however, requires a series of adjustments, made possible by the margins of flexibility they preserve. For this reason, studying how they are used makes it possible to highlight a series of decisive aspects in the development of infrastructural processes, primarily being a significant work of localization and coordination. An in-depth analysis of the laboratory kit therefore provides the opportunity for a better understanding of scientific infrastructures.

The work presented here is based on several interviews and observations conducted by the authors in a laboratory of cellular biotechnology at the University of Milan. The interviews carried out in the laboratory were with both junior and senior researchers who use these kits in their experimental practices for diverse reasons, particularly in experiments where it is necessary to extract DNA or purify cell lines.

Considering the laboratory to be an automated space (Keating, Limoges and Cambrosio 1999) within which an ever greater standardization of processes can be found, thanks to technologies capable of processing an array of materials and information, is not a new idea. The kits, however, do not appear to have attracted much attention, and their contribution to the automation and standardization of scientific processes remains unexplored.

These devices favor the articulation of diverse phases of internal and external laboratory work, therefore taking part in the “alignment of several levels of work organization” while remaining a hidden aspect of daily research work together with other “taken-for-granted tasks”, such as “washing pipettes and signing up to use the ultra-centrifuge”(Fujimura 1987, 258). They thus make an important contribution to the construction of “doable problems” (Fujimura 1987), through which appropriate research questions are posed and designs and experimental practices oriented. The processes of articulation and alignment will be analyzed later in the text by introducing the concept of "standardized packages" in relation to that of the "boundary objects" of Star and Griesemer, as well as the stabilization of scientific facts by Latour (Fujimura 1992).

Focalizing on the kits allows us to further develop our analysis by observing how these devices participate in the complex processes of standardization and classification (Star 1999; Star and Bowker 1999; Star and Bowker 2002). The kits thus contribute to the socio-material production of the infrastructures of scientific research which connect experiments, researchers, instruments, genes, computers, data, protocols, bio-technological companies, and so on. In fact, this work hypothesizes that the kits used in the laboratory are part of a complex circuit and intends to analyze how they are conceived, utilized and problematized (or not). We shall also analyze how they take part in experimental practice and contribute to maintaining the activity of networks of practice within and without the laboratory.

The chapter is articulated in four paragraphs. The first describes the kits we encountered in the course of our investigation through the descriptions given by the researchers we interviewed. The second defines the problem posed in the centre of the chapter – whether to evaluate the kits as isolated objects, boundary objects, or research infrastructures themselves - empirically and analytically. Given that the kits can only be considered to be part of such infrastructures, the third paragraph highlights their contribution to their maintenance. We take the opportunity of investigating them as learning devices in the fourth paragraph. The conclusions pose some final considerations, further focalizing on the contribution offered by the kits at local level in sustaining and nurturing the more extensive infrastructures of scientific research.

1. How and where we found them

The social life of a laboratory teems with different kinds of objects. Due to the increasing presence of technologies and databases, laboratories can be regarded as *technologically dense environments* (Bruni 2005; Bruni, Pinch and Schubert 2013) interconnected with such an extended network of social and technological worlds as to put their boundaries into question. As the place *par excellence* where knowledge, instruments and procedures converge, the laboratory is so populated with technical objects as to become an interactive space that takes part in technological and social ecologies – by which it is simultaneously incorporated and constructed – among heterogeneous elements including materialized, ready-to-hand knowledge and knowledge in the making. An example of material knowledge is provided by kits, which are both technical objects and a vehicle for cooperative social activities. Increasingly used in laboratory practice, kits are devices that allow researchers to organize and combine not only elements such as primers, antibodies and reagents, but also time schedules and working practices necessary to successfully performing experiments. They are standard systems, rather than mere agglomerates of elements, which make it possible to routinize and speed up some steps of the experimental practice.

> "A kit is not just one single thing. Otherwise it wouldn't be a kit but a single reagent. It is rather a collection of reagents, solutions and small tools. For example, in DNA extraction, there is a matrix packed into a small column for cells. There are also more complex sets containing a column, a column holder and a magnet. Kits are always made of different objects, often related to each other, and therefore include instruments, reagents and solutions." (Marco, Senior Researcher)

> "A kit is a system that allows you to carry out an experiment in an extremely standard condition, which is quicker than the traditional laboratory practices that have led to the identification and development of this experimental process. (…) It is a closed box, because you don't know the components it is made of. You just have to follow the instructions. It makes your life much simpler, but it probably makes you lose something in terms of knowledge." (Chiara, Senior Researcher)

A kit is therefore a set of elements that have been optimized, collected and organized for the execution of a particular step in the experimental process. Although kits appear to be "ready-to-hand" devices, their users must integrate them into their specific experiment, exploiting their margins of flexibility. A kit needs to be assembled; it is a system conceived and designed to be employed in a local environment, a miniaturized form of a socio-technical system connecting manufacturing companies and laboratories, making different aspects and expectations converge (Mongili 2007).

A standard kit is, for example, the one used for DNA extraction.

> "An example of super-standard routine tool is the DNA extraction kit (…). It was one of the first compounds to be standardized and sold as a kit (…). It is easy to use. You don't have to ask questions. It is a set of solutions that allows you to isolate the DNA from a tissue or cells (…). When you open the box, you find bottles, solutions and a booklet with protocol instructions, which tell you exactly what to do. You open it and go step-by-step, one, two, three, and so forth. The approach is very simple. All items contained in the kit are related to each other. You first use the buffer to

> isolate the cells, so, if you have a tissue with cells and you want to get the DNA and break them, this is the first instrument you have to use. Then you go to the next step and so on. To make it clearer, I can tell you that now there are kits for DNA extraction for which we just discuss price. You say to the company: "Look, we need this for five years", so we take it and we have ten kits every year…" (Chiara, Researcher, Laboratory Technician)

This has a high level of stabilization and low interpretative flexibility (Bijker 1995), an instrument that has reached high stability, with low margins of uncertainty and ambiguity. As we will see later on, a kit can be regarded as stable when it becomes a reliable and secure tool to be used for measurements and employment of already combined, ready-to-hand elements in the experimental phases where they are required.

In laboratories, however, it is also possible to find low stability in locally used kits: on the one hand, they maintain a high level of standardization; on the other, they show a higher interpretative flexibility. The kit used for the purification of neural cells is an example of one that still needs to be adapted and translated into everyday work practices in order to become naturalized.

> "In the case of the purification kit, there was a congress in Florence. At congresses there are often stands with representatives from various companies advertising their products, among them new kits. I had some spare time and there was this stand (…) where they were showing some advanced techniques [and], since we needed to isolate and purify our neurons, [I thought] I should get some information. We had to purify cells; we needed to differentiate them in order to isolate and transplant them or have a more homogeneous in vitro population. They gave me a brochure with their products and we talked about technical features, purity, the various parameters to use this kit, costs, etc. I knew this system only by hearsay. Seeing it and talking with a specialist surely helped. (…) This kit is actually sort of plastic, modular, because you can have the column, the solutions, the column holder, and so on, which is standard, but, depending on your specific case, you can choose the molecule you want to use to separate your cells. They have a full book. There might be a hundred different molecules, so I asked for the specific one we wanted to use. I knew that [in another laboratory] they were using that system and were very happy about it, they had bought a 20 reaction-kit and asked us if we wanted to try it (…), so even the direct experience of a collaborator counts." (Marco, Senior Researcher)

A kit that is not yet stabilized in everyday practices requires the presence and participation of a wide range of heterogeneous actors: the fair, congress, company, company experts (sales managers and technicians), laboratory, expert project researcher, the young apprentice who is learning the technique, and researchers from other laboratories. To become effective in practice, the kit is tested within a network of actions and actors (both human and non-human). Only after its capacity to integrate itself into the local situation is verified can it be included in the laboratory's relationship network. A kit that is not yet stabilized, therefore, activates a network of different sets of knowledge, both inside and outside the laboratory, and involves local adjustments where the standard created elsewhere by other experts must demonstrate its functionality and adaptability.

> "You need a kit to carry out a certain analysis of cerebrospinal fluid. You Google it and start searching. There might be the super-famous and reliable company selling it or the Chinese company you don't know anything about, so you look at the websites, see if the people from these companies make any reference to individuals, groups or studies that have already used these tools. Then you examine their works and understand their level, whether they are rubbish or important research teams. Otherwise, if you have a difficult test to make and you know that there is a guy at Columbia University who makes it, you send him an email and say: "Excuse me, I've seen that there are 35 kits for this test, what do you recommend?" So, to find the right one, you have to combine different kinds of research… Once you have collected some basic information, you look for the people who might have used it and ask them." (Chiara, Senior Researcher)

To find and use a kit, it is necessary to enter a network of relationships-the Internet, the manufacturing company, a laboratory, an online catalog, someone who has already used it-based on exchange and learning processes, made up of heterogeneous actors and interests.

> "Another option is to find the reference to a certain kit in an article. This sometimes happens with single reagents, and surely for antibodies. For example, for the RNA analysis I went through the literature to find out what kit had been used and I discovered that one had given significant results. You can see from the quality of the publication of results and the number of recent publications focusing on this kit instead of others and confirming the quality of results. It is a sort of index. I remember we made our choice on the basis of this parameter." (Marco, Senior Researcher)

> "The history of a kit can also start within a laboratory. Each of us, in our personal history, has used various kinds. If you know that someone has already used one, you ask for opinions, otherwise you go on Google and type in the name of what you have to do, for example 'DNA extraction'. Another option is to turn to the literature – that is, papers by someone who

> has already used it for specific purposes and you take the information you need from the sections about materials and methods. It doesn't mean it will work in your case, but you can start from there." (Paola, Senior Researcher)

A kit requires and allows the coordination of a multitude of actors and seems to assume the characteristics of a *boundary object* (Star and Griesemer 1989), because, despite its ability to incorporate standards and instructions for use, it appears as a plastic object which maintains a certain integrity and lends itself to a variety of uses, needs and interpretations. It becomes an object that enters the laboratory with a set of stabilized meanings, connecting different worlds and requiring further interpretations and translations. These can also diverge from the indications inscribed in the object. Users may also happen to attach themselves to manufacturers and become co-authors and co-producers, constructing the object by combining local and commercial needs. A kit is indeed a complex object, mediating various competences and actions: it emerges from and, at the same time, promotes interdisciplinary networks. When it enters the laboratory practice, it tends to become a naturalized object, which is inscribed in and adapts itself to a network of actors and processes.

> "A less stable kit is the one used for the macro-array analysis, a large-scale gene analysis of cells, tissues or other things, for which a series of kits are now available (…). These kits are very expensive; they cost more than 2,000 euros for one or more reactions, compared to 200 euros for the DNA kit… For some kits, such as the macro-array, there is a constant innovation also related to the codified genes… The database you find on the Internet is constantly updated. So you say, I used this system with 47,000 probes, and another says, I used it with 52, 57 or 60 thousand, with coding or non-coding functions. So, the complexity of the system gradually increases, as does the stability and variety of the various kits. In this case, it is necessary to have different techniques, with large patents and tens of bioinformaticians working on them. There has been much more research work and, above all, there's a constant evolution." (Marco, Senior Researcher)

Every kind of kit intertwines with different competences. While the most standard involve routine competences, more complex and less stable ones require highly qualified expertise from users and a greater ability to connect with the research world and the companies, which, in turn, use laboratory experiences to produce more differentiated and specialized kits.

While kits in research work are based on networking processes that can be regarded as infrastructures, their contribution to these processes, their construction and dynamics, still requires clarification.

2. Kits as infrastructures?

> "There are so many kits now. Everything is kitted! Yes, kitted. In the past, everything was handcrafted, it was more fun."
> (Luciano, Senior Researcher)

If we wanted to give a linear description of what happens in a laboratory, we could represent it as a process that goes through different steps, characterized by the increasing relevance of standardized activities embodied in kits and tools, along with the progressive reduction of the handcraft dimension entrusted to the researcher's manual ability. Thus, if the exploration of new ideas is successful, part of the activities carried out in a disorderly way in this first stage is codified into protocols, allowing other laboratories to use and adapt them to their needs. In turn, most standard protocols used in a wide range of experimental projects can further stabilize into kits and/or other new tools.

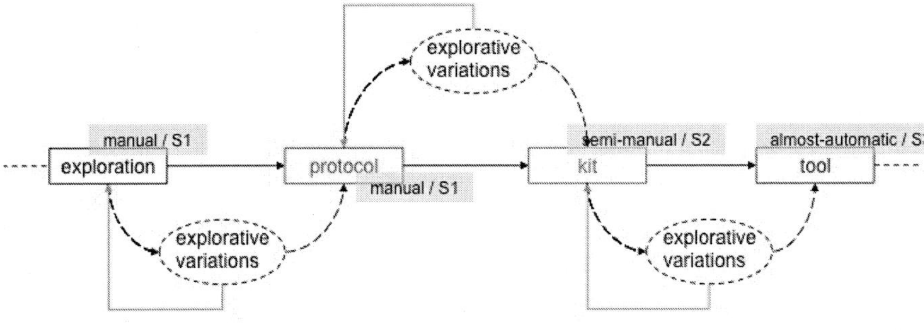

Fig. 5-1: the laboratory as a linear process.

The transition from the first to the second stage brings to mind the "creation of order out of disorder", referred to by Latour and Woolgar in 1979, a distinguishing trait of the activity of a research laboratory; successive passages correspond to the construction of increasingly standard *immutable mobiles* (Latour 1987). Certainly, this representation of research activity is not only reductive but also inaccurate, if not misleading. In fact, exploration, creation of protocols and usage of kits and tools simultaneously coexist within laboratories: an experiment is the combination of various activities, each referable to one of the above-mentioned "classes" of activities. Each activity becomes meaningful and can only be performed within this specific combination, corresponding to the specific network of heterogeneous actors that makes it possible and is, conversely, made possible by its presence.

In this perspective, it is not possible to consider kits as isolated, separate objects:

> "Basically, machines and Kits do a number of things; they allow you to do a number of things… Tools in general are associated with kits." (Marco, Senior Researcher)

Kits are pieces of research activities standard enough to be boxed up, mass-produced and used by many. They are part of heterogeneous networks allowing a number of activities and contributing to their coordination, even though their structure and functioning are opaque to most of the actors involved.

> "Technology has become more refined. There are kits, and people use them even without knowing their technical specifications or how they have been developed… So you use the technique, but you don't really understand its basis and you don't know the limits they can have in terms of results. Scientists specializing in that particular technical field instead know the technique very well; they know what the limits are, but they are confined to that research field and cannot carry out a multidisciplinary study. That's why it is important to create these networks." (Luciano, Senior Researcher)

Due to their characteristics, kits can be compared to "infrastructures". In this regard, it is worth dwelling briefly on the definition of infrastructure given by Star and Ruhleder (1996). First, it must be regarded as a process; it exists because it is in action, and continues to solve the tension between the rules and standards making technologies available on a large scale and local needs, which must be met by these technologies. Infrastructures are so inherently dynamic that it is more difficult to understand *what* they are than *when* they are. According to Star and Ruhleder:

> "An infrastructure occurs when local practices are afforded by a larger-scale technology, which can then be used in a natural, ready-to-hand fashion. It becomes transparent as local variations are folded into organizational changes, and becomes an unambiguous home – for somebody. This is not a physical location nor a permanent one, but a working relation – since no home is universal" (1996, 114).

An infrastructure therefore emerges when a large-scale technology is available and naturally ready to be used in local environments, where it is perfectly embedded into the organization and everyday practice of its users. If we leave out the reference to the "larger-scale technology" for a moment, there is no doubt that kits share some of the characteristics of infrastructures. They occur within networks of heterogeneous actors, which make them socially available; they activate processes requiring standard practices and coordination of actors involved in their use; they are tools that, although produced on a large scale, are adapted to local needs. Similar to large networks and technological systems, kits often appear quite opaque to their local users and only become visible and problematic when their limits emerge (preventing researchers from doing what they would like to do) or they do not work.

> "A less stable kit, which still shows some limits, is for example the one used to perform chromatin immunoprecipitation, which is called ChIP. There are a small number of manufacturing companies, so you have to have clearer ideas. The ChIP kit is designed to measure to what extent a transcription factor is bound to a gene promoter. I used it, but I didn't feel confident about it; it didn't seem reliable enough to me. I once obtained contrasting results, so I went back to the old way and did everything by myself. It took more time, but I couldn't risk it." (Paola, Laboratory Technician)

It is quite evident at this point that kits are not infrastructures as such, but represent a significant part of them. Being disposable, they cannot be identified *tout court* with an infrastructure. While true that the networks in which kits are embedded are extended and heterogeneous, their extension and heterogeneity is certainly very different from those of infrastructures; however flexible and versatile, a kit will never be used by as many different actors and for as many different activities as those made possible by other infrastructures (for example, a database such as GenBank or a computing platform that offers data processing services to a large number of laboratories and experiments). Nevertheless, it cannot be considered as a mere device operating as an isolated item, closed in itself, but instead a component of a larger and articulated system – precisely the characteristic that makes it a valuable research tool.

Because they are integral parts of technoscientific infrastructures, they share some of their most significant characteristics such as opacity, standardization, coordination and networking. In this sense, kits are *boundary objects*, pre-packaged portions of ready-to-hand scientific practices, which, although highly standardized, lend themselves to a variety of uses. It is worth pointing out that this multiplicity is not so much related to the capacity of the kits to adapt to different needs and produce different outputs as it is to the possibility of using their standard outputs – which can be subject to small adjustments – in various experimental contexts.

Although, in molecular biology research, kits often include organic components, they cannot be regarded as bio-objects – that is, a new class of objects made up of organic (vegetal, animal, human) and inorganic materials, which have entered everyday practices as a result of scientific research, such as GMOs. They are instead part of bio-objectification processes, meaning those processes through which "different life forms are created and are given life, and perhaps, multiple lives" (Webster 2011, 2). Within these processes, scientific research plays a leading role and kits, while their contribution is still largely unexplored, are an integral part of it.

3. The contribution of kits to the construction and maintenance of a scientific infrastructure

Kits represent a typical element of contemporary scientific research, which is particularly eager for speed and efficiency. These two aspects are clearly related: the need to conduct experiments in the shortest time possible in order to be competitive in the publishing market urges researchers to adopt increasingly efficient tools. Conversely, the availability of increasingly efficient research tools – which make it possible to obtain results in little time and at relatively low cost – has sped up laboratory activities and enhanced competition. In this scenario, the use of kits has become more and more frequent:

> "Machines, instruments, including kits, greatly reduce the time for analysis. It doesn't take one second per cell, but one microsecond. So in one minute you have analyzed 20 million cells, and so on… many companies now tend to propose kits, instruments and reagents, allowing for much shorter preparation times." (Chiara, Senior Researcher)

> "Kits are used more frequently now. In the space of five years, it has become evident. Even more, whereas five years ago a kit required ten passages, now it requires three. So there has also been an evolution in this. For example, take our DNA extraction kit. I've been here for eight years now, but this kit was even used before. This is to say that it has entered our everyday life, is an integral part of it. So there are things we won't change. We know that with this kit we are pretty sure about results; it is reliable, we don't waste ten days or a week to have a genotype. So, it is hard to change

something that produces data in such a safe and quick way. If you perform the same practice over and over again and the efficiency of your kit is still high, clearly you are not willing to change it so easily." (Paola, Senior Researcher)

Although a researcher's life has apparently become simpler, even compared to the recent past, it is accompanied by increasing levels of stress, due to faster-paced work rhythms. The advent of increasingly efficient technologies has not resulted in more free time and energy to be used for other things, but has multiplied the number of simplified activities to be performed. In short, the rapid conduction of experiments, which would provide energy to other fundamental aspects – such as, reflection, review of literature, discussion with colleagues in and out of the laboratory, writing, checking results – has instead merely led to the conduction of a larger number of experiments.

This is neither good nor bad in itself, but simply means a whole different scenario exists compared to the past. Researchers carry out an increasing number of experiments, and therefore have more occasions to check and analyze results in-depth, as they are able to explore many new possibilities simultaneously. At the same time, the same researchers seem to have less time to reflect on the results they obtain, being absorbed both by the possibility of performing new experiments and the necessity of rapidly publishing their results, so as to be the first to place a missing piece in the great puzzle of science.

As we have seen, the participation of kits in the research process allows the coordination of numerous actors: the researchers in that specific laboratory or others, editors and reviewers of scientific journals, manufacturing and distribution companies, and materials required for their functioning. The networks from which kits are generated are extensive (Latour 1991): if the kit has proved useful, it will be mentioned and recognized in the publication illustrating the research work.

"The kit will be mentioned among reagents or in the dedicated acknowledgements section to 'the specialist who assisted us.' The company can also be cited among the authors – it means a lot to them. In Italy this is perhaps more difficult, but in the United States there are companies paying PhDs or research grants to have specific publications about their kits. In this case, they won't be cited among the authors, and the kit is a standard one." (Marco, Senior Researcher)

"Articles contain detailed descriptions of experimental protocols, where it is said: to perform this part of the experiment we used the Qiagen's Maxi Prep kit. The catalog number can either be included or not, but there is always a reference to the kit. You exactly follow the instructions contained in the kit, so in the paper you write: 'according to the manufacturer's instructions…' or you describe any change you made." (Chiara, Senior Researcher)

By mentioning the kit used in an experiment in a publication, authors can easily provide many pieces of information at the same time. If the readers – whether fellow researchers, editors or reviewers – know the specific kit in question, they can easily understand how a specific experimental passage has been dealt with; otherwise it is equally easy for them to get all information about its functioning through the description provided by the manufacturer, or they can just continue to read and do so only if they need further information. In this way, kits contribute to the standardization and naturalization of data and procedures by stabilizing thresholds, defining values, quantities and compositions, thus reducing the numerous contingent aspects of the scientific activity related to the specificities of a laboratory and the researchers who work in it. At the same time, of course, this reduction also implies a loss of information and the transferring of part of the responsibilities for the results obtained to the manufacturing companies, which therefore become *a priori* guarantors of activities apparently only "performed" by the researchers who use that particular kit. It is not difficult to identify here a profound analogy between kits and scientific instruments whose role in the processes of scientific knowledge production has been studied in STS more deeply.

References to kits made in scientific articles or other communications circulating within the scientific community allow them to be placed in the category referred to by computer scientists as "meta-data" – that is, a set of second-level pieces of information that makes first-level ones comprehensible, providing a suitable interpretation framework.

By observing research work in laboratories, it is possible to see that the participation of kits in scientific research processes exposes them to a double shift, simultaneously both a requirement and proof of their importance.

The first shift is identified by a specific name: "consumption". Being the scientific equivalent of the numerous disposable objects that populate our everyday life, kits have a very clear consumption trajectory: from the researcher's bench to the garbage can or, more often, the toxic waste bin, which involves a further extension of the network of actors involved in scientific research. Toxic wastes are indeed disposed of by specialized companies, following specific procedures and expert competences. As happens in any family, where waste management is based on agreements often difficult to negotiate and subject to periodical conflicts, waste disposal regulations within laboratories (such as putting waste in the right bin) and the measures to make them effective (for instance, the periodical emptying of bins) are a relevant part of the research activity and can become a source of disagreement.

> "In the end, work becomes wearying also for those little things that you have to do every day, such as: who empties the bin?" (Giulia, Junior Researcher)

Kits are not, of course, the only items that fall into the "consumables" category – which is so relevant as to be specifically included in the definition of the budget destined to research projects and in financial reports. Research activity is also made up of pipettes, reagents, organic materials and slides, just to mention a few. Kits, however, have a distinguishing trait: compared to other consumables, they are more – if not the most – complex, and therefore expensive. At the same time, their consumable nature makes them different from other traditional scientific instruments, which are part of the laboratory scene for longer and follow different trajectories when they are discarded. Instruments are not consumed, but worn out; they require maintenance rather than being constantly replaced; they are not disposed of, but disused.

As for the second shift, we have already referred to it when we defined kits as meta-data. After being used in experiments, they are mentioned in scientific articles, catalogs from manufacturing and distribution companies, and databases consulted by researchers to select those that suit them best. Alternatively, they are deemed insufficiently reliable and therefore it becomes necessary to perform that part of the experimental work manually. This shift can be called "de-materialization" or "symbolization", and this transformation is as necessary as the consumption itself. Being able to move from the laboratory bench to the scientific article, the kit is different from many other consumables, which can hardly aspire to be mentioned in the latter.

4. Kits as a learning device

What kits "make it possible to do" in scientific research practice is related to processes of coordination of heterogeneous actors (human and non-human), mainly having a cognitive nature, where cognitive – according to the situated action paradigm (Lave 1988; Lave and Wenger 1991) – does not refer to "mental" activities but expert material actions anchored to local contexts. The introduction, adoption and translation of kits into laboratory life therefore entails a number of processes, which can be generically defined as learning.

First of all, their use makes it necessary to recur to a certain dose of what in literature is called *relationship work* (Gherardi 1990), primarily aimed at maintaining and reproducing everyday social practices. Because kits do not subvert the organization of practices, but gradually inscribe themselves in them, they introduce new forms of action, progressively becoming transparent objects, integral parts of routines intertwining tacit knowledge and previous experience. This does not mean they remain static, but quite the opposite: to become everyday objects, kits require a *work of re-articulation* (Corbin and Strauss 1993). Their introduction necessarily involves a new organization of work, which also occurs through the supervised training of young researchers, exploration of new experimental practices, adoption of new languages and a knowledge of new procedures and protocols. Of course, this process becomes particularly relevant when kits are insufficiently stable, and therefore have a higher interpretative flexibility, making them problematic while, at the same time, more adaptable to the specificities of the local environment.

As components of larger infrastructures – that is, systems that take on shape and life by integrating themselves into extensive networks both in and outside the laboratory – these devices are charged with what Engeström (1999) referred to as *knotworking*, a type of knowledge in action able to "tie separate threads of activity". Their participation in the life of the scientific laboratory is therefore not one of passive dependence, but active contribution to network construction and development. In this sense, they can be regarded as actors in a true Actor-Network Theory perspective.

As objects *of* and *for* learning, kits become a central element in the process of "legitimate peripheral participation", which brings less expert researchers from the periphery to the center of laboratory practice (Lave and Wenger 1991). This characteristic is particularly evident when kits are used by younger researchers, who tend to *take them for granted* as natural components of research practice. Conversely, this process of naturalization is indicated as a problem by senior researchers, their main users being young researchers who recur to them because they simplify procedures, protect them from individual variability and make them quicker and more reliable:

> "Yes, they are especially used by newcomers; they immediately get accustomed to using kits. Of course, it helps, but it is also risky, because they have no idea about their content and, unlike us seniors, if they had to, they wouldn't even know how to perform an extraction of genomic DNA. They just use the kit and that's it (…). So I say to my students: be careful

when you open it, because a kit is a closed box, you don't know the components it contains, you have very simple instructions to follow, it makes your life simpler, but it probably makes you lose something in terms of knowledge." (Chiara, Senior Researcher)

Senior researchers' awareness of the risks implied in the use of the kit as a "closed box" makes them see it as problematic, making them wonder how it can be transformed from a *matter of fact* to a *matter of concern* (Latour 2004). Their suggestion is to open this closed box and explore it through a learning process that should make researchers more aware of the implications of usage:

"Of course, we seniors use them, too. There's no doubt they're useful, but I often use them with younger researchers to teach them how to use, treat and handle them; we try to understand together how they work, what they are made of. Yes, it may take time in the beginning, and young researchers can be quite impatient, but it is important to understand; you have to read instructions, you have to know them, you have to learn how to use them, how to turn them into a resource of the laboratory, rather than just a blind-box solution." (Paola, Senior Technician)

"For every student or anyone entering the lab for the first time, there should be at least one person providing explanations. Once you've examined it the first time together, then you can do it yourself." (Chiara, Senior Researcher)

Whether it is taken as a natural object or a problematic device, the kit *enrolls allies and domesticates bodies and practices* (Callon 1986; Viteritti 2013). In this sense, the enrolment process activated by the kit can be defined as a network-based mechanism of learning; all negotiating actions – from the less problematic adoption to the more active, which transforms the kit into an object of learning – are part of this mechanism. The same happens with the *alignment* process, which implies a translation that involves a number of actors, all converging towards the introduction and incorporation of the new object in the research.

Kits can also be regarded as *programs of actions* (Akrich 1990), since their original design becomes the script that directs practice. Those inscribed in the kit, however, require a local training process, which is activated when the object is used and involves both actors in and outside the laboratory, including the manufacturing companies.

"The people from the company from which we wanted to buy the kit wanted to know exactly the characteristics of our experiment, because they

> had sold this kit also for other purposes. We wanted to use it for a precise application, so, together we said: let's try it. Each of us wanted feedback from the other. So the company gave us a reaction for free; the sales manager [...] and the specialist came to the laboratory and we tried it together. They had never worked on these cells, human neurons, and we had never used a kit for this. Alessia, our researcher, tried it with me and the people from the company. The test worked. They gave us lots of explanations and we decided to buy it and refine it with a certain number of cells. Alessia used it twice and it worked, but it is necessary to take into account the specific parameters we set up. End of the story: we bought it [...], but now that we know how to use it, it's in our hands." (Marco, Senior Researcher)

In a sense, we can say that the kit prescribes the ways to involve the various actors who design and adopt it, making it necessary to establish alliances and agreements, and becoming a partner of scientific practice. Starting from the explicit information stage – conducted with the company's technicians or made available through the instructions for use – the object is transformed into a sort of pre-cooked, ready-to-hand knowledge, progressively incorporated in the practice through processes of learning and appropriation.

It shapes its environment of usage, inscribes specific organizational forms and social practices (Akrich 1990), presents itself as a closed, "turnkey" device with a defined and stable technical content and a set of well-packaged prescriptions. However, to become part of that specific research activity, it must be opened, understood, inscribed, adapted, questioned, acquired and consumed in practice. To exist as a technical object, it must dissolve itself, and its disappearance shows a total symmetry with its users who, together with it, contribute to the experimental process.

Because it is designed and adopted in the presence of numerous specialists who establish alliances with the object and incorporate it in their practice as an everyday object, it requires a multidisciplinary knowledge set.

> "If we take the macro-array kit, for example, it can be used by any average researcher, but to conduct analyses you have to be supported by specific competences. You need the statistician, the bioinformatician, the one who knows the machine, etc." (Marco, Senior Researcher)

The kit therefore becomes a *plastic training object* regulating action, connecting different knowledge sets that require trained (or trainable) users – it necessitates not only skilled and competent users, but needs to be questioned and put to test by expert users. Although in the laboratory it sometimes seems natural to take it for granted, especially when it is in the hands of junior researchers, the kit needs to bring out the network to which it belongs to become part of a specific environment: it must be opened and put into context; it must lose its transparency, its being taken for granted, and must become a problem. The kit can be considered as a training device able to translate networks, shape its context of use, create individual and collective competences as well as languages and practices, becoming a relevant part of the situated socio-material learning (Viteritti 2012).

5. Conclusions: small processes for big infrastructures

Within molecular biology research, kits present themselves as the vehicle for a series of adjustments, which make it possible to mediate between standardization and flexibility, simplicity of execution and specialized competences, rapidity and awareness, creation of infrastructures and mobility/circulation. This mediation would be possible neither in general nor in the abstract, but can only occur in actual practices and situated actions.

Kits are embedded in networks with variable degrees of articulation and extension – from the smallest, related to those more established, to the largest, required by those less so – which are however able to involve a number of different actors, mostly located outside the laboratory. In this way, they participate in the research activity and contribute to its organization, mediating between the necessity to overcome local specificities and the need to not sacrifice them entirely. As boundary objects, they are sufficiently stable to allow the experimental activity conducted in a laboratory to be communicated to other researchers – for example through the references made in scientific publications – but are nevertheless able to adapt to the specific purposes and needs of the laboratory. This is evident in the training of young researchers, who are taught to use kits "cleverly", or in the availability and interest of companies in developing new versions able to meet the requirements of new experiments. In this sense the kits play a relevant and underestimated role in aligning the tasks at diverse organizational levels. It is almost as if the articulation processes necessary in connecting the tasks relative to practices at experimental, laboratory and socio-material world levels find a form of standardization-which renders such processes stable in taken-for-granted routines-in the kits. In this way, rather than carrying out the purification of DNA manually every time (at experimental level), establishing how to proceed, acquiring the various substances and materials (at laboratory level), locating the various companies or other laboratories that produce these substances, negotiating for their supply and convincing colleagues and funders that the right choices have been made (extra-laboratory or socio-material world level), the kit aligns and stabilizes all these tasks and all the actions necessary for their completion at a stroke. At the same time, this stabilization consents those degrees of flexibility that make the kits adaptable to the specific needs of that particular experiment or laboratory.

By observing the laboratory from the perspective of the kits used in it, it is possible to highlight aspects of the research activity in general, and its infrastructural processes in particular, that have so far mostly been neglected. From this perspective, it becomes clear that the distinguishing features of scientific infrastructures – standardization, flexibility and adaptability to local needs, naturalization and opacity – do not magically appear as a product of the assemblage of extensive networks of heterogeneous actors, but are rather the result of the interactions occurring in a situated way between their constituting elements. More than just adapting to local practices by offering services and imposing constraints, scientific infrastructures – similar to databases or technological platforms equipped with research tools able to serve various laboratories and research teams – become the product of an unrelenting work of mediation between standardization and flexibility, made possible and promoted by disposable systems such as kits. In this respect, their consumption can be compared to that of motor fuel: it's not possible to do without it. In the case of kits, however, the comparison might be misleading; their consumption does not in fact merely support the experimental activity, but bears with it the constant combination between local adaptability and general standardization without which scientific research could not go on.

References

Akrich, Madeleine. 1990. "De la sociologie des techniques à une sociologie des usages: l'impossible intégration du magnétoscope dans les réseaux câblés de première génération." In *Techniques et Culture*, (16): 83-110.

Bijker Wiebe E. 1995. *On Bicycles, Bakelites, and Bulbs*. Cambridge, MA: The MIT Press.

Bijker Wiebe E., Thomas P. Hughes and Trevor Pinch (edited by). 1987. *The Social Construction of Technological Systems: New Directions in the Sociology and History of Technology*. Cambridge, MA: The MIT Press.

Bowker, Geoffrey C. and Susan L. Star. 1999. *Sorting Things Out: Classification and Its Consequences*. Cambridge, MA: The MIT Press.

Bruni, Attila. 2005. "La socialità degli oggetti e la materialità dell'organizzare: umani e non-umani nei contesti lavorativi." In *Studi Organizzativi*, 1: 113-129.

Bruni, Attila, Trevor Pinch and Cornelius Schubert. 2013. "Technologically Dense Environments: What For? What Next?" In *Tecnoscienza. Italian Journal of Science & Technology Studies*, 4(2): 51-72.

Callon, Michel. 1986. "Some Elements of a Sociology of Translation: Domestication of the Scallops and the Fishermen of St. Brieuc Bay." In *Power, Action and Belief: A New Sociology of Knowledge*, edited by John Law, 196–233, London: Routledge & Kegan Paul.

Corbin, Juliet and Anselm Strauss. 1993. "The Articulation of Work Through Interaction." In *Sociological Quarterly*, 1: 71-83.

Edwards, Paul N., Steven J. Jackson, Geoffrey C. Bowker and Cary Knobel. 2007. *Understanding Infrastructure: Dynamics, Tensions, and Design*. National Science Foundation Report.

Engeström, Yrjö, R. Engeström and T. Vähäaho. 1999. "When the center does not hold: The importance of knotworking." In *Activity theory and social practice: Cultural-historical approaches*, edited by Seth Chaiklin, Mariane Hedegaard and Uffe Juul Jensen, Aarhus: Aarhus University Press.

Fujimura, Joan H. 1987. "Constructing 'Do-Able' Problems in Cancer Research: Articulating Alignment." In *Social Studies of Science*, 17(2): 257-293.

—. 1992. "Crafting Science: Standardized Packages, Boundary Objects, and 'Translation'." In *Science as practice and culture*, edited by Andrew Pickering, 168-211, Chicago: The University of Chicago Press.

Gherardi, Silvia and Antonio Strati. 1990. "The Texture of Organizing an Italian University Department." In *Journal of Management Studies*, 27(6): 605-618.

Hutchins, Edwin. 1995. *Cognition in the Wild*. Cambridge, MA: The MIT Press.

Karasti, Helena, Karen S. Baker and Florence Millerand. 2010. "Infrastructure Time: Long-term Matters in Collaborative Development." In *Computer Supported Cooperative Work*, 19: 377-415.

Keating, Peter, Camille Limoges and Alberto Cambrosio. 1999. "The Automated Laboratory: Generation and Replication of Work in

Molecular Genetics." In *The Practices of Human Genetics*, edited by Michael Fortun and Everett Mendelsohn, 125-142, Dordrecht: Kluwer.
Latour, Bruno. 1987. *Science in Action. How to follow Scientists and Engineers through Society*. Milton Keynes: Open University Press.
—. 1991. *Nous n'avons jamais été modernes*. Paris: La Découverte.
—. 2004. "Why Has Critique Run out of Steam? From Matters of Fact to Matters of Concern." In *Critical Inquiry*, 30: 225-248.
Latour, Bruno and Steve Woolgar. 1979. *Laboratory Life: the Social Construction of Scientific Facts*. Princeton: Princeton University Press.
Lave, Jean. 1988. *Cognition in Practice: Mind, Mathematics and Culture in Everyday Life*. Cambridge: Cambridge University Press.
Lave, Jean and Etienne Wenger. 1991. *Situated Learning: Legitimate Peripheral Participation*. Cambridge: Cambridge University Press.
Mongili, Alessandro. 2007. *Tecnologia e Società*. Roma: Carocci.
Pellegrino, Giuseppina. 2012. "Beyond an IT-driven Knowledge Society: Knowledge as Intertwined Sociotechnical Circulation." In *European Review - Academia Europæa*, 20(2): 164-172.
Star, Susan L. 1999. "The Ethnography of Infrastructure." In *American Behavioral Scientist*, 43(3): 377-391.
Star, Susan L. and Geoffrey C. Bowker. 2002. "How to Infrastructure." In *Handbook of the New Media*, edited by Leah A. Lievrouw and Sonia L. Livingstone, 230-245, London: Sage.
Star, Susan L. and James R. Griesemer. 1989. "Institutional Ecology, "Translations", and Coherence: Amateurs and Professionals in Berkeley's Museum of Vertebrate Zoology, 1907–1939." In *Social Studies of Science*, 19: 387–420.
Star, Susan L. and Karen Ruhleder. 1996. "Steps toward an ecology of infrastructure: Design and access for large information spaces." In *Information Systems Research*, 7: 111–34.
Viteritti, Assunta. 2012. "Sociomaterial Assemblages in learning scientific practice: Margherita's first PCR." In *Tecnoscienza: Italian Journal of Science & Technology Studies*, 1: 29-48.
Viteritti, Assunta. (2013), "It's the body (that does it)! The production of knowledge through body in scientific learning practice." In *Scandinavian Journal of Management*, Special Issue "*Body, Senses and Knowing in Organization*", 29(4): 367-376.
Webster, Andrew. 2012. "Introduction: Bio-Objects: Exploring the Boundaries of Life." In *Bio-Objects. Life in the 21st Century*, edited by Niki Vermeulen, Sakari Tamminen and Andrew Webster, 1-10, Farnham: Ashgate.

Acknowledgements

The two authors jointly worked and discussed on the chapter. In particular, Federico Neresini is the author of sections 2, 3 and the conclusions. Assunta Viteritti wrote the introduction, sections 1 and 4.

CHAPTER SIX

GIVE US A PROTOCOL AND WE WILL RISE A LAB: THE SHAPING OF INFRA-STRUCTURING OBJECTS

STEFANO CRABU

What can be studied is always a relationship or an infinite regress of relationships. Never a "thing".
(Bateson 1978, 249)

1. Mapping the problem

The attention paid by Science and Technology Studies (STS) to the relationship between human actors and technological artifacts leads us to consider social practices and situated settings of interaction as an emerging outcome of a network of socio-material relationships (Law 1994; Mol 2002; Barad 2003; Orlikowski 2007). As a whole, these relationships involve heterogeneous processes of knowledge constructions, scientific facts, routines and technologies. Particularly, in its ecological acceptation (Star and Greisemer 1989; Collins and Yearly 1992; Fujimura 1995; Suchman 2000), this perspective illustrates how social contexts are generated and permanently reshaped through complex socio-material relationships that involve classifications, conventions and standards, which are sometimes invisible to the same human actors (Mongili 2007).

As pointed out in a seminal article by Star and Ruhleder (1996), infrastructures as well as technical objects are based on specific relational ecologies and are built around particular works and social practices. The definition of infrastructure later proposed by Star (1999) seems particularly evocative and permits us to clarify the analytical position, inspired by the ecological perspective, which will be adopted in this chapter.

> "[...] infrastructure is a fundamentally relational concept, becoming real infrastructure in relation to organised practices (see also Jewett and Kling, 1991). So, within a given cultural context, the cook considers the water system as working infrastructure integral to making dinner." (Star 1999, 380).

Following the suggestion of Star, no material entity or technical object exists for itself in the form of an immanent social fact or fixed context containing entities and actions. On the other hand, infrastructures – and in a broader sense "technologically dense environments" (Bruni 2005; Bruni *et al.* 2013) – represent a relational concept since they only occur within a plot of ecologies of action. As a result, an object becomes sociologically relevant as long as it can be considered a manifestation of organizational properties rather than something purified and separated from its context of creation and use (Star and Greisemer 1989; Star and Ruhleder 1996; Star 2010).

This debate has been particularly relevant for those scholars interested in the sociomaterial processes of production and sharing of scientific knowledge, who set themselves in line with a tradition of thought that has superseded the standard positivistic view that used to assign sociological immunity to scientific knowledge (Latour and Woolgar 1979; Knorr-Cetina 1981). In this respect, far from being a mere objective representation of nature, scientific knowledge has been described as an outcome of the cooperation between human actors, material contexts and interdependent technical entities in a relationship of mutual configuration (Latour 1990; Rabinow 1996; Cambrosio and Keating 1998; Keating and Cambrosio 2003, 2012).

Starting from the previously mentioned theoretical assumptions and based on the data collected during an ethnographic research carried out in a biomedical research centre in Northern Italy, this chapter explores in depth the processes of creation, use and *mise-en-contexte* (Latour 1992, 89) of a specific technoscientific object. Research on infrastructures and technological artifacts in general has traditionally focused on the work of coordination and cooperation made possible by the classifications, standards and protocols incorporated therein (Star and Griesemer 1989; Bowker and Star 1999). In this respect, infrastructures were analysed by looking at the people who took them for granted or the practices of design and development.

In this contribution, however, we will focus on one particular artifact, whose infra-structuring character emerges from the situated activities of use and *mise-en-contexte.* One of the most interesting aspects, as will be shown, lies in the fact that the use and *mise-en-contexte* involve the construction and manipulation of the artifact itself. We will see how a professional community requires not only the skills to use artifacts, but also the knowledge necessary to manipulate and recreate them in everyday activities. In particular, this chapter will focus on the protocols adopted by molecular biology laboratories, which seem to be artifacts of particular relevance for the implementation and positive results of experimental practices. It will be shown how this specific component of laboratories may represent an emblematic case study, which can help thematise the infra-structuring role of technical objects as relational entities supporting the production of scientific knowledge and cooperation between different human subjects.

Assuming this case study as a starting point, we will discuss a number of theoretical issues that will be used to propose and investigate the notion of an *infra-structuring object* as a conceptual device useful for studying the processes of production and sharing of expert knowledge. In particular, assuming the concept of an *infra-structuring object*, we would like to stress how the relationship of mutual generation between technologies, human subjects and contexts implies the creation and management of particular objects that can incorporate and make transparent several elements, such as standards, regulations, pedagogies, routines, conventions and power relationships. The prefix "infra" refers to the possibility that these elements might become transparent (but not less important) during the implementation of work or experimental practices, since they are used in a routine, "natural" and consistent way within the context (except when there are unexpected ruptures), whereas the suffix "structuring" focuses on how objects are allowed to shape and re-shape the local setting, as well as how the ongoing practices may open up new potential uses of such objects. In line with an ecological perspective, the suffix "structuring" serves the need to emphasize the level of "naturalisation" (Bowker and Star 1999) rather than "stabilization" (Bijker 1995), evoking a semantic field that is akin to "plasticity" and the opposite of "rigidity", which is rather suggested by the concept of "structural".

2. Investigating protocols: research context and method

The term *protocol* can be traced back to the mediaeval Latin word *protocollum* and refers to the first sheet (*proto*) glued (*collum*) to the front page of the official records of a transaction (see Oxford English Dictionary). This document was used to specify the total value of a transaction and could also include the name of the notary who drew up the papyrus. From an etymological point of view, the term "protocol" shows an interesting material-discursive juxtaposition, denoting a document that is ancillary to a "main text", which is likewise necessary and contextual to the former. The juxtaposition of the technology of writing and its medium (papyrus) can be considered a precursor of modularity. The two texts could not be separated: the *protocol* was only meaningful in relation to the accompanying main text. In general, the artifact "papyrus" could circulate between a number of different contexts and places thanks to the presence of the *protocol* on the front page.

Following a longitudinal perspective, the protocols have survived to the present day, pervading the clinical world and life sciences[1]. Within the field of biomedicine, they accompany and regulate the experimental practices of every researcher and express the formal image of the expected outcome of the standardization and dissemination of best practices. In clinical and experimental contexts, we may find different types of protocols aimed at regulating the use of personal protective equipment, access to laboratories, extraction of human DNA, culture of cell populations, animal testing and other activities relating to diagnostic practices. As far as molecular biology is concerned, the protocol describes the set of instructions, methods, "ingredients"[2], materials and procedures that make up an experiment.

In spite of the key role played by this object within laboratories and clinical areas, STS have paid little attention to the devious processes of development and situated use of these protocols. Whilst deeply involved in the investigation of research activities within a number of laboratories operating in various fields of natural sciences – such as biochemistry (Latour and Woolgar 1979; Knorr-Cetina 1981), neuroscience (Lynch 1985) or physics (Collins 1975; Collins and Harrison 1975) – works falling within the domain of *Laboratory Studies,* both classical and more recent ones (Latour 1983; Scott 1991; Knorr-Cetina 1995; Neresini 2008; Doing 2008; Viteritti 2012), have so far omitted a detailed analysis of the processes of protocol construction in relation to scientific practices.

Major contributions to the study of protocols come from the field of *Sociology of Health and Illness*, which focuses on the relationship between medical knowledge and materiality within clinical practice (Löwy 1995;

---

[1] See Lynch (2002) for a detailed discussion on the different types of protocols from an ethnomethodological perspective.

[2] Based on my fieldwork, the vast majority of practitioners consider laboratory protocols as a sort of cookbook, and the laboratory is described as if it were a domestic kitchen.

Timmermans and Berg 1997, 2003; Berg 1998). These last contributions, however, fail to point out the protocol in its processual and relational components. The above authors rather preferred to adopt an analytical approach that only considers the protocol as an organizational device consisting of formal elements aimed at regulating and standardizing medical activities.

In light of these considerations, it should be noted how the protocol, even in its formal aspects, displays a relational nature that might be worth investigating. This artifact, as shown later in this chapter, becomes the generator of a heterogeneous techno-scientific context where data is to be produced by all actors in a relatively unambiguous way. Within the context of molecular biology, the need to explain natural phenomena in formal terms – in order to make knowledge available to the relevant scientific community – reveals a very strong tension between local knowledge, tacit knowledge and public knowledge (Knorr-Cetina 1981). Along with the implementation of information infrastructures (Star 1999), this tension was resolved through the systematic and consistent use of protocols, which can lead to the standardization of methods and the construction of knowledge and data that can be made publicly available. However, even in this case, the implementation of protocols may be a complex and difficult task since they require articulation in a local setting and always in different ways (Lynch 2002). For this reason, it would be interesting to investigate the ecology of actions oriented towards the construction, implementation and situated use of protocols, rather than simply focusing on the protocol in itself.

This contribution will focus on two particular types of protocols: the first describes how molecular biology articulates cell culture activities, whereas the second relates to the activities of purification of the plasmid DNA produced within the laboratory. The following considerations are based on a broader five-month ethnographic research that was carried out within a leading institute of molecular oncology located in Northern Italy. The empirical material discussed here was collected during a 12-week ethnographic observation (Silverman 1997) of a team of molecular biologists. At the time of the research, the team was in the process of forming an independent research unit consisting of three junior researchers assisted by a senior biologist, a laboratory technician and occasionally other practitioners with suitable experience in laboratory activities[3]. From a methodological point of view, great attention has been devoted to

[3] For the purpose of privacy and confidentiality, practitioners are hereinafter referred to by fictitious names. All names and sensitive information were also removed from the empirical material used in this chapter.

everyday activities conducted by the three junior researchers. By following a team of novice researchers, it became possible to investigate the activities of construction and use of protocols along with the learning practices of scientific work. More precisely, some ethnographers believe that investigating a team of novices may be a good strategy for becoming aware of a number of problems, doubts and difficulties typical of those who, lacking an in-depth knowledge of a specific practice, do not necessarily take existing data for granted (Schwartz and Jacobs 1979; Angrosino 2005). Novices are defined as such since they need to follow an apprenticeship scheme, which allows ethnographers to collect valuable data they could not normally access. To some extent, it will be the novice – in the guise of the ethnographer – who will reasonably ask questions to experienced researchers and stimulate detailed and valuable reports. By implicitly delegating the task of asking questions, the ethnographer may obtain detailed information on experimental practices without the need to disrupt everyday activities within the laboratory. On the other hand, following the senior biologist and laboratory technician presents the ethnographer with an opportunity to observe the styles and practices for the transmission of expert knowledge, which will then be stratified within the protocol.

The ethnographic fieldnotes, including the dialogues intercourse among actors, have been fully transcribed in digital format. The coding of the transcribed material was carried out through a qualitative analysis software, following the principles of constructivist grounded theory (Charmaz 2006, 2009). More precisely, the adoption of a Grounded Theory-driven process of data coding initially enabled the emergence of descriptive labels. At a later time, this process was accompanied by the development of more theoretical labels oriented by an ecological approach in order to consider all the elements involved in the creation, use and contextualization of the protocols.

3. Protocols in action: the shaping of an infra-structuring object

Starting from two emblematic ethnographic accounts, we will now examine the ecology of actions oriented towards the construction and *mise-en-contexte* of two different protocols. In the first account, the protocol relating to the culture of cell lines was "hand-built" by the practitioners themselves, whereas in the second account, the practitioners were involved in the "personalisation" and contextualisation of a protocol supplied to the laboratory by a Biotech Company operating on a global scale. This last protocol relates to the purification of the plasmid DNA.

3.1 Cell culture and the handmade protocol

In this section we will see biologists getting involved in the development of an experimental method aimed at identifying the sequence of actions necessary to ensure that an immortal cell line – known as HeLa[4] – reproduces itself in a suitable environment. These cells offer an experimental biological model, which will be used for the implementation of biomedical research activities within the laboratory, such as cytoxicity testing of molecules that are potentially therapeutic for human beings. The results should be published in specialised international journals. For this reason, it is extremely important that the practice of cell culture is suitably learned by all staff members.

> "At 10:20 AM, Luigi, the senior biologist, interrupts the three young biologists, who were busy writing some notes, and says: "If you write the protocols in pdf format, preferably in English, they can become part of the laboratory internal documentation, which can then be used as reference by other practitioners and foreigners who will join the lab in the future". Luigi then starts describing in great detail the set of activities to be carried out. Chiara, Pippo and Leila – the three young biologists – carefully take notes using pen and paper. After a few minutes, Luigi stops and says: "Let's start the test, come on! Ask Colombo for the cells. He will lend us a flask. If you experience any difficulties, you can find me at the director's office. After a few minutes I follow Chiara along the corridor and, somehow irritated, she says: "When Luigi speaks and gives instructions... I mean, it is something that requires concentration. He says things in a rush, without giving you the time to... What are you doing? You have to take note immediately so that you can start working and testing. Under the bio-safety cabinet, I say. Because you will have to write a description of the methods. In other words the protocol needs to be accurate."

This episode shows how the protocol goes through a number of steps during its development, before it can reach a suitable configuration for circulation within the laboratory itself. As in the situation described above, the process involves moving from one text to another, from oral exchanges to written drafts (Figure 6-1), until an object is developed that will become

[4] HeLa cells consist of a type of immortal cell line widely used in cancer research. See Landecker (2007) and Skloot (2010) for further details about the generation of this cell line. For a discussion on the technoscientific implications of HeLa cell line, refer to Casati *et al.* (2012).

collectively relevant when it is shared.

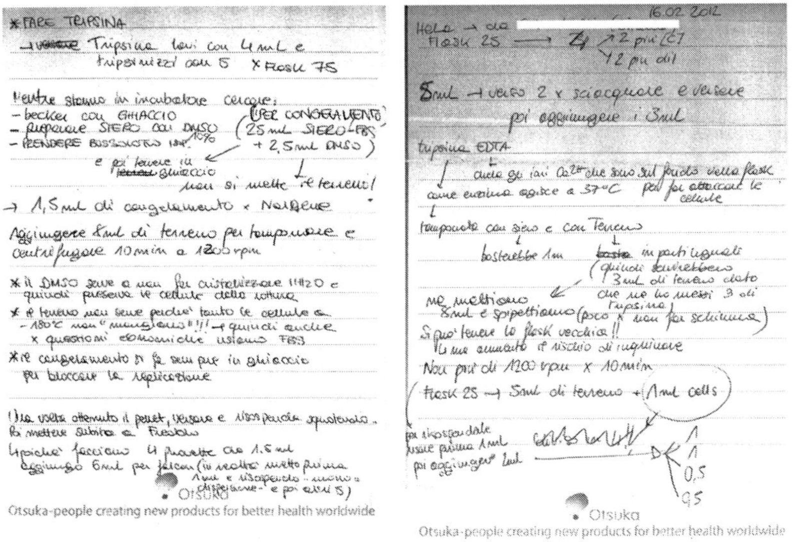

Fig. 6-1: laboratory handwritten draft.

Based on the description provided by Luigi, the three junior biologists share ideas and compare their notes for consistency, before starting the experimental activity under the biosafety cabinet with the help of additional artifacts and technologies.

In their rough draft notes, as shown in Figure 6-1, the junior biologists write down any information they may find relevant for the construction of the final protocol. These notes also include a number of particularly colourful expressions coming directly from the jargon of experimental practice, for example: "agitare un po' la fiaschetta" (lightly shake the flask), "spipettare almeno dieci volte per staccare bene le cellule" (pipette at least 10 times to detach the cells), "dare colpetti al fondo della fiaschetta" (tap the bottom of the flask), etc. These expressions will then be removed from the finalised protocol, which will be prepared in electronic format, printed and distributed in electrostatic copy. The construction of the protocol requires ongoing negotiation and careful consideration on the part of Chiara, Pippo and Leila during the first experimental practices. This stage basically includes the initial testing of the protocol for usability, where an attempt is made to align, in a meaningful way, orality (oral description), writing (notes), technologies (biosafety cabinets, pipettes, centrifuges, reagents) and biological entities (cells).

The use of pen and paper, and eventually a word processing tool, shows how the process of protocol construction takes place through the mediation of different technological genealogies, which should be aligned, layered and made mutually compatible. These technologies enable the senior researcher to translate the smooth, approximate and narrative description of experimental activities into usable documentation. Junior researchers, for their part, have the opportunity to learn specific relational skills that allow them to note down in a didactic and sufficiently clear way the basic concepts for the design and implementation of experimental activities: technologies to be used, places to cross, reagents and time scales. The translation of the oral description into written notes, and eventually a computer-processed protocol, should be accurate, concise and consistent. The finalised and ready-to-use protocol (Figure 6-2) must be capable of circulating within the laboratory and the drafter should have in mind different recipients and users other than himself. In this respect, the style and register of the finalised protocol will be aimed at a greater formality, behind which lie routine, skills and knowledge taken for granted and treated as common heritage by the professional community.

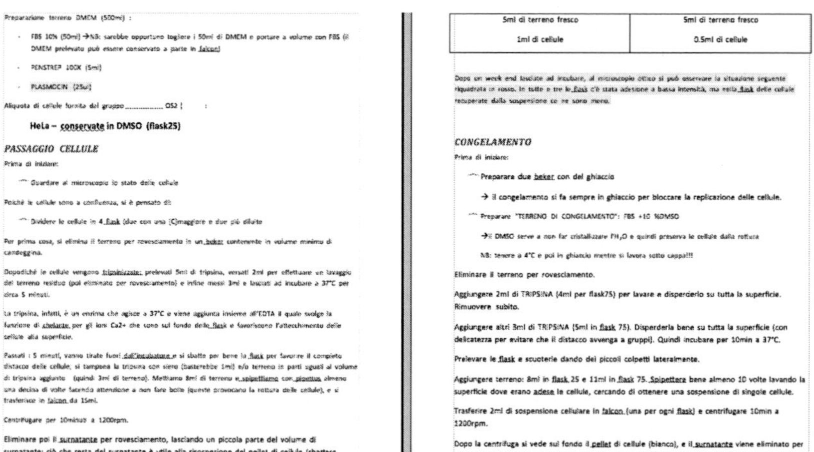
Preparazione terreno DMEM (500ml) :

- FBS 10% (50ml) →NB: sarebbe opportuno togliere i 50ml di DMEM e portare a volume con FBS (il DMEM prelevato può essere conservato a parte in falcon)
- PENSTREP 100X (5ml)
- PLASMOCIN (25ul)

Aliquota di cellule fornita dal gruppo................. OS2 ;

HeLa – conservate in DMSO (flask25)

PASSAGGIO CELLULE

Prima di iniziare:

- Guardare al microscopio lo stato delle cellule

Poiché le cellule sono a confluenza, si è pensato di:

- Dividere le cellule in 4 flask (due con una [C]maggiore e due più diluite

Per prima cosa, si elimina il terreno per rovesciamento in un beker contenente in volume minimo di candeggina.

Dopodiché le cellule vengono tripsinizzate: prelevati 5ml di tripsina, versati 2ml per effettuare un lavaggio del terreno residuo (poi eliminato per rovesciamento) e infine messi 3ml e lasciati ad incubare a 37°C per circa 5 minuti.

La tripsina, infatti, è un enzima che agisce a 37°C e viene aggiunta insieme all'EDTA il quale svolge la funzione di chelante per gli ioni Ca2+ che sono sul fondo delle flask e favoriscono l'attecchimento delle cellule alla superficie.

Passati i 5 minuti, vanno tirate fuori dall'incubatore e si sbatte per bene la flask per favorire il completo distacco delle cellule; si tampona la tripsina con siero (basterebbe 1ml) e/o terreno in parti uguali al volume di tripsina aggiunto (quindi 3ml di terreno). Mettiamo 8ml di terreno e spipettiamo con pipettus almeno una decina di volte facendo attenzione a non fare bolle (queste provocano la rottura delle cellule), e si trasferisce in falcon da 15ml.

Centrifugare per 10minuti a 1200rpm.

Eliminare poi il surnatante per rovesciamento, lasciando un piccola parte del volume di surnatante: ciò che resta del surnatante è utile alla risospesione del pellet di cellule (sbattere delicatamente la falcon). Alla risospensione vengono aggiunti 3ml di terreno fresco e si spipetta con pipetta da 2ml; viene prelevato tutto e distribuito aggiungendo a 5ml di terreno 1ml in due flask25 e 0.5ml nelle altre due flask25. Abbiamo usato anche la flask vecchia (anche se facendo

5ml di terreno fresco 1ml di cellule	5ml di terreno fresco 0.5ml di cellule

Dopo un week end lasciate ad incubare, al microscopio ottico si può osservare la situazione seguente riquadrata in rosso. In tutte e tre le flask c'è stata adesione a bassa intensità, ma nella flask delle cellule recuperate dalla suspensione ce ne sono meno.

CONGELAMENTO

Prima di iniziare:

- Preparare due beker con del ghiaccio

→ il congelamento si fa sempre in ghiaccio per bloccare la replicazione delle cellule.

- Preparare "TERRENO DI CONGELAMENTO": FBS +10 %DMSO

→il DMSO serve a non far cristallizzare l'H_2O e quindi preserva le cellule dalla rottura

NB: tenere a 4°C e poi in ghiaccio mentre si lavora sotto cappa!!!

Eliminare il terreno per rovesciamento.

Aggiungere 2ml di TRIPSINA (4ml per flask75) per lavare e disperderlo su tutta la superficie. Rimuovere subito.

Aggiungere altri 3ml di TRIPSINA (5ml in flask 75). Disperderla bene su tutta la superficie (con delicatezza per evitare che il distacco avvenga a gruppi). Quindi incubare per 10min a 37°C.

Prelevare le flask e scuoterle dando dei piccoli colpetti lateralmente.

Aggiungere terreno: 8ml in flask 25 e 11ml in flask 75. Spipettare bene almeno 10 volte lavando la superficie dove erano adese le cellule, cercando di ottenere una sospensione di singole cellule.

Trasferire 2ml di sospensione cellulare in falcon (una per ogni flask) e centrifugare 10min a 1200rpm.

Dopo la centrifuga si vede sul fondo il pellet di cellule (bianco), e il surnatante viene eliminato per rovesciamento lasciando una piccola parte di volume al fine di non rischiare di perdere le cellule e per favorirne l'immediata rispspensione. La falcon viene messa subito in ghiaccio.

Il congelamento delle cellule va fatto in criovials (NALGENE) conteneti 1.5ml di cellule risospese in

Fig. 6-2: ready-to-use protocol.

The three junior researchers actively cooperate and share ideas during experimental practice. They discuss and negotiate in order to identify the best procedure to be used for the development of a "scientifically sound and functional" method. This becomes clear when the three inexperienced biologists are assisted by Gina, an expert laboratory technician.

> **Leila:** "Gina, how long do you leave HeLa cells in the incubator? Ten minutes?"
> **Gina:** "What? Ten minutes is too long. Come on, that is way too long."
> **Leila:** "The other cells, I used to incubate them for five minutes."
> **Gina:** "Do you mean exactly 5 minutes? Well, roughly... I never take the time. At least that's how I do it... in a while you are going to take them off and observe them through the microscope to make sure that everything is OK. My dear, you just have to get used to it. I have been working on cells and toxicity testing of drugs for many years. And in that case you have to be extremely careful. You cannot make mistakes when preparing toxicity bioassays [assessment experiments]. But do not worry, you are just starting out" [laughter] While we are waiting for the incubator, I'll go get a becher with some ice. Working on cells requires low temperatures, but you know that, don't you? And I cannot see any ice round here!"
> **Leila:** "Well, I don't know. But Luigi told me that I should follow you and do as you do. He always tells me that I should do as Gina does."
> **Gina:** "Well, everyone knows that. And now you know it too." [laughter]

Based on the above excerpt, we are able to highlight the key role played by discursive practices in the process of protocol configuration. As in other productive organisations, even in the world of science, narratives should be treated as a work tool (Gherardi 2012). The words of Gina, a particularly influential practitioner within the laboratory, transmit local knowledge that resists formalisation, while denoting technical mastery and professional community membership. It is a narrative knowledge (Bruner 1990) that supports both the construction of the protocol and the learning of scientific work. While hearing these stories, apprentices endorse them, experience them first-hand and begin to learn in a competent way the meaning of "being an expert biologist". In this case we are not in front of an emic representation of science as knowledge oriented by the principle of falsifiability (Popper 1992). The words of Gina rather show a sort of narrative knowledge (Bruner 1990) oriented by the principle of plausibility, which contributes to the structuring of the experimental style and skills of junior researchers. These skills will be gradually acquired by apprentices and will become part of the protocol, thus becoming (inter)textual, invisible and indisputable, as long as that protocol is needed to keep the cell line alive.

A second remark concerns the microphysics of power (Foucault 1972) that lies under the process of protocol configuration. In this case, what confers power and authority to Gina's utterances is not so much her formal status, but rather her experience and the ability to convey professional knowledge that requires considerable communication skills. These attributes subvert the formal hierarchical order to the extent that it is Luigi himself, a senior researcher, who asks the apprentices to scrupulously observe Gina's instructions. Despite Gina's formal educational background, which is definitely less qualified than the apprentices, she is able to manage the knowledge and ability to mediate between technologies, biological entities and human subjects, and this confers her a key role within the laboratory. The practices of expert knowledge transmission, in this case, show how power is not distributed in a dichotomous and formal way, which introduces a clear distinction between dominated and dominants. Power is not localised, unilaterally exercised and discretionarily held by someone. Power, as a relational concept, is rather exercised through a changing, ubiquitous and widespread relational ecology. During this first ethnographic account, the construction of a ready-to-use protocol called for an ecology of actions that might have been particularly relevant for the learning activities of scientific work. At this stage, we are able to identify a number of features that can help clarify the nature of the protocol as an infra-structuring object.

First of all, the setup of methods highlighted its "structuring" dimension. As a result, the laboratory has been processually reconfigured, not only as a place of knowledge production, but also as a pedagogical educational setting (Fenwick and Edwards 2012; Viteritti 2012), where knowledge acquisition and sharing take place in a considerably different way as compared to higher education systems. In this respect, the protocol triggered a number of subjectivation processes that can affect the identity development of the actors involved. Gina is not simply a lab technician, she is a "teacher" and custodian of expert knowledge. Novices, on the other hand, will learn new skills and techniques and will reconfigure their identity towards the profile of the "professional biologist".

In addition, the method setup and development ensure incorporation into the protocol of the (inter)textual dimensions (knowledge, pedagogies, power relationships, routines and conventions) which will, through use, assume an invisible and transparent cogency.

3.2 DNA purification and profaned protocol

"Gina is preparing the list of orders. I was surprised by the cost of some products. Gina explains to me: "KITs are very expensive. Some of them may cost up to 10,000 EUR. They should bear the CE marking, should be standardized and so on... I mean, for certain things you would only use KITs now. As explained by the directors, you need to specify what KIT you use every time you publish an article."

In the last few decades, scientific laboratories have increasingly been using standardized instruments called KITs. They include a number of heterogeneous elements (reagents, solutions, buffers, pipettes, instructions for use, and so on) with nomenclature and standard features for the optimisation of experimental practice. These biotechnological entities are produced by Biotech Company and are always accompanied by a protocol containing the relevant operating instructions. By way of introduction, a KIT can therefore be understood as a tool that incorporates a reified and standardized form of knowledge. In this section, we will see what happens when experimental activities require the use of a KIT and a protocol that is produced by an external organisation.

In this case, biologists will be engaged in the *mise-en-contexte* of a standardized method for the purification of the plasmid DNA[5] previously produced *in vitro* within the laboratory itself. Once ready to use, the KIT protocol raises questions typical of technologies that require articulation within a local setting (Latour 1993). Following the ecology of actions that are put in place "to accommodate" the KIT within the laboratory will help us better understand the infra-structuring nature of protocols.

"Pippo is carefully writing down some notes on the KIT protocol. He lays down his pen on the counter and says to me: "I have added Luigi's instructions. Look, there are two techniques to purify DNA. The double centrifuge, also included in the KIT, and the other one with the sterile gauze. Here we use the one with the sterile gauze. Our centrifuge does not seem to be suitable for this KIT. That's how we do it in our lab. I mean, if you have laboratory skills, you can also do it your way. But only because you have experience. For instance, Luigi, who has ten years' laboratory experience, may slightly deviate from the KIT protocol."

The situation just presented suggests very clearly how a standardized protocol requires a process of translation within local settings (Latour

[5] In molecular biology, the use of plasmids (small DNA molecules) is very popular since it allows biologists to multiply specific genes of interest by means of bacterial colonies. Purification of plasmid DNA is an operation of genetic manipulation that is rather important in biomedical research, and the ability to perform it further, strengthens the sense of belonging to the professional community of molecular biologists.

1987, 1993). In this case, it is necessary to transpose other artifacts, procedures, conventions and standards from one context to another through local adaptations. Adaptation processes are complex and operate at different levels. They can be mainly found in the official protocol. This text is transformed and customised (Figure 6-3) by junior biologists in relation to their situated learning trajectory and any technologies they use that are not covered by the official document. This happens because the laboratory, which was expected and inscribed in the KIT by Biotech Company, does not match the experimental setting at Pippo's disposal. Through its situated articulation, the official protocol is corrected and amended based on local technologies (e.g. the centrifuge), conventions and situated experimental practices.

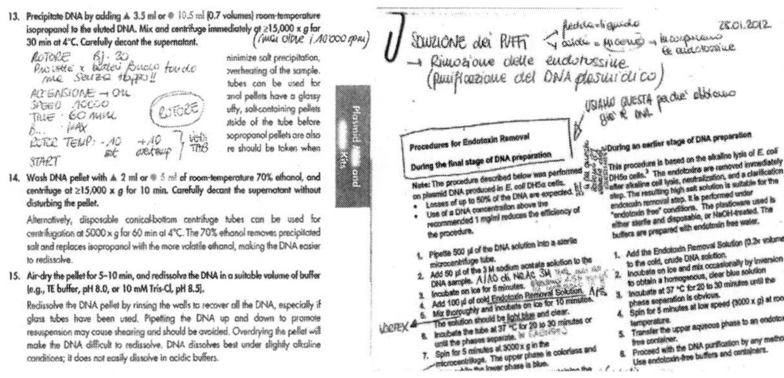

13. Precipitate DNA by adding 3.5 ml or 10.5 ml (0.7 volumes) room-temperature isopropanol to the eluted DNA. Mix and centrifuge immediately at ≥15,000 x g for 30 min at 4°C. Carefully decant the supernatant.

…inimize salt precipitation, …overheating of the sample. …tubes can be used for …anol pellets have a glossy …uffy, salt-containing pellets …tside of the tube before …sopropanol pellets are also …re should be taken when

14. Wash DNA pellet with 2 ml or 5 ml of room-temperature 70% ethanol, and centrifuge at ≥15,000 x g for 10 min. Carefully decant the supernatant without disturbing the pellet.

Alternatively, disposable conical-bottom centrifuge tubes can be used for centrifugation at 5000 x g for 60 min at 4°C. The 70% ethanol removes precipitated salt and replaces isopropanol with the more volatile ethanol, making the DNA easier to redissolve.

15. Air-dry the pellet for 5–10 min, and redissolve the DNA in a suitable volume of buffer (e.g., TE buffer, pH 8.0, or 10 mM Tris-Cl, pH 8.5).

Redissolve the DNA pellet by rinsing the walls to recover all the DNA, especially if glass tubes have been used. Pipetting the DNA up and down to promote resuspension may cause shearing and should be avoided. Overdrying the pellet will make the DNA difficult to redissolve. DNA dissolves best under slightly alkaline conditions; it does not easily dissolve in acidic buffers.

Plasmid and Kits

28.01.2012

Procedures for Endotoxin Removal

During the final stage of DNA preparation

Note: The procedure described below was performed on plasmid DNA produced in E. coli DH5α cells.
- Losses of up to 50% of the DNA are expected.
- Use of a DNA concentration above the recommended 1 mg/ml reduces the efficiency of the procedure.

1. Pipette 500 µl of the DNA solution into a sterile microcentrifuge tube.
2. Add 50 µl of the 3 M sodium acetate solution to the DNA sample.
3. Incubate on ice for 5 minutes.
4. Add 100 µl of cold Endotoxin Removal Solution.
5. Mix thoroughly and incubate on ice for 10 minutes. The solution should be light blue and clear.
6. Incubate the tube at 37 °C for 20 to 30 minutes or until the phases separate.
7. Spin for 5 minutes at 3000 x g in the microcentrifuge. The upper phase is colorless and … the lower phase is blue.

During an earlier stage of DNA preparation

This procedure is based on the alkaline lysis of E. coli DH5α cells.[3] The endotoxins are removed immediately after alkaline cell lysis, neutralization, and a clarification step. The resulting high salt solution is suitable for the endotoxin removal step. It is performed under "endotoxin free" conditions. The plasticware used is either sterile and disposable, or NaOH-treated. The buffers are prepared with endotoxin free water.

1. Add the Endotoxin Removal Solution (0.2x volume) to the cold, crude DNA solution.
2. Incubate on ice and mix occasionally by inversion to obtain a homogenous, clear blue solution.
3. Incubate at 37 °C for 20 to 30 minutes until the phase separation is obvious.
4. Spin for 5 minutes at low speed (3000 x g) at room temperature.
5. Transfer the upper aqueous phase to an endotoxin free container.
6. Proceed with the DNA purification by any method. Use endotoxin-free buffers and containers.

Fig. 6-3: additional text used to contextualize the kit.

Moreover, it may happen that the official protocol is not sufficiently clear and requires additional interpretation. In this case, junior researchers, assisted by an expert biologist, refer to other texts (Fig. 6-3) – often found on the web – that will complement the official method included in the KIT. These activities of textual stratification, as in the case of the *handmade* protocol, are closely intertwined with learning practices. Junior researchers working on the *mise-en-contexte* of the KIT acquire new techniques and skills together with the ability to carry out protocol usability testing. In this case it seems appropriate to use the image of "flirt" (Bruni 2011) as a metaphor for the interpretation of the relationship between human subjects and heterogeneous materials. Through the situated use of KITs, biologists establish an endless game of sociomaterial relationships so that the KIT and method can interact with the local experimental setting. This triggers an ecology of actions showing the attempt to redefine and re-contextualize official practices. Through the metaphor of "flirt" a set of local knowledge, conventions, habits and uses (Knorr-Cetina 1981) are juxtaposed and layered as (inter)texts in the official protocol. Through the daily use of the KIT, this knowledge becomes routine, ends up being taken for granted and is eventually naturalized (Latour 1993), if only temporarily.

At this stage, it would be interesting to further investigate what happens within the context of a laboratory when working with the KIT and its protocol.

> "While giving instructions to Chiara and Pippo, Pamela (senior biologist) says: "You see, we work more and more often with KITs. Before we used to do everything by hand. We used to prepare the KIT ourselves. And then, you know, there was no official document or hardly any. All you had was your own notes. But it was by oral transmission. You had to work side by side with someone more experienced than you and you just had to get used to it. Now, you see, everything is standardized and you end up losing a number of skills and abilities. The funny thing, though, is that the KIT always shows the word "PROB"[6] in plain view, and that's how they cover their back. But why? Because, as I usually remind young people, you also need to be able to operate the KIT. I mean, there is a set of margins to be taken into account when using it."

Pamela's words suggest how the process of KIT articulation results not only in a transformation of the official methods, but also in the reconfiguration of the local experimental setting. This is possible because the KIT incorporates knowledge, standards and conventions that may affect established practices or make them obsolete and transform the way of learning and teaching scientific research practices. The *mise-en-contexte* and naturalisation of a KIT protocol can therefore be interpreted as the emerging outcome of the conflict between formal requirements and situativity. This conflict finally requires a process of articulation involving the structuring of local settings. Within the situated setting, the initial object is modified and the process stimulates the transformation of the setting itself.

Focusing on the *mise-en-contexte* of the protocol we were able to analyse the ecology of relationships that come to be established between actors (not only humans) shaping experimental practices. However, another important point needs to be made. Thanks to their incorporation of

---

[6] The "PROB" marking indicates that analyses performed using a KIT are probabilistic and are always subject to a margin of error. This allows Biotech Company to decline any responsibilities towards individual laboratories in case of error.

standards – officially recognised by the international scientific community – the KIT protocols always bear a "solid" component (e.g. the standards relating to reagents). This component never dissolves within a plurality of contexts. This feature enables the production of scientific data that can be circulated on a global basis through publications in international journals. However, further adjustments are always possible, and sometimes even necessary, in order to keep the dialogue between ready-to-use methods and local settings alive.

> "Gina is explaining to Giovanna (a trainee) the need to constantly update the laboratory logbook. With an assertive tone, Gina says: "The logbook should document everything, especially when there is something new. This is because the operators in charge of the experiment can clearly see what you did. For instance, it may happen that one of the reagents included in the KIT might have been replaced and you wouldn't even realize because you know the protocol by heart and nobody noted this down in the laboratory logbook. We have already experienced similar situations in the past."

This conversation reveals the involvement of practitioners in an endless process of maintenance of the protocol in relation to its context of use. Once again, this can be achieved through the (inter)textual juxtaposition and the introduction of other objects (e.g. the laboratory logbook), the emergence of which was stimulated by the method included in the KIT. In this respect, the activity of *mise-en-contexte* should be interpreted as an outcome rather than a precondition dictated by the alleged cogency of the organisational structures within the laboratory. The relevant protocol enables a broader techno-scientific project requiring continuous adaptation between protocol and context, and between pre-inscription and use.

In this section, the analysis of the processes of *mise-en-contexte* allowed us to better understand the key role played by infra-structuring objects in shaping the relationship between human actors, technologies and experimental contexts. We have seen how the protocol may lead to the creation of new texts that will be circulated within the laboratory. At the same time, the protocol has come to incorporate a heterogeneous set of conventions, routines, habits and uses that were stimulated by the relationship between KIT components, human subjects and local setting.

4. The protocol is not only a data-shaping device...

Based on some distinct ethnographic accounts, we have seen how the protocols, unlike their apparent rigidity, may be open to potentially new and subversive uses of the action plans included therein through their situated use (Suchman 1987). Similarly, when methods are locally constructed and/or articulated, they enter daily practices, becoming part of the techno-scientific background to whose generation they contributed. Having (inter)textually incorporated procedures, classifications, standards, technical convergences, organisational processes and power relationships, protocols may be considered socio-material entities that turn scientific research practices and their organisation into a visible and justifiable disciplinary regime of truth. In this way, scientific knowledge can be communicated to the whole scientific community.

Both ethnographic events show a common tendency to consider protocols as infra-structuring objects. This suggests the focus should be on the ecology of actions that come to be established between knowledge, scientific practices, human subjects and technologies. On the basis of the above considerations, we are now able to identify more precisely the infra-structuring nature of protocols.

A first dimension of the protocol that allows us to conceptualise it as an infra-structuring object lies in its modular character. In particular, this refers to the "infra-" dimension of the object. We have seen how the protocol incorporates a number of different elements in a stratified way. The concept of modularity is particularly relevant for the identification of the different layers forming the infra-structuring object, including certifications (e.g. nomenclature of reagents and biological entities), standards (e.g. reaction time, concentration of reagents, solutions and buffers), pedagogies, technical negotiations and different technological genealogies (written notes, digitisation, computer processing), which become a transparent and naturalised routine through use. In this case, the most prominent feature is the fact that the infra-structuring object cannot be modified in full following its construction. Each change may involve a specific level and should be integrated with other levels through little adjustments. Moreover, given their stratified nature, infra-structuring objects are never created *ex novo,* but rely on an existing infrastructural basis (e.g. the centrifuge available in the laboratory), which shapes their qualities and attributes.

A second feature, which is closely connected to the "structuring" dimension, is the ability of infra-structuring objects to transpose artifacts, information, standards and conventions from one context to another.

Practitioners attempt to contextualise these elements by "flirting" (Bruni 2011) with the protocol. To do so, it becomes necessary to develop organisational routines and innovate experimental practices through the cooperation between human subject, technological devices and biological entities (e.g. cells, plasmids and DNA). This characteristic contributes to the generation of the techno-scientific context, which should not be treated as a pre-existing and unchangeable social fact.

Finally, the third feature that allows us to establish a connection between the two dimensions described above refers to learning and knowledge. As we have seen through the analysis of scientific laboratory protocols, infra-structuring objects are complex and heterogeneous sets of elements, which, once naturalised, become part of the current scientific practices now taken for granted and institutionalised. As a result, the ability to use protocols allows novices to become part of an institutionally recognised professional community. This latter feature, being based on work practices reproduced and supported collectively, is the main carrier of the socialisation processes of novices. The protocol can therefore be considered an anchor for learning and is able to transform the laboratory into a teaching setting. Ultimately, this aspect implies the ability to manage the different knowledge regimes incorporated into infra-structuring objects. The creation and *mise-en-contexte* of protocols is therefore intertwined with a set of situated abilities (Suchman 1987) that exceed the formal and paradigmatic level. These skills allow actors to carry out their daily activities and are particularly relevant in the socialisation processes of novices.

The above considerations show the ambivalent nature of protocols, which cannot simply be considered as devices to "extract order out of disorder" (Latour and Woolgar 1976, 36-37) and produce scientific data. The interpretative lens offered by the ecological approach shows a much more complex set of elements, which can guide us through a broader final analysis of the concept of infra-structuring objects.

Skeptical observers may question if the protocol can be considered, more simply, just a boundary object (Star and Griesemer 1989; Star 2010). Star and Griesemer developed their concept to take into account the coordination and management of work through multiple and divergent actors. More precisely, the contribution of Star and Griesemer is, therefore, related to the problem of how members of different social worlds interact, and they argue that a boundary object can facilitate the multiple translation of knowledge to arrange a settlement between multiple social worlds. My work, in contrast, did not limit its focus on cooperation between actors from different social worlds. An infra-structuring object differs from a boundary object, because it is used by scientists of the same professional community to define a space of technical work. In this sense, the concept proposed here is less abstract, and it helps to understand how a technoscientific context becomes less ambiguous and less amorphous. In this sense, an infra-structuring object does not serve only as a communication interface between different social worlds, but rather as a device of dialogue between human actors who have to share the same line of work and resources (e.g. materials, roles, skills and tools).

5. ... it is an infra-structuring object!

Following the daily activities of creation and contextualisation of the protocols used in molecular biology, the aim of this chapter was to show how the infra-structuring object is the emerging outcome of an ecology of actions in the form of written notes, narratives, scientific experimentation and so on. Of particular interest is the type of ecology implemented for/by the *infra-structuring object* itself. In this respect, what is more interesting is not so much the purposeful evolution of these objects, but rather the ongoing naturalisation of conventions, routines, practices and knowledge through/in the protocol. Ultimately, the infra-structuring object accompanies human subjects into everyday practices, proving to be an individual and collective resource allowing the interaction between knowledge, artifacts and technologies within a space that covers the molecular (e.g. *in vitro* molecular activities) and global levels (e.g. publication of results in international journals).

In light of these considerations, we are able to identify three main features that could allow the extensive use of the concept of an infra-structuring object as a conceptual device useful for the study of the contemporary processes of knowledge creation:

1. **Heterogeneity**: the infra-structuring object is the multidimensional and reticulated expression of a set of heterogeneous elements. These elements include discursive and non-discursive practices, narratives, technologies, biological entities, conventions, routines, standards and classifications;
2. **Situativity**: the infra-structuring object is sufficiently plastic to undergo modifications, specific transformations and local articulations. From an analytical perspective, local articulations activate situated relationships of power between the actors involved in the management, sharing and

acquisition of a set of abilities oriented towards the use and *mise-en-contexte* of the object itself;

3. **Generativity**: the infra-structuring object includes a number of sufficiently solid components that can generate transformations of the local settings of use. These elements also enable the mediation and interaction between different practices distributed on a molecular, local and global level.

As a result, infra-structuring objects manifest a hybrid character, which allows for the joint analysis of human actors, technologies, practices of knowledge production and power relationships that hold these elements together. In this respect, infra-structuring objects stimulate the combined action of human subjects, technologies and knowledge. In particular, they support collective action, allowing the interaction between human subjects and other technological objects. From a theoretical point of view, this concept consents to focus on the practices of creation and use of specific technological artifacts in order to identify the ecology of actions needed to build and use them.

Finally, infra-structuring objects also grant mediation between the material and social dimension. Potentially, any configuration between social and material dimensions could be naturalised. However, this naturalisation always represents a plastic outcome that is open to change.

References

Angrosino, Michael. 2005. "Recontextualizing observation: ethnography, pedagogy, and the prospects for a progressive political agenda." In *The Sage Handbook of Qualitative Research,* edited by Norman K. Denzin, and Yvonna S. Lincoln, 729–745, London: Sage.

Barad, Karen. 2003. "Posthumanist performativity: Toward an understanding of how matter comes to matter." In *Signs*, 28: 801–31.

Bateson, Gregory. 1978. *Steps to an Ecology of Mind*. New York: Ballantine.

Berg, Marc. 1998. "Order(s) and Disorder(s): Of Protocols and Medical Practices." In *Differences in Medicine: Unraveling Practices, Techniques, and Bodies*, edited by Annemarie Mol and Marc Berg, 226–49. Durham, NC and London: Duke University Press.

Bijker, Wiebe E. 1995. *Of Bicycles, Bakelites and Bulbs: Toward a Theory of Sociotechnical Change*. Cambridge, MA: The MIT Press.

Bowker, Geoffrey C. and Susan L. Star, 1999. *Sorting Things Out: Classification and Its Consequences*. Cambridge, MA: The MIT Press.

Bruner, Jerome. 1990. *Acts of Meaning*. Cambridge, MA: Harvard University Press.

Bruni, Attila. 2005. "Shadowing Software and Clinical Records: On the Ethnography of Non–Humans and Heterogeneous Context." In *Organization*, 12: 357–78.

—. 2011. "Lavorare in sala operatoria, flirtare con la materialità del mondo." In *Etnografia e Ricerca Qualitativa*, 4: 83–97.

Bruni, Attila, Trevor Pinch and Cornelius Schubert. 2013. "Technologically Dense Environments: What For? What Next?" In *Tecnoscienza. Italian Journal of Science & Technology Studies*, 4: 51–72.

Cambrosio, Alberto and Peter Keating. 1998. ""Going Monoclonal": Art, Science and Magic in the day–to–day use of hybridoma technology." In *Social Problems*, 35: 244–60.

Casati, Sara, Stefano Crabu, Marialuisa Lavitrano and Mauro Turrini. 2012. "HeLa. Reconstructing an Immortal Bio." In *Tecnoscienza. Italian Journal of Science & Technology Studies*, 3: 147–62.

Charmaz, Kathy. 2006. *Constructing Grounded Theory. A practical Guide through Qualitative Analysis*. London: Sage.

—. 2009. "Shifting the grounds: Constructivist grounded theory methods." In *Developing grounded theory: The second generation,* edited by Janice M. Morse, Phyllis Noerager Stern, Juliet Corbin, Barbara Bowers, Adele E. Clarke and Kathy Charmaz, 127–54. Walnut Creek: University of Arizona Press.

Collins, Harry M. 1975. "The seven sexes: A study of a phenomenon of replication of experiments in physics." In *Sociology*, 9: 205–24.

Collins, Harry M. and Robert G. Harrison. 1975. "Building a TEA laser: The caprices of communication." In *Social Studies of Science*, 5: 441–50.

Collins, Harry and Steven Yearly. 1992. "Epistemological Chicken." In *Science as Practice and Culture*, edited by Andrew Pickering, 301–26, Chicago: University of Chicago Press.

Doing, Park. 2008. "Give Me a Laboratory and I Will Raise a Discipline: The Past, Present, and Future Politics of Laboratory Studies in STS." In *The Handbook of Science and Technology Studies. Third Edition*, edited by Edward J. Hackett, Olga Amsterdamska, Michael E. Lynch and Judy Wajcman, 279–317, Cambridge, MA: The MIT Press.

Fenwick, Tara, and Richard Edwards (edited by). 2012. *Researching Education through Actor-Network Theory*. London: Wiley-Blackwell.

Foucault, Michel, trans. 1972. *The Archaeology of Knowledge*. New York: Pantheon Books.

Fujimura, Joan. 1995. "Ecologies of Action: Recombining Genes, Molecularizing Cancer, and Transforming Biology." In *Ecologies of Knowledge*, edited by Susan L. Star, 302–46, Albany: State University of New York Press.

Gherardi, Silvia. 2012. *How to Conduct a Practice-Based Study: Problems and Methods*. Northampton: Edward Elgar Publishing.

Keating, Peter, and Alberto Cambrosio. 2003. *Biomedical Platforms: Realigning the Normal and the Pathological in Late-Twentieth-Century Medicine*. Cambridge, MA: The MIT Press.

—. 2012. *Cancer on Trial: Oncology as a New Style of Practice*. Chicago: University of Chicago Press.

Knorr-Cetina, Karin. 1981. *The Manufacture of Knowledge: An Essay on the Constructivist and Contextual Nature of Science*. Oxford: Pergamon Press.

—. 1995. "Laboratory Studies: The Cultural Approach to the Study of Science." In *Handbook of Science and Technology Studies*, edited by Sheila Jasanoff, Gerald E. Markle, James C. Peterson, and Trevor Pinch, 140–66, London: Sage.

Landecker, Hannah. 2007. *Culturing life: how cells became technologies*. Cambridge, MA/London: Harvard University Press.

Latour, Bruno. 1983. "Give Me a Laboratory and I will Raise the World." In *Science observed, Perspectives on the Study of Science*, edited by Mike Mulkay and Karin Knorr–Cetina, 141–70, London: Sage.

—. 1987. *Science in Action: How to Follow Scientists and Engineers through Society*. Harvard: Harvard University Press.

—. 1990. "Drawing things together." In *Representation in Scientific Practice*, edited by Michael Lynch and Steve Woolgar, 19–68. Cambridge, MA: MIT Press.

—. 1992. *Aramis ou l'amour des techniques*. Paris: La Découverte.

—. 1993. *La clef de Berlin et autres leçons d'un amateur de sciences*. Paris: La Découverte.

Latour, Bruno and Steve Woolgar. 1979. *Laboratory Life: The Social Construction of Scientific Facts*. Princeton: Princeton University Press.

Law, John. 1994. *Organizing Modernity*. Oxford: Blackwell.

Löwy, Ilana. 1995. "La standardisation de l'inconnu : les protocoles thérapeutiques en cancérologie." In *Techniques & Culture*, 25-26: 73–108.

Lynch, Michael. 1985. *Art and Artifact in Laboratory Science: A study of shop work and shop talk in a research laboratory*. London: Routledge.

—. 2002. "Protocols, practices, and the reproduction of technique in molecular biology." In *British Journal of Sociology*, 53: 203–20.

Mol, Annemarie. 2002. *The Body Multiple: Ontology in Medical Practice*. Durham, NC: Duke University Press.

Mongili, Alessandro. 2007. *Tecnologia e società*. Roma:Carocci.

Neresini, Federico. 2008. "Dentro e fuori dal laboratorio. Trasformazioni della tecnoscienza e analisi sociologica." In *Rassegna Italiana di Sociologia*, 49: 349–76.

Orlikowski, Wanda J. 2007. "Sociomaterial practices: Exploring technology at work." In *Organization Studies*, 28: 1435–48.

Popper, Karl R. 1992. *The Logic of Scientific Discovery*. London: Routledge.

Rabinow, Paul. 1996. *Making PCR: A Story of Biotechnology*. Chicago: University of Chicago Press.

Schwartz, Howard and Jerry Jacobs. 1979. *Qualitative Sociology. A Method to the Madness*. New York: The Free Press.

Scott, Pam. 1991. "Levers and Counterweights: A Laboratory That Failed to Raise the World." In *Social Studies of Science*, 21:7–35.

Silverman, David. 1997. *Qualitative Research: Theory, Method and Practice*. London: Sage.

Skloot, Rebecca. 2010. *The Immortal Life of Henrietta Lacks*. New York: Crown Publishers.

Star, Susan L. 1999. "The Ethnography of Infrastructure." In *American Behavioral Scientist*, 43: 377–91.

—. 2010. "This is Not a Boundary Object: Reflections on the Origin of a Concept." In *Science, Technology, & Human Values*, 35: 601–17

Star, Susan L. and James R. Griesemer. 1989. "Institutional Ecology, "Translations", and Coherence: Amateurs and Professionals in Berkeley's Museum of Vertebrate Zoology, 1907–1939." In *Social Studies of Science*, 19: 387–420.

Star, Susan L. and Karen Ruhleder. 1996. "Steps toward an ecology of infrastructure: Design and access for large information spaces." In *Information Systems Research*, 7: 111–34.

Suchman, Lucy. 1987. *Plans and Situated Action: The Problem of Human–machine Communication*. Cambridge: Cambridge University Press.

—. 2000. "Organizing Alignment: A Case of Bridge–Building." *Organization*, 7: 311–29.

Timmermans, Stefan and Marc Berg. 1997. "Standardization in Action: Achieving Local Universality through Medical Protocols." In *Social Studies of Science*, 26:769–99.

—. 2003. *The Gold Standard: The Challenge of Evidence–Based Medicine and Standardization in Health Care*. Philadelphia: Temple University Press.

Viteritti, Assunta. 2012. "Sociomaterial Assemblages in Learning Scientific Practice: Margherita's First PCR." In *Tecnoscienza. Italian Journal of Science & Technology Studies,* 3: 29–48.

Acknowledgments

I would like to thank my colleagues at the *Pa.S.T.I.S. Research Unit* (Padova Science Technology and Innovation Studies) for their intellectual insights provided during the writing of this chapter. I am grateful to the anonymous reviewers for their critical and detailed feedback. Finally, I wish to thank Paolo Magaudda, Francesco Miele and Manuela Perrotta for their fruitful comments on a previous draft of this chapter.

III. Users, Information Infrastructures and Mobilities

CHAPTER SEVEN

PARASITE USERS? THE VOLUNTEER MAPPING OF CYCLING INFRASTRUCTURES

JÉRÔME DENIS AND DAVID PONTILLE

Introduction

Throughout the past decade, cities around the world have invested in "green" transportation policy, and bicycle-related policy has ranked highest on the list. As such, bikes have seriously transformed the appearance of urban spaces. Linked to renewed forms of political ecology (Tironi 2012), bicycle facilities have skyrocketed and cycling practices themselves have grown considerably since public bicycle sharing systems have dramatically increased the cycler population. Much research, partly initiated by the first bicycle messenger movements, has focused on these changes (Jones 2005; Blickstein 2010; Forsyth and Kryzek 2011). Drawing on Lefebvre's (1991) notion of lived space or *espace vécu* and on Borden's (2001) work on the practice of skateboarding, this research has shown how cycling may constitute a form of urban space appropriation (Kidder 2008). Often down on the pavement themselves, scholars have studied cyclists' experiences firsthand.

This appropriation and use by the masses nevertheless suffers from a lack of overall intelligibility for urban infrastructures dedicated to bicycles. Since Latour and Hermant's (1998) exploration of the backrooms of Paris, we know there are no instruments or sites that allow for a city – and certainly not for a major city – to access a panoramic, exhaustive vision of the objects and people inside it. Though relatively up-to-date maps showing road infrastructures classed and identified by stabilized categories do exist, maps illustrating bicycle infrastructures available to users are judged as largely incomplete and not useful, whether produced by local governments, states, or regions. For the last several years, as online geographical tools have developed and new digital cartographies have appeared (Abrams and Hall 2006; Zook and Graham 2007), parallel initiatives have emerged in attempts to classify and map these infrastructures. Among these initiatives, which differ in their programs as well as in the tools they use, we propose here to focus on OpenStreetMap (OSM), an Open Data Commons Open Database License of geographical information used to group user-generated maps and hand the task of spatial representation over to the inhabitants (here, cyclists). From swamps to school bus stops, from coastlines to ski trails, the individual databases inventorying objects on OSM are varied and numerous. The bike facilities database is significant, though, since aside from being available for integration via an API (Application Programming Interface) into appropriate rendering software, it may also be visualized as a dedicated layer on the map service's main website (www.openstreetmap.org). In one click, a city map or a section of a city map reveals the lanes, paths, and other elements that organize and facilitate cycling. These elements have been inventoried by volunteer cartographers during dedicated riding tours, with the help of geo-localization devices, and integrated in the collaborative map thanks to editing software that allows the volunteers to directly edit road sections at home.

In this chapter, we focus on collective discussions that punctuate the elaboration of the database, insisting on the place that is given to users. We distinguish three types of manifestations of users in debates, each with their own specific issues: 1) Users of the geographical database 2) Users of urban infrastructures such as they have been framed and defined by the law, and 3) "Flesh and blood" users of the bike facilities. In their own way, these means of apprehending users reveal both a different part of the challenge of creating an "open" database and the elusive nature of bicycle infrastructures, much less simple to describe and inventory than meets the eye. But first, we specify our position within the main issues of database studies.

1. A sociology of databases

OSM is one of the cornerstone tools in the new forms of technical democracy (Callon *et al.* 2001) appearing in the early 90s. Amongst others, these devices have participated in transforming the forms of amateur-led actions that, in "field" sciences such as geography, have long held an important role (Ellis and Waterton 2004; Chilvers 2007). From the point of view of geography practices, OSM has taken on considerable importance. Along with other tools for volunteered geography, OSM cleared the way for important changes within the discipline. Some scholars go so far as to speak of the emergence of neo-geography (Turner 2006; Mericskay and Roche 2011), and some emphasize the now inevitable accumulation of millions of volunteer-produced geographical data (Goodchild 2007; Elwood 2008).

Such cartographical activities are doubly interesting for infrastructure studies. On the one hand, they create the rules and conditions for understanding previously obscure and even invisible objects. In this sense they offer a perfect site for exploring map performativity (Kitchin and Dodge 2007; Wood and Fels 2008). At the heart of the OSM contributors' bike facilities inventorying practices, the city as a *cyclable* place comes to the surface. On the other hand, the cartographical activities bear the traces of the conversations taking place about them. These conversations or debates exist in the form of a mailing list, archived online as a "talk" and as a Wiki page that consolidates the various recommendations. To create a map that shows a city as bicyclable, relevant describable entities must be progressively stabilized along with shared categories, scales of detail, and exchange formats, and all of these elements must be discussed. The result is a more or less unending mass of debates, more or less riddled with conflict[1]. Examining these debates provides privileged access to the conditions under which the database is elaborated, and namely to what Bowker and Star (1999) call "categorical work", that is, collaborative activity via which agreements are made on frameworks for descriptive reference. It is largely through this work that participants build up the OSM database and largely by this process that they bear witness to the efforts required to assemble a collective and necessarily performative description of the world (Law 2009).

Three main aspects distinguish bike facilities mapping activities from those usually examined in infrastructure studies. First, social studies of databases tend to focus on infrastructures with historical weight, often based on other infrastructures that have largely been solidified (Star and Ruhleder 1996). Certainly this is also the case with amateur bike cartographers, whatever their focus area may be: the vocabulary they use is always partly stabilized and shared. The attempt, however, is particularly fresh. In France, "OSM Talk" took off only in 2006. Topics remain very open and work on the different categories is colossal. Participants ask very basic questions, leading to discussions that touch upon general problems. How best to differentiate between roads – by size? By type (highway, main road, side street)? The task is further complicated

[1] This question of description troubles is analyzed in Denis and Pontille (2013).

by the constant task of translating from English-language nomenclature to French.

Next, the information infrastructure the amateur cartographers look to build is based on urban infrastructure that is itself in perpetual evolution. As we suggested in the introduction, urban spaces are subject to endless transformations and are gradually filled with new equipment. Far from once-and-for-all stabilization, objects such as these are difficult to qualify and in some ways, it is impossible to stabilize vocabulary used for naming. Infrastructure studies tend to underplay this question, focusing instead on ways of ordering scientific work (Hine 2006) or equipping communities of practice (Turner *et al.* 2006). Amateur cartographers deal mainly with issues in advance of mapping itself, centered on the correct methods for describing the world and proposing adjusted representations of it (Lynch and Woolgar 1990).

Finally, databases are generally considered as "boundary infrastructures" (Bowker and Star 1999) working to facilitate coordination between actors who belong to previously identified social worlds, even if they are partly redefined as databases are configured. This is the case, for example, of scientists, amateur naturalists, and trappers in Star and Greisemer's canonical 1989 article. In our case, information infrastructures may potentially be used by everyone. Participants and their interests are heterogeneous. Enthusiastic lone cyclists find themselves side by side with creative entrepreneurs looking to create a commercial application based on the OSM database, with members of an association of practitioners close by. The OSM database is under a Creative Commons Attribution-ShareAlike 2.0 license meaning that it is open for use by actors likely to appropriate it for diverse projects. As such it is difficult to define, even gradually, the scope of concerned users – and this is not the case for most other databases (Millerand and Baker 2010).

In the present chapter we will focus on this last point. Much ink has been spilt on the crucial role of users in innovation processes. Within STS, fascinating studies have shown how users played an important role in stabilizing technical objects, describing processes of co-configuration by which collectives of users and technologies mutually shaped one another (Pinch and Bijker 1987; MacKenzie and Wajcman 1999; Oudshoorn and Pinch 2003). Akrich (1992) shed light upon the processes of inscription of frameworks of action in technical objects, which explicitly or implicitly define the skills and appropriate positions of users. Volunteer cartographers' categorical work is also subject to these kinds of processes: users are constantly discussed in online discussions. But they are present in a very particular way. In most cases, users are treated as entities to be underplayed as much as possible. It would appear that in order to create a neat, clean database that responds to the criteria of coherence necessary to bring it to a point of stabilization, choices made for the sake of users should be avoided.

Our study is drawn from a qualitative analysis of 692 messages dealing with bike facilities in the "OSM Talk" archive site for the French-language discussion boards[2] between August 2006 (opening month) and April 2012. The messages make up sixty discussion threads and were identified by three word searches: "cycl-", "vélo", and "vélib."[3] Our analysis of these threads uses the principles of grounded theory (Glaser and Strauss 1967), meaning that we had no pre-established research hypotheses insofar as the dynamics and objects concerned. In this way we were able to favor unexpected results instead of looking to make confirmations. Via an initial chronological reading of all of the messages, we were able to familiarize ourselves with a mixed bag of technical vocabulary. We then proceeded to identify the messages' different themes, locating throughout repeated readings the contributors' problems and the solutions devised as the database was gradually developed. There is no quantification involved in the method, and no significance given to questions that appear with a certain frequency or receive a certain number of responses. Our goal was rather to remain attentive to the different forms of expression and questions that would go unnoticed in a statistical analysis. However brief they may be, the messages amount to arguments – they call upon certain groups, they formulate judgments, and they bring into prominence various entities we wish to include in our purview. In this exploration process, users struck us as being particularly interesting entities, both omnipresent and ambiguous. Rather than resources used in debates or end results, more or less ideal, toward which agreements could aim, users appear indeed as a source of difficulty and concern in the categorical work.

2. Database users: boundaries

There is one recurring formula in the contributor discussions that is particularly striking for the outside observer: "We don't tag for the rendering" ("*on ne taggue pas pour le rendu*"). The phrase is often addressed to novices arriving on the forum boards, or by experienced

---

[2] http://lists.openstreetmap.org/pipermail/talk-fr/. The contributor names have been changed, but the messages have not been altered. The English translations attempt to best render the original tone, spelling, and grammar.

[3] Vélib is the name of Paris's bicycle sharing system.

participants as disclaimers to a question ("I know, no tagging for rendering") and serves as a reminder for all that categorical work should not be motivated by a desire to produce an attractive map (a rendering). The phrase marks the limits of a collective activity for which users (assimilated here in the context of the online map, that is, the final product of the database) are far from the realm of relevancy in the actual creation of the database. As such, users appear as "parasites" that should be disregarded since they only hinder the process of good bicycle pathway description.

But parasites are never easy to get rid of. Though they are often dismissed with recurring formulas, users still haunt the debates and decisions of contributors, seeming to go out one door and come back in through another. The "rendering" is an initial foil that reveals strong tension between two activities. One is intended to produce a "neat" map and the other is meant to create a database with "coherent" data. This tension between esthetic quality and logical principles is particularly salient in the conversations dealing with pathway representation. Is it better to draw just one path and use several categories to describe it (one for cars, one for bicycles…) or is it preferable to indicate separate paths for each means of transportation? The example of bridges highlights the different results of these two options.

> *Bertrand*
> *Fri. Dec. 4 19:25:17 GMT 2009*
> On this map of Toulouse, you can see two bridges that have bike lanes.
> http://www.openstreetmap.org/?lat=43.58801&lon=1.43681&zoom=16&layers=B000FTF
> The southern bridge is a highway=secondary with cycleway=track. The northern bridge is represented by three ways (a highway=secondary and two highways=cycleway). At first, the southern bridge used the same representation.
> I edited it because the first way strikes me as more coherent. Physically, there's only one bridge whereas with three ways, there are three bridges (you see that when you zoom).
> But the Wiki says it's best to "draw a separate way tagged with highway=cycleway (see http://wiki.openstreetmap.org/wiki/FR:Map_Features#Voie_cyclable_.28cycleway.29). So I guess I messed up. For me, the advantage of using a separate way is that the rendering is better. But I often read that it's better to have coherent data than high-quality rendering. What do you think? [p. 176-177]

For OSM contributors, calling the same lane by different names provides a correct map, but without high-quality data. They suggest representing as many separate lanes as there are ways to travel them (car, bike…), a method that emphasizes data coherence on the map rendering. The difficulties contributors have in following this rule come to the fore with objects whose cartographic representation is particularly delicate. This is the case in this example with bridges, but we came across fascinating discussions, for example, on traffic circles. In debating the often-hazy limits of categories, the contributors experience the same kinds of tensions that were present in epistemological controversies in geography when geographical information systems were first developed (Schuurman 2000) and must deal with the challenges of producing a database as a finished product (Bowker 2000).

Who, exactly, are the users being dismissed when contributors look beyond the map's rendering in the context of their collective activity? We have located two types, closely linked. The first are *machines*: the software that produces the rendering. When discussed, these automated users are given minimal attention, and certain categories remain active so that they could still work out.

> *Nicolas*
> *Wed. Dec. 30 17:24:58 GMT 2009*
> The idea of adding an additional segregated=yes|kerb key for "cycleway=lane" seems better to me because the software keeps interpreting cycleway=lane and we can give additional useful information. [p. 176]

At other moments, the software is dealt with more harshly. The "no tagging for rendering" leitmotiv is accompanied by a remark suggesting the software should deal with issues itself, that with time and new versions, they will be able to deal better with the categories in the midst of implementation, in other words, they will conform to the database.

The second type of users associated with "rendering" are the software's direct human partners, the *programmers.* This includes programmers who create services, free or commercial, linked to OSM, whether simple maps, itinerary calculators, or geo-localization systems that indicate nearby tourist destinations. The way in which these second users are disregarded is in many ways identical to the first. But from time to time, there is a more explicit form: a balancing act that must occur between various actors by which not tagging to produce rendering becomes not putting oneself in the programmer's shoes, or, more specifically, not accepting to do favors (for example, systematically identifying pathways according to certain criteria) for people who will eventually benefit from the final product or service and who should be the ones doing the job in the first place. In speaking of the programmers as users who should not play into the concerns of the categorization process, the participants attempt to clarify the respective positions of all actors (including themselves) in a socio-technical network in which interdependence is strong and the boundaries remain indistinct.

In contrast to this constant yet never fully successful rejection, there is one figure representative of a relevant user, though rarely explicitly present: the *OSM contributor*. Inventing the acceptable means by which to categorize bike facilities, and identifying outside elements in which they have no real stake, mailing list members implicitly define a collective of volunteer cartographers who are able, little by little and as the approved categories are stabilized, to produce a shared map. However, the existence of this community remains unstable. There is no actual, concrete group of users for whom identifying certain shared characteristics would make it easier to create the categories they will end up using. As such it is not a boundary object, or even a boundary infrastructure that the list's contributors are creating. For them, the goal is not to come up with a way to align, as flexibly as possible, different, pre-defined social worlds (Bowker and Star 1999). For other databases, the question of user identification and adequate spokesperson is often complex, since users evolve throughout the course of the project and the configuration of various categories is synonymous with the progressive emergence of a "web of users" that is never completely stable (Millerand and Baker 2010). In the case of OSM, the uncertainty is even greater. Collectives likely to use the database are open, constantly emerging, and making reference to the users they are made up of does not serve to reduce any doubt, or to make choices between different possible options any easier.

However, despite regular invitations to forget the "rendering" and disregard a number of potential users, the recurring theme is sometimes questioned. Even though categories should not conform to criteria for cartographic visibility and intelligibility, and even though they do not need to respond to the needs of map users, they must, somehow, "be good for something."

> *Gérard*
> *Tues Sept. 8 09:17:21 BST 2009*
> > Careful not to tag for the rendering ;-)
> OK, but shouldn't we tag so that people can get some use of this? [p. 90]

For some, the creation of database categories should nevertheless be oriented toward a kind of possible use, not just of the geographical database, but of the bike facilities themselves. From this standpoint, cyclists are users that should, at least, be taken into consideration.

3. Uses in law

During the processes of categorizing, describing, and inventorying bicycle infrastructures, cyclists generally appear via the law. Many legal qualifications define pathways in regard to specific uses, authorizations and prohibitions. A section of a road may be closed off to cars, another reserved for pedestrians, etc. One might think such legal qualifications would bring few debates, remaining a solid reference for descriptive work on the database and a transparent resource used to integrate different uses to the database. Just as in everyday situations, though, the collaborative collection of geographical data creates contrasting, even disputed, links to the law. Ways of invoking law, and the situations that the law is intended to cover, are neither univocal, nor transparent (Ewick and Silbey 1998). In the mailing list debates, the law appears under various forms; its position and its power fluctuate from one occurrence to another.

Law appears first in the discussions as parts of legal code, used as stabilized references to guide part of a description.

> *Ernst*
> *Mon April 19 22:43:02 BST 2012*
> I'm starting to doubt something… according to article L362-1 of the Environmental Code, "motor vehicle traffic is prohibited off of roads not part of the public road domain at the State, department, and local levels, as well as on rural roads and private paths open to public motor vehicle traffic." There are more details over here: http://www.franche-comte.developpement-durable.gouv.fr/article.php3?id_article=96
> So for me motor traffic is prohibited by default on all trailways. [p. 215]

But as the first part of this quote shows, even if texts appear in total firmness, they still do not suffice to automatically clarify situations or produce unanimous descriptions. In this way, the law is just as liable to be a guiding force as it is a source of further doubt. Aside from the fact that they have yet to be interpreted in a collective fashion, the texts remain abstract, detached from the observations made by cartographers on site. Indeed they reveal the links between facilities and different user types, but in too vague a fashion for them to constitute a resource sufficient enough to produce descriptive categories.

When it is present on site, where OSM contributors make their explorations, the Law appears most easily used in descriptive work. Rather than being reproduced in citable texts, the law in this case is instated via everyday graphic ordering objects – sign posts or ground markings.

> *Owen*
> *Fri June 12 00:38:03 BST 2009*
> Highway=footway should only be used for lanes reserved for pedestrians and usually indicated by the sign B22b.
> (http://wiki.openstreetmap.org/wiki/Image:Fr-B22b-Obligatoire_pour_les_pietons.gif)
> It's true that highway=footway was used pretty much everywhere, but that was before highway=path was created. [p. 72]

This is no doubt the most easily-manipulated reference for contributors. As soon as a solid link is established between certain signs and categories within the growing database, the description work is made considerably easier. This is due to the performativity of these types of signs: their mere presence constitutes a legal qualification (Silbey and Cavicchi 2005). The urban planners and, in the end, the sign installers assumed the responsibility of aligning legal categories and sections of the road when each sign was put into place (Denis and Pontille 2011). As such, passersby could benefit from an *in situ* classification. With signs, the legal inclusion and exclusion of users occur on the level of infrastructure itself in an extremely stabilized, intelligible manner. The strength and simplicity of signs are notably revealed when they are missing.

> *Zebulon*
> *Wed Aug 19 09:04:35 BST 2009*
> Something I've been wondering for a while? How do you tag paths that bikes can travel on but for which no signage has been put in place (no bike path sign or bikes prohibited sign). [p. 80]

This kind of questioning shows the extent to which the law is both important and difficult to apprehend in the context of the contributors' description work. How should one best describe a cyclist's potential to ride in a certain place when it is known that cyclists are authorized but there is no proof of such a fact, no text, no sign? In this specific case, the solution discussed online ended up deciding on a generic category – path – from which cars would be excluded but which leaves open the possibility for travel by all non-motorized vehicles as long as no other signs (marking pedestrian-only access) are present:

> *René*
> *Wed Aug 19 09:25:17 BST 2009*
> Hello!
> For me, as long as this sign is not present http://wiki.openstreetmap.org/wiki/Image:CH_fussweg.svg (or any related sign) pathways should be tagged highway=path instead of highway=footway. So bikes are allowed to ride (http://wiki.openstreetmap.org/wiki/OSM_tags_for_routing/Access-Restrictions). [p. 80]

In referencing signs and their absence, not only are pathways identified but the categories themselves are specified. *Footway* and *cycleway* are linked to paths whose legal qualification is clear, whereas *path* is used for describing paths that have been observed as being shared. The legal qualifications therefore appear to be unstable descriptive elements. Though signage makes it possible to ensure legal conformity when entering sections of pathways into the database in the context of describing their uses, it is also true that when signage is absent, many questions are left unanswered.

Besides, the presence of cyclists on a given pathway is itself a cause of doubt in terms of the law. Rules are easily skirted around and even flouted by the "real" uses observed by the contributors. Then, classifying and taking users into consideration becomes a question of morals. The use of generic categories makes it possible to ensure that no categories of the database will contradict the law. But it tends to weaken the utility of a tool for whose practical worth descriptive precision is paramount. This is why cyclists and their uses are no small matter in the mailing list discussions.

4. Flesh and blood users

Must "real" uses be accounted for? Can contributors really be expected to develop a normalized descriptive system entirely built around the cyclists? In the introduction to the English-language wiki, which contains an inventory of categories and a list of good cartographic practices, there is a clearly negative answer to this question:

> Important principle: no two cyclists are the same - OSM should concentrate on recording the facts so that map users/client software can make up its own mind, rather than imposing an arbitrary general-purpose definition of "cyclability".
> (http://wiki.openstreetmap.org/wiki/Cycle_routes/cyclability)

This declaration is a crystallization of the problems raised when uses are integrated into the database construction process. "Flesh and blood" users are here clearly shown in their capacity as parasites in the context of the contributors' category development.

Again, though, the situation is not so simple. In the course of the mailing list discussions, the cyclists are constantly brought up. The French version of the introductory text raises the issue of this difficulty, but remains much more ambiguous as to the solution.

> Two cyclists will never agree upon the definition of a "good" bike route. Some look only for off-road bike lanes and others prefer the shortest route possible even if it means riding on national roads. And still others prefer flat routes, and others want hills. But this is why OSM is a powerful tool. It is able to capture all of the data that cyclists might want, and each is then free to choose the route he or she prefers.
> (http://wiki.openstreetmap.org/wiki/FR:Bicycle)

Whether one aims at all users, individual ones, or none at all, figures of speech are well-known (Oudshoorn *et al.* 2004). And more so than just in terms of a mission statement, potential uses are difficult *not* to evoke. But it is precisely because they seem for some contributors to be intimately tied to facilities that users end up adding a specific level of description to the bicycle infrastructure. In bringing them up, even in a context of doubt, volunteer cartographers open up a large swath of new descriptive possibilities that they have difficulty dismissing with a simple movement.

Bike facilities users are discussed in the mailing list in one of two main ways. They appear as being relatively fixed, linked always to a specific representative: *their material*, that is, the type of bicycle they ride. In this case, bike infrastructure is described in terms of whether it is suitable or not for a certain bicycle: mountain bike, street bike, racing bike… Upon first glance, such a discussion would appear to simplify the task of description since it allows already largely stabilized distinctions to be drawn on. But it remains a sensitive area of conversation: when contributors raise the issue, others respond that entering into such a categorical realm is problematic because it evokes the possibility of limiting the list to certain objects. The issue is not so much the differences between equipment as the numerous "cyclist objects" on the market. Such a proliferation makes it tempting to disregard all of these distinctions, especially since nobody is immune to the possible appearance of a new, popular bicycle that would make it necessary to re-label all of the pathways already identified within the database.

Zoo-York
Tues 3 Nov. 17:49:00 GMT 2009
As Lamarck said, "The use creates the function"!
Gael, we're going to need to add the following cases to the site:
Bikemobile
Side-car bike
Trailer bike
Delivery tricycle
Bike with baby seat
;-)
No joke, there are several pathways that restrict use for certain cyclists (it's never fun to carry your bike on your shoulder, but if you can get 1 kilometer ahead that way, it's definitely worth it)
These types of obstacles could never be managed by the bikes in the above list, though.
I think the question of width dealt with in the discussion was most relevant for traffics (ways) with the security implications that result rather than barricades or something of the sort (nodes).
Both aspects are accounted for on the map and if they're taken into consideration for route calculations that's all the better. [p. 145]

Users appear, secondly, as elements causing much more complex issues for volunteer cartographers: *their practices.* How does one make sure the descriptions included in the database are in keeping with "real" cyclist practices? The question is not a minor one. It shows once again that OSM contributors are involved in a meticulous inquiry, conducted to carefully explore the diversity of cycling experiences.

As the previous excerpt suggested, one of the main areas concerned by this practical dimension is safety. Much of the user-centered discussion originates with safety concerns, for the sake of which users are not seen as inert figures divided up into distinct types, but rather spring to life in situation, riding their bicycles.

Koala2000
Mon Nov 2 10:58:33 GMT 2009
lane_in_the_middle: this is a tag I made up to describe a cyclable lane between traffic lanes and parking lots (for example, Boulevard Excelmans, Paris 16th arrondissement). It seems to me that this kind of configuration is pretty dangerous because drivers could exit from their cars without paying attention, thinking that the space between the traffic area and the parking lot is big enough (for him) and then forget the cyclists. I know what I'm talking about, I almost paid the price. Two things: either you crash into the guy and his open car door or you swerve over into the traffic and you're crushed. So for me this tag indicates a dangerous area. (And if my memory serves me correctly, the bike lane is more or less painted with skinny white

> lanes all smushed together, so yeah, really does a good job setting it apart.) [p. 136]

In this case, a category is invented by a volunteer cartographer, justified by the danger of the situation the type of lane places cyclists in. In other cases, the issue of safety encourages contributors to choose an existing category that leaves the least ambiguity insofar as danger levels are concerned. But the criteria are not simple to manipulate. Safety is a vague term and creates indecision with regards to coding between two opposite types of categorization. One is acceptable, and even defended: the objective categorization of concrete elements that serves to increase cyclist safety or, on the contrary, serves to represent clearly identified risks. The other is grounded in a banished concept: subjectivity. Contributors are regularly asked to make the distinction between the inventorying of objects and "material" elements that may be inventoried and measured in terms of safety, and impressions of or judgments on the danger of certain mapped roads.

There is another dimension at stake for the contributors, an even more complex source of grueling debates and never-ending questioning. In developing the database, should one take into consideration the viable (or unviable) nature of the paths? Should the database account for degrees of "practicability", *cyclability*? And if so, how?

> *Denis*
> *Mon Oct 26 16:00:33 GMT 2009*
> I put every trailway in the highway=path category and that covers the fact that there's no rule saying bikes can't ride down them. Still, that doesn't mean they're easy to ride. Most of the time the paths aren't paved, or worse… I would not be happy if a route planner sent me down one of these on a regular street bike. [p. 20]

In opting for an element of practicability in their descriptions, some contributors use a coding method that directly serves users. And in dealing with user practices in such a way as to consider their diversity and not neglect them, the contributors recognize, in a way, the appropriation the cyclist makes in performing his or her cycling activity (Kidder 2008). Indeed, taking care to account for cyclists in practice when defining the relevant OSM categories is equivalent to recognizing the deeply relational character of the cyclable infrastructure. It comes down to fully accepting the difficulty of separating uses from their objects; to adopting a pragmatic attitude and remaining attentive to the ways in which urban reality is enacted.

Upon studying the discussion dynamics of OSM contributors, several elements may be gathered with regards to the confrontation between users and their practices. First, the users manifest themselves in a variety of forms throughout the debates. The two previous excerpts showed that the users are sometimes called to the fore via their equipment, sometimes via their preferences, and, finally, sometimes via their general practices. Their spokespersons are equally varied: from relatives to bicycling associations up to and including the individual contributors putting themselves in a generic user's shoes.

Next, it is important to stress the contributors' repeated refusals to insist upon the concrete uses of the bicycle infrastructures. Such refusal suggests that too much insistence on uses would not only be placing too much weight on the shaky viability of judgments suspected of subjectivity, but would also mean confronting a diversity of uses that cannot be reduced to the single measure of a cyclability level, or a even a simple recommendation.

> *Iann*
> *Wed May 11 00:16:21 BST 2011*
> On 10/05/2011 23:29, Gérard wrote:
> -what would you say to bicycle=advisable?
> Recommendable… For my mother who can barely ride a bike, for me, used to riding on the national roads dragging a carrier, for an all-suspension mountain biker, for a speeder riding a total carbon bone-shaker, for someone like me who loves riding in traffic, for the handicapped bikers' association, in order to enjoy the scenery, or in order to ride as smoothly as possible? Because you're being told all of this is subjective…
> Let different algorithms work differently using basic information, taking care to consider the specific weight of each user type – a synthetic indicator calculated in a dynamic way will be much more useful than a subjective synthetic indicator that's static and polemical. [p. 250]

In other words, going full force ahead with a description of the "practicability" of infrastructures within a database would dramatically multiply the registers used for describing the infrastructure. In addition to the legal definitions an indefinite number of new versions would be imposed, products of one-time, unique practical uses, each time with a different combination of material, user competences, preferences and the physical properties of the roads.

Finally, the previous excerpts also reveal the figure of a user with a potentially renewed shape. The cyclists who use bike facilities are presented from time to time as users of "rendering" software and online services that latch onto OSM and propose their own algorithms, maps, route calculation tools, etc. In the contributors' conversations about the place they should carve out for cycling practices, these intermediary technologies, situated between the geographical database and the cyclists, are important. The presence of such technologies paves the way for a distribution of roles in the consideration of users, and therefore in the ways multiple versions of cycling infrastructures can be represented. The refusal to include cyclability criteria in the database is shared by some, repeated throughout the discussions, and proves particularly resonant when one of the contributors, presenting himself as a programmer, suggests that he will deal with the subject and use the criteria in a special database. The database belongs to his association (which later becomes a company) and will be used as support for an algorithm he is in the process of developing. The database will constitute the added value for an online multi-criteria (most comfortable, fastest, etc.) route calculation service. In so doing, he is in line with the English-language site's mission statement, cited earlier, when he suggests that OSM should focus on the most "objective" possible descriptions of bicycle infrastructures.

> *FabriceRSF*
> *Tues Nov 3 14:18:58 GMT 2009*
> Like with what we did in Tours, we plan to have the association note each road depending on the:
> - auto traffic
> - road width
> - general safety
> - and also lots of more or less objective elements!
>
> So in the end we're able to say that it's more like "suggestions" on which roads to use.
> So I was thinking of building a database that would be connected to OSM (based on the identification of ways and nodes) and then I would make available the different tools necessary to bring it up to date. [p. 142]

We won't go into detail here on the numerous discussions sparked on the mailing list on the subject of this software. However, it is important to keep in mind that the concrete user practice approach as a possible way of describing infrastructures is an important crystallizing factor for general conversation on OSM's own status. Debates about the best way to consider cyclist practices lead to the identification of different complementarities between an open software and an innovative service, or between the role of a contributor and that of a programmer (in this example, both roles are played by the same person).

In the end, through all these cases, it is the nature of the contributors' description work itself that is called into question. When these issues of practices appear – and in particular the debates on the necessity of coding infrastructure cyclability – there is proof of a kind of tension present throughout the collective inquiry conducted by the volunteer cartographers: tension, that is, between the "factual" description of infrastructures and the temptation to evaluate them.

5. An elusive object and unstable users

What can be learned from this exploratory plunge into volunteer cartographers' categorical work? First, though urban infrastructures such as bicycle facilities should in theory go without saying for most users, unless they break down (Star and Ruhleder 1996; Graham 2010), the creation of a database designed to produce a complete description is a challenge that leads to questions about its everyday intelligibility. We are also led to consider the difficulties involved in creating an open, collaborative database wherein the boundaries between contributors and users remain uncertain.

In an attempt to describe objects for which traditional descriptions do not suffice, different authors have made efforts to renew the vocabulary used in STS. "Open" innovations do not jive well, for example, with the terminology used in describing immutability and black boxes, whereas the notion of fluidity is better suited for grasping its modes of existence (Mol and Law 1994 ; de Laet and Mol 2000). Analyzing discussions on the categorical work for the OSM database is an opportunity to make the same kind of adjustment. In creating an information infrastructure dedicated to a physical infrastructure, volunteer cartographers apprehend the second not as a fixed assemblage, identifiable in a nearly transparent manner, but rather as an elusive object (Law 2004) sometimes having clear boundaries and at other times not. These difficulties illustrate that the *taken-for-grantedness* of bicycle infrastructures is a matter of tacit knowledge (Collins 2001), informed by corporal urban experiences full of affects (Jones 2005), which is particularly difficult to share and even more difficult to translate into a series of normalized categories (Bowker 2000). In trying to put bicycle facilities "into" a database, volunteers experience practically the issues that have been emerging in urban studies. Numerous works have foregrounded the heterogeneity of the urban reality, which should be analyzed as an assemblage, rather than a material or cultural monolith (Farías and Bender 2010 ; McFarlane 2011). In the mapping process, the volunteers discover the changing shapes and edges of bicycle facilities assemblages.

Moving from the concrete uses of urban infrastructure to the production of its intelligibility in a database is thus a challenge for practitioners' knowledge. It foregrounds an elusive side of the infrastructure, gradually refined through categorical work. From this point of view, the amateurs' position – that is to say, the users or even experts on the cyclable city - does not appear as a privilege but rather as a form of attachment to an infrastructure from which one must learn, in part, to separate.

In addition to this elusiveness of urban infrastructure, we must add the elusive nature of the collective made up of the OSM database contributors and mailing list participants. On first glance, one might consider that the mailing list is a space where a community of practice comes together (Lave and Wenger 1991) and via which cartographic skills and an understanding of cycling infrastructures are, little by little, consolidated. However, as we have seen in various excerpts in this chapter, though cartographic practices are discussed, it is also the actual use of bicycles and urban infrastructures facilitating or hindering this use that comes up in the discussions. In this way, the mailing list contributors also constitute a community of experience in Akrich's sense (2010) whereby practices in a cyclable city are narrated and shared throughout often-technical discussions.

Yet it would appear that there are major differences between the communities described in the specialized literature and the contributors working together on the OSM database. Aside from the fact that they make up a volatile population, far from being stabilized, the volunteer cartographers working to make cycling facilities more intelligible do not actually bear witness to any desire to become a united group. Contrary to certain participants working on lists dedicated to health issues, these cartographers' participation does not tend toward creating an epistemic community (Akrich 2010), whose collective identity would be based on new forms of knowledge that might enter into competition with other, more established expertise. Aside from several "old timer" participants who provide advice on various categories and how to use them, the participants intervene only from time to time, and their contributions are limited to mapping elements of the infrastructure on sections of the territory they regularly travel on and know well. As we have seen here, in addition, there is the fact that it remains impossible, even throughout the course of consolidating categories, to designate more or less stabilized groups of potential users. This other dimension of the endeavor's elusiveness makes it nearly impossible for contributors to draw upon the demands of coordination between communities (here, practitioners, developers, final users…), even though STS have described these demands as the traditional fuel of database construction in cases of scientific research.

The elusiveness of the database and the elusiveness of the collective working to develop and discuss it are, without a doubt, partly linked to the specific case of OSM and bicycle infrastructures. But with the rise of participative data collection tools and the multiplication of "opendata" initiatives in various sectors, we can hypothesize that these two dimensions will play an increasingly important role in other fields.

References

Abrams, Janet and Peter Hall. 2006. *Else/Where: Mapping — New Cartographies of Networks and Territories*. University of Minnesota Design Institute.

Akrich, Madeleine. 1992. "The De-scription of Technical Objects." In *Shaping Technology/Building Society. Studies in Sociotechnical Change*, edited by Wiebe Bijker and John Law, 205-224, Cambridge, MA: The MIT Press.

—. 2010. "From Comunities of Practice to Epistemic Communities: Health Mobilizations on the Internet." In *Sociological Research Online*, 15(2).

Blickstein, Susan G. 2010. "Automobility and the Politics of Bicycling in New York City." In *International Journal of Urban and Regional Research*, 34(4): 886–905.

Bowker, Geoffrey C. 2000. "Biodiversity Datadiversity." In *Social Studies of Science*, 30(5): 643–683.

Bowker, Geoffrey C. and Susan L. Star. 1999. *Sorting Things Out: Classification and Its Consequences*. Cambridge, MA: The MIT Press.

Callon, Michel, Pierre Lascoumes and Yannick Barthe. 2009. *Acting in an Uncertain World. An Essay on Technical Democracy*. New York: The MIT Press.

Chilvers, Jason. 2007. "Deliberating Competence: Theoretical and Practitioner Perspectives on Effective Participatory Appraisal Practice." In *Science, Technology, & Human Values* , 33(2): 155–185.

Collins, Harry M. 2001. "Tacit Knowledge, Trust and the Q of Sapphire." In *Social Studies of Science* 31(1): 71–85.

de Laet, Marianne and Mol Annemarie. 2000. "The Zimbabwe Bush Pump: Mechanics of a Fluid Technology." In *Social Studies of Science*, 30(2): 225–263.

Denis, Jérôme and David Pontille. 2011. "Aménager des espaces circulables. La dynamique des déictiques." In *Sciences de la Société*, 80: 177–192.
—. 2013. "Une infrastructure élusive. Aménagements cyclables et troubles de la description dans OpenStreetMap." In *Réseaux*, 178-179: 92-125.
Ellis, Rebecca and Claire Waterton. 2004. "Environmental citizenship in the making: The participation of volunteer naturalists in UK biological recording and biodiversity policy." In *Science and Public Policy*, 31(2): 95–101.
Elwood, Sarah. 2008. "Volunteered geographic information: future research directions motivated by critical, participatory, and feminist GIS." In *GeoJournal*, 72(3-4): 173–183.
Ewick, Patricia and Susan S. Silbey. 2008, *The Common Place of Law: Stories from Everyday Life*. Chicago: University of Chicago Press.
Farías, Ignacio and Thomas Bender. 2010. *Urban Assemblages. How Actor-Network Changes Urban Studies*. New York: Routledge.
Forsyth, Ann and Kevin Krizek. 2011. "Urban Design: Is there a Distinctive View from the Bicycle?" In *Journal of Urban Design*, 16(4): 531–549.
Glaser, Barney G. and Anselm Strauss. 1967. *The Discovery of Grounded Theory: Strategies for Qualitative Research*. Chicago: Aldine Transaction.
Goodchild, Michael F. 2007. "Citizens as sensors: the world of volunteered geography." In *GeoJournal*, 69(4): 211–221.
Graham, Stephen (edited by). 2010. *Disrupted Cities*. New York: Routledge.
Hine, Christine. 2006. "Databases as Scientific Instruments and Their Role in the Ordering of Scientific Work." In *Social Studies of Science*, 36(2): 269–298.
Jones, Phil. 2005. "Performing the city: a body and a bicycle take on Birmingham, UK." In *Social & Cultural Geography*, 6(6): 813–830.
Kidder, Jeffrey L. 2008. "Appropriating the city: space, theory, and bike messengers." In *Theory and Society*, 38(3): 307–328.
Kitchin, Rob and Martin Dodge. 2007. "Rethinking maps." In *Progress in Human Geography*, 31(3): 331–344.
Latour, Bruno and Emilie Hermant. 1998. *Paris ville invisible*. Paris: Les empêcheurs de penser en rond/La Découverte.
Lave, Jean and Etienne Wenger. 1991. *Situated Learning: Legitimate Peripheral Participation*. Cambridge: Cambridge University Press.
Law, John. 2004. *After Method: Mess in Social Science Research*. London: Routledge.

Law John. 2009. "Seeing Like a Survey." In *Cultural Sociology*, 3(2): 239–256.

Lefebvre, Henri. 1991. *The Production of Space*. Blackwell Publishers.

Lynch, Michael and Steve Woolgar (edited by). 1990. *Representation in Scientific Practice*. Cambridge, MA: The MIT Press.

MacKenzie, Donald and Judy Wajcman (edited by). 1999. *The Social Shaping of Technology*. Manchester: Open University Press.

McFarlane, Colin. 2011. "The city as assemblage: dwelling and urban space." In *Environment and Planning D: Society and Space*, 29(4): 649–671.

Mericskay, Boris and Stephàne Roche. 2011. "Cartographie 2.0 : le grand public, producteur de contenus et de savoirs géographiques avec le web 2.0." In *Cybergeo: European Journal of Geography*, document 552. Accessed May 2012.

Millerand, Florence and Karen S. Baker. 2010. "Who are the users? Who are the developers? Webs of users and developers in the development process of a technical standard." In *Information Systems Journal*, 20(2): 137–161.

Mol, Annemarie and John Law. 1994. "Regions, Networks and Fluids: Anaemia and Social Topology." In *Social Studies of Science*, 24(4): 641–671.

Oudshoorn, Nelly and Trevor Pinch (edited by). 2003. *How Users Matter. The Co-construction of Users and Technology*. Cambridge, MA: The MIT Press.

Oudshoorn, Nelly, Els Rommes and Marcelle Stienstra. 2004. "Configuring the User as Everybody: Gender and Design Cultures in Information and Communication Technologies." In *Science, Technology, & Human Values*, 29(1): 30–63.

Pinch, Trevor and Wiebe E. Bijker. 1987. "The Social Construction of Facts and Artefacts: or How the Sociology of Science and the Sociology of Technology might Benefit Each Other." In *The Social Construction of Technological Systems. New Directions in the Sociology and History of Technology*, edited by Wiebe E. Bijker, Thomas P. Hughes and Trevor J. Pinch, 17–50, Cambridge, MA: The MIT Press.

Schuurman, Nadine. 2000. "Trouble in the heartland: GIS and its critics in the 1990s." In *Progress in Human Geography*, 24(4): 569–590.

Silbey, Susan S. and Ayn Cavicchi. 2005. "The Common Place of Law. Transforming Matters of Concern Into the Objects of Everyday Life." In *Making Things Public: Atmospheres of Democracy*, edited by Bruno Latour and Peter Weibel, 556–564, Cambridge, MA: The MIT Press.

Star, Susan L. and Karen Ruhleder. 1996. "Steps toward an ecology of infrastructure: Design and access for large information spaces." In *Information Systems Research*, 7:111–34.

Tironi, Manuel. 2012. "Mettre l'écologie en mouvement. Les controverses aux origines du projet Vélib'." In *Tracés*, 22(1): 65-83.

Turner, Andrew. 2006. *Introduction to Neogeography*. Sebastopol, CA: O'Reilly Media.

Turner, William, Geoffrey C. Bowker, Les Gasser and Manuel Zacklad. 2006. "Information Infrastructures for Distributed Collective Practices." In *Computer Supported Cooperative Work*, 15(2-3): 93–110.

Wood, Denis and John Fels. 2009. *The Natures of Maps: Cartographic Constructions of the Natural World*. Chicago: University of Chicago Press.

Zook Mattew and Mark Graham. 2007. "Mapping DigiPlace: geocoded Internet data and the representation of place." In *Environment and Planning B: Planning and Design*, 34(2): 466-483.

Acknowledgments

This article was translated by Jill McCoy. We are very grateful to her professional commitment and her useful comments on a previous version of this chapter.

CHAPTER EIGHT

DIGITAL TEXTS: READING AND WRITING PRACTICES AT THE DIGITAL TURNING POINT

GIAN PAOLO LAZZER AND PAOLO GIARDULLO

Introduction

This contribution will take into consideration the development of new trajectories concerning writing and reading practices, focusing on text digitalisation and, in particular, on e-book technology. E-books, in all their forms, including PDF, ePub, Kindle and so on, are one of the most popular supports for the circulation of digital texts.

E-books have gained great success in recent years: in 2010, Amazon declared that the percentage of e-book sales had overtaken hardcover sales, and further market research confirms this data, showing that the number of e-book readers is growing and also that the number of websites offering services linked to e-book distribution is expanding (Santino 2011; Rainie *et al.* 2012). In the United States, 21% of the citizens have an e-book reader device, while in European countries (e.g. Germany), the share of e-book penetration, even if much lower, doubled in a single year from 2010 to 2011.

The unexpected success of some previously unknown authors such as Riccardo Bruni, who self-published his book with more than 3,000 Kindle copies sold in three months between 2012 and 2013[1], or Amanda Hocking, who reached the top of the book chart in the United States thanks to her self-published e-book on Amazon[2], indicate that e-books are probably fostering different kinds of practices.

---

[1] http://iltirreno.gelocal.it/regione/2013/05/15/news/autori-in-vetta-alle-classifiche-l-ebook-si-addice-alla-toscana-1.7070759

[2] http://www.theguardian.com/books/2012/jan/12/amanda-hocking-self-publishing

Indeed, the heterogeneous elements we have listed—new editors, e-book readers and writers—seem to suggest that a unique way to interact with e-book technology has been stabilised once and for all, but some other elements hint that this does not merely have to do with a matter of quantity. E-books are part of a "digital turning point" in which old practices appear reshaped through processes of multiplication and diversification.

In the discussion below, new trajectories will be described, depicting how those trajectories are emerging from the interactions between the users and the Internet. In fact, most of the practices linked to the e-book need the support of the Internet, because that is the environment in which the digitalisation of texts takes place.

In this chapter, we will first clarify the theoretical basis of our work, considering theories of practice (Schatzki *et al.* 2001) and the STS approach to information infrastructure (Bowker and Star 1999) as useful keys for the interpretation of our object of inquiry. Then we will summarise the methodological scheme that we have followed: a mixed method that uses both network analysis and direct observation of websites. In the second part of the chapter, a deep analysis of the graphs produced will be combined with the data collected during the direct observation. In particular, we will focus on two cases that have been recognised as especiallly significant as examples of these new phenomena.

1. Two complementary approaches for the study of practices on the Web

We are assuming writing and reading practices as our units of analysis. In this section, we will clarify the concept of practice, according to the recent turning point claim by Schatzki *et al.* (2001). We think that the theory of practices is the proper lens through which to observe our object of inquiry for many reasons: firstly, it gives us strong analytical elements useful for identifying what a practice is, and secondly, it is able to describe changes and developments. Last, but not least, it recognises the importance of the material side of the issue.

In particular, we will describe the trajectories (Southerton *et al.* 2012) that these practices are outlining, focusing our attention on the relationship between users and the information infrastructure that lies under the digitalisation processes of texts.

Practices have been imported from Wittgensteinian philosophy into social theory debate since the contributions of Anthony Giddens (1984), with the dual concept of structuration, and Pierre Bourdieu (1977, 1990), with the proposal of *habitus*; this so-called first generation of social theorists and researchers (Postill 2010) assigned to practices the role of useful lenses for understanding societal reproduction processes. More recently, a so-called second generation of scholars (*Id.*), among its other contributions, has individuated in techno-scientific material objects an important role in the studies of practices that are also able to identify innovation. In particular, the contribution of Science and Technology Studies to social theory debate, in proposing innovative concepts such as actant (Latour 1987), socio-material relationship (Law 1994), and socio-technical *agencement* (Callon 2005), has been recognised as extremely significant for social change and transformation. Practice theory, in fact, has recognised the role of material objects, and their agency has been clearly expressed in the key work *The Practice Turn in Contemporary Theory* (Schatzki *et al.* 2001). The importance of material objects, and by extension, of techno-scientific objects, has been stated, considering that practices are:

> "[socially] generally construed as materially mediated nexuses of activity" (Schatzki *et al.* 2001, 20)

On the basis of this statement, e-book technology assumes the role of socio-material nexus for the shaping of the trajectories of writing and reading practices. We think that the analysis of e-books is the best way to achieve our purposes, because e-books are the most evident and concrete output of the practices we are considering. E-books render practices identifiable and, at the same time, they configure the performances that are able to push for the diversification of the practices themselves. Indeed, e-book technology is a socio-technical object that unveils the different elements that concur in the development of the practices linked to digital texts and, at the same time, it shows the reciprocal influence between the practitioners and the information infrastructure.

Maybe some clarifications are now necessary. We are not taking into consideration single actions or performances—a practice isn't just doing or saying something—but we consider practices to be the results of acts performed by groups which are deeply intertwined in heterogeneous environments made both by social and material elements. As Schatzki explains, *"nonhuman entities help constitute human sociality"* (2001, 20).

According to recent contributions on theories of practice, we point to some analytical elements that may help in identifying a practice and, in doing so, may also identify its trajectories.

It may seem tautological, but the most important thing is the presence of practitioners, people who share a common background of knowledge, scopes and emotions (Warde 2005). Therefore, the constitution of a community is another element useful for identifying a practice. A community has its own rules, habits or norms that must be followed, and these kinds of procedures are frequently standardized through handbooks or codes that explain *what* to do and *how* to do it. A practice must be a routine, a series of performances that also needs commitment in terms of time.

Practice has clear scopes, projects and goals (Schatzki 1996) that are linked to the distribution of rewards such as social honour, prizes or economic advantage, but which are also linked to punishments.

Practices also require the use of special tools and proper equipment, as for instance objects and material artefacts.

To summarise, in Alan Warde's words, a practice:

> "requires that competent practitioners will avail themselves of the requisite services, possess and command the capability to manipulate the appropriate tools, and devote a suitable level of attention to the conduct of the practice. This is, of course, in addition to exhibiting common understanding, know-how, and commitment to the value of the practice." (Warde 2005, 145)

The environment where practices take place is obviously important for their diversification and multiplication.

In speaking about digital texts, the Internet can be considered the basis for the shaping of reading and writing practices; the central element that characterises the Web is exactly the interaction between heterogeneous entities (data and users, just to split the complexity into two categories) and the exchange of information through particular standardized packages (Borgman 2003). For this reason, we consider it of much interest to conceive the Internet as an informational infrastructure.

The theoretical approach to the study of informational infrastructure, developed inside the framework of Science and Technology Studies, here assumes a crucial role. As pointed out by Star and Ruhleder (1996), "infrastructure appears only as a relational property", and later on, Star (1999) reaffirmed that "infrastructure is a relational concept, becoming real infrastructure in relation to organised practices". Typical elements of sociological inquiry mentioned in these extracts, such as "[social] relation" and "[social] practice", have prompted a renewed interest by scholars because of their translation on the Web through computer-mediated communication (CMC) and, more recently, through mobile devices. The Internet has fostered the development of diverse and multiple forms of social practices, for instance, bottom-up knowledge production processes, shown by the blooming of Wiki platforms. (For a complete review, see Consalvo and Ess 2011).

Hence, different kinds of uses are performed according to the different social worlds in which they are promoted, defined and regulated. Looking at these kinds of practices, the concept of "naturalised use" (Bowker and Star 1999) seems particularly effective in explaining the variety of actual practices visible on the Internet. For two reasons, the concept of "naturalisation" becomes extremely useful in approaching the study of the bonding of practices to the Internet. First, if compared to other approaches of socio-technical innovation, as for instance the Social Construction of Technology (SCOT), naturalised use allows us to consider the plasticity and openness of the objects. In fact, while SCOT has allowed researchers to better recognise the role of users in shaping socio-technical objects (Oudshoorn and Pinch 2003), it has promoted the idea of rigid outcomes of trajectories for socio-technical evolution through the concepts of "closure" and "stabilisation" (Bijker 1995). These concepts are part of the same process, assuming that practices' trajectories, disciplined by socio-technical objects, are meant to reach a conclusion. Then, according to SCOT, after a phase of negotiation (the so called "interpretative flexibility", Bijker 1992), relevant social groups impose a rigid adoption of the acceptable, plausible and possible practices linked to a material object. This process embodies a sort of finalistic idea of socio-technical development, which is, even if not in absolute terms (Bijker 1995, 291), irreversible once it is fixed. The multiplicity of the examples about the Internet disregard this vision, showing that the naturalisation approach can better intercept multiple, heterogeneous and diverse practices that actually take place.

Second, naturalised use has been conceived for the studies of informational infrastructure, and the Internet, as hinted above, is exactly an informational infrastructure. More analytically, the Internet fulfils all the features individuated by Star and Ruhleder (1996). It is embedded in ordinary life, and its embeddedness is becoming more and more transparent in people's daily activities. Consequently, it shares many links with different kinds of practices. For these reasons, we argue that the Internet can be studied as an informational infrastructure, adopting the lens of naturalisation.

2. The infrastructure's representation

Therefore, following the concepts offered by this integrated theoretical approach, we have analysed the network of websites supporting e-book technology, the environment where writing and reading practices are outlying new trajectories through the naturalised use of the Net.

In order to approach the practices, we concentrate our inquiry on the Web, with the aim of directly observing the traces of web users. For this purpose, we merged the pros and cons of two different research techniques: network analysis and direct observation[3]. The former gave us the opportunity to define where the practices are performed; the latter allowed us to inquire into user interactions within the infrastructure.

As for the network analysis, we recorded a session of web surfing, starting from the simple query "e-book download" on an Italian version of Google[4]. We used a web tool called NaviCrawler[5], which has been implemented as a recorder for navigation sessions, and which is able to collect raw data for network analysis (Marin and Wellman 2011). In this way, we collected measurements from each website visited during the session; then we represented the network obtained using Gephi[6], a specific graphical software for network analysis.

At the end of this procedure, we gained a clear definition of the field to explore, based on quantitative criteria useful for a narrow observation of the websites. The graphs we obtained represent nodes (255 websites) and hedges (280 links between the websites), so they describe the network in which we surfed. Then we proceeded with a more in-depth observation according to the hubs (main websites) and the cliques (websites with the same colour) represented in the graphs.

---

[3] This is the typical pragmatic approach of a mixed methods research program (Morgan 2007; Lieberman 2005).

[4] Hence, all the results we will show will necessarily refer to Italian sites.

[5] http://webatlas.fr/wp/navicrawler/

[6] https://gephi.org

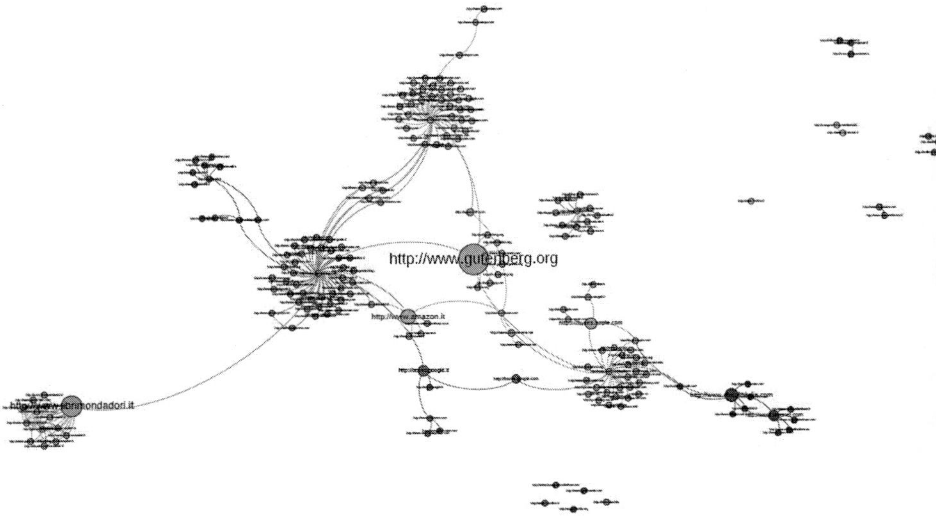

Fig. 8-1: the graph obtained through NaviCrawler Data.

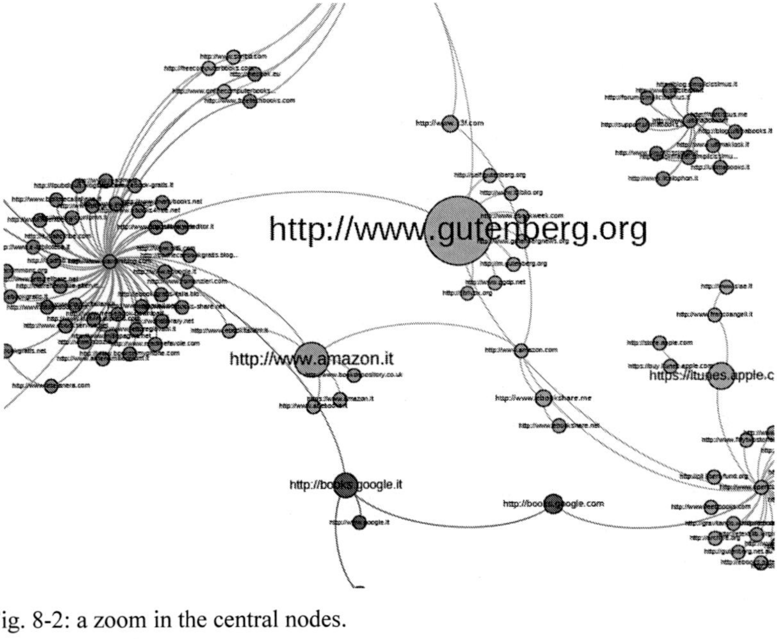

Fig. 8-2: a zoom in the central nodes.

3. Project Gutenberg and LaTelaNera: two trajectories

In figure 8-1, we have (part of) the scenario we obtained through our recorded session of navigation; here we represented the data, preferring the measure of betweenness centrality[7]. The higher the value of the betweenness centrality, the larger the size of the node in the graph.

We can easily individuate four sites with a high degree of betweenness (fig. 8-2): Gutenberg.org (hereafter Project Gutenberg or PG), Amazon.it (Amazon), Itunes.apple.com (iTunes) and Books.google.it (Google Books). These websites are the most easily reachable in the network; in fact, they are the most important hubs in the Italian network for getting e-books.

The different nuances of black and grey represented on the graph indicate different cliques of websites. As we can see, the four main sites do not share connections between them. This is due to the fact that they embody at least two different approaches to e-book technology. While Amazon, iTunes and Google Books are mainly commercial platforms for e-book selling—they are competitors, so they do not link to each other—PG has a completely different aim, because it is a free access library.

PG is also the biggest node of the net, taking up the most relevant position. For these reasons, scope and position, we observed the clique it forms and the extracts coming from its Wiki pages that give a general presentation of the website.

PG was born in 1971, with the aim of becoming a free access library, transforming the highest number of books into digital texts. When Michael S. Hart (1947–2011), the founder of PG, conceived the original idea of digitising texts, he was moved by the intention of spreading knowledge as much as was possible:

> "Everyone in the world, or even not in this world (given satellite transmission) can have a copy of a book that has been entered into a computer."[8]

The key element of Hart's philosophy was to consider, as the actual value of computers, the storage capacity rather than the computing. As can be read directly on the site:

---

[7] Network analysis assumes that centrality is a measure of interpersonal influence (Freeman 1979).

[8] http://www.gutenberg.org/wiki/Gutenberg:The_History_and_Philosophy_of_Project_Gutenberg_by_Michael_Hart#The_Project_Gutenberg_Philosophy

"This was totally serendipitous, as it turned out that two of a four operator crew happened to be the best friend of Michael's (Hart, A/N) and the best friend of his brother. Michael just happened "to be at the right place at the right time" at the time there was more computer time than people knew what to do with, and those operators were encouraged to do whatever they wanted with that fortune in "spare time" in the hopes they would learn more for their job proficiency.
At any rate, Michael decided there was nothing he could do, in the way of "normal computing," that would repay the huge value of the computer time he had been given [...] An hour and 47 minutes later, he announced that the greatest value created by computers would not be computing, but would be the storage, retrieval, and searching of what was stored in our libraries.
He then proceeded to type in the "Declaration of Independence" and tried to send it to everyone on the networks... which can only be described today as a not so narrow miss at creating an early version of what was later called the "Internet Virus."[9]"

The first history e-book was the United States Declaration of Independence, which Hart converted (by typing it) on the Xerox Sigma V mainframe at the Materials Research Lab at the University of Illinois. This mainframe was one of the nodes of ARPANET, the first actual attempt to connect different computing machines through landlines, the ancestor of the Internet.

Hart's purpose was to multiply opportunities for reading through the computer; this meant that the project started to reshape reading practice, taking advantage of the opportunity of a rising informational infrastructure.

Presently, as we have had the opportunity to observe, PG's website is a portal both for reading free e-books and for uploading new contributed e-books[10]. In the Mission Statement of PG the elements that configure the submission of e-books to its website and the guidelines for the community are reported:

"The mission of Project Gutenberg is simple:
To encourage the creation and distribution of eBooks.
This mission is, as much as possible, to encourage all those who are interested in making eBooks and helping to give them away.

---

[9] http://www.gutenberg.org/wiki/Gutenberg:The_History_and_Philosophy_of_Project_Gutenberg_by_Michael_Hart#The_Project_Gutenberg_Philosophy

[10] We have to point out that PG has never published texts violating authors' copyright; as PG declares, "Copyright for most of these books has expired in the United States. (They may still be copyrighted in other countries). So anybody may make verbatim or non-verbatim copies of those works".
(http://www.gutenberg.org/wiki/Gutenberg:No_Cost_or_Freedom%3F)

> In fact, Project Gutenberg approves about 99% of all requests from those who would like to make our eBooks and give them away, within their various local copyright limitations.
> Project Gutenberg is powered by ideas, ideals, and by idealism.
> Project Gutenberg is not powered by financial or political power.
> Therefore Project Gutenberg is powered totally by volunteers.
> Because we are totally powered by volunteers we are hesitant to be very bossy about what our volunteers should do, or how to do it.
> We offer as many freedoms to our volunteers as possible, in choices of what books to do, what formats to do them in, or any other ideas they may have concerning "the creation and distribution of eBooks."
> Project Gutenberg is not in the business of establishing standards. If we were, we would have gladly accepted the request to convert an exemplary portion of our eBooks into HTML when World Wide Web was a brand new idea in 1993; we are happy to bring eBooks to our readers in as many formats as our volunteers wish to make."

PG, from what we read in this extract from the Mission Statement of the community, is not completely lenient, even if it may appear extremely free. The community is not *"bossy about what our volunteers should do, or how to do it*" but at the same time is very sharp, considering itself to be outside of the "*business of establishing standards*". It has declined exterior requests to convert e-books into a single format. This is a key element which animates the community, the general ideal of free access to knowledge. PG offers, and requests of its volunteers, e-books in all the formats available. PG remains neutral about the technological format, hence not imposing the adoption of a given device. This lack of standardization is one of the key rules which opens the readers to the opportunity to be free in adopting the reading devices they prefer. Devices such as e-book readers are very important elements in this sense, and they are the "battlefield" on which the commercial firms, the other three central nodes of the network, engage their counterparts.

Twenty-five years after the foundation of PG, in 1996, the idea of the accumulation of written texts on hard disks met the trajectory of portable technology. This shift in the original trajectory was promoted by the intervention of different kinds of actant. While PG continued digitalising books, electronic corporations started producing portable devices for e-book reading, as for instance, the Electronic Paper developed by E Ink Corporation, in 1997. This was a key step for the beginning of commercialisation; just one year later, in 1998, e-books became available on the first commercial web platforms such as eReaders.com. Google Books started in 2004, and Amazon proposed its standardized e-book format, flanked by its own e-book reader device (Kindle) in 2007, while iTunes announced its own e-book storage app in 2010. The firms involved in digital commercialisation of text, *de facto*, diversified the opportunities for reading. But this run for diverse standards also implied processes of closure. Closure is exactly what PG wants to avoid.

> "You may convert to and distribute this work in any binary, compressed, marked up, non-proprietary or proprietary form, including any word processing or hypertext form. However, if you provide access to or distribute copies of a Project Gutenberg-tm work in a format other than "Plain Vanilla ASCII" [N/A ASCII is the American Standard Code for Information Interchange, established in 1968] or other format used in the official version posted on the official Project Gutenberg-tm web site (www.gutenberg.org), you must, at no additional cost, fee or expense to the user, provide a copy, a means of exporting a copy, or a means of obtaining a copy upon request, of the work in its original "Plain Vanilla ASCII" or other form. Any alternate format must include the full Project Gutenberg-tm License as specified in paragraph 1.E.1.[11]"

The commitment of using basic coding such as ASCII could be seen as a strategy used to maintain completely open access to the e-books collected in the PG database. This obligation, required of all the contributors, is also a key element for readers; it enables them to benefit from open access e-books without any specific technical requirement, just a connection and a screen. This opportunity multiplies reading practices without any cost to the reader and without any restriction due to the copyright. More recently, this "open access" approach to digital texts has also been translated to writing practices within the PG community.

[11] http://www.gutenberg.org/wiki/Gutenberg:The_Project_Gutenberg_License

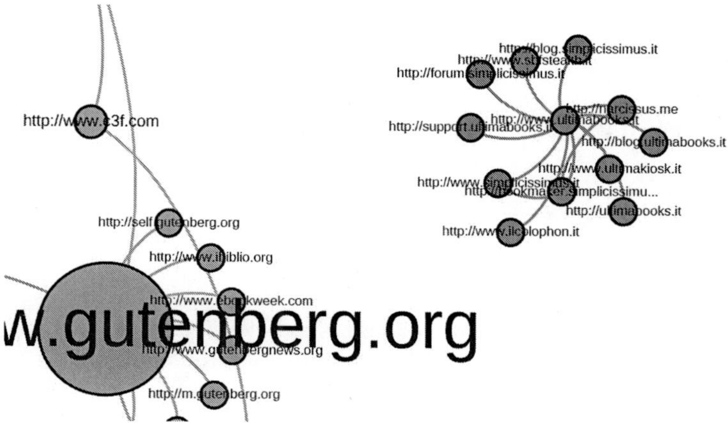

Fig. 8-3: a zoom in on PG's clique (light grey) and Narcissus' clique (dark grey).

In the clique formed by PG, we individuated a website called self.gutenberg.org (fig. 8-3). This website is the online platform for free self-publishing services. As we can read from the related blog:

> "The generous souls at Project Gutenberg opened a new venue supplementing their traditional collection of public domain classics. (And, by the way, that collection of usually all the works, even minor ones, of major authors and the major works of minor authors now runs more than 40,000). But what about contemporary works, particularly by people who have not found or perhaps do not want to go with traditional publishers? Now Project Gutenberg has opened a separate collection and discovery area for such works referred to as a Self-Publishing Portal. Anyone can access and search the site to view and/or download documents. If the reader chooses to register with Project Gutenberg, as all the authors have, they can participate in the Authors Community Cloud Library and post comments, feedback, ratings, and reviews. A book details page and Wall will attach this social network style input for other readers to see.
>
> According to John Guagliardo, long-time volunteer at Project Gutenberg and a director of the Gutenberg Consortia Center, as well as of the World Public Library Association, the new service stems from years of receiving "tons of requests" from unpublished authors."[12]

Often, self-publishing is a strategy for *wannabe* writers who are disposed to pay the publisher directly. This phenomenon is also called "vanity press," and it has been well-known in the publishing industry since the

[12] http://self.gutenberg.org/SiteMaintenance.htm?aspxerrorpath=/Blog/default.aspx

1950s (Sullivan 1958). However, as mentioned in the Introduction, the phenomenon is now becoming more visible. Our observations show that self-publishing is now becoming less expensive for authors, thanks to particular initiatives like self.gutenberg[13].

> "Authors interested in posting to the Self-Publishing Portal upload documents in PDF format. This is the only format that the new service will use. The announcement of the new service alludes to the compatibility of PDFs with iPad, Kindle, and other e-readers. […]"[14]

The community of PG clearly expresses its rules in a handbook, as the excerpt reports. In this case, PDF format is the only format permitted for authors. But the purpose, once again, is to use the most common format in order to facilitate the circulation of digital texts, regardless of the device used by readers.

> "Authors who contribute to the new repository retain all their rights to their works. Guagliardo repeatedly stated that Project Gutenberg had no ownership nor any transfer of ownership (e.g., through a perpetual license). The content is not public domain, nor even necessarily offered under a Creative Commons license. If an author chooses to submit their ebook to another repository, that is their choice. If an author finds a commercial publisher and chooses to withdraw their ebook from the Project Gutenberg site, that too is their choice. If an author wants to revise their work and submit a new version, Project Gutenberg will comply. But whatever resides on the Project Gutenberg Self-Publishing Portal is there to be shared with readers "for personal study and non-commercial sharing."[15]"

For the same reason, easing e-book circulation, the community has established that it does not impose any copyright or any restriction for the authors.

While PG is totally free, we found other websites that offer similar services for self-publishing with fees, such as simplicissimus.it (the dark grey clique in fig. 8-3).

> "Narcissus retains 40% of the cover price, without any constraint on the cover price, size of the file (up to 40MB) and sales.
> You can freely choose the cover price and change it at any time.

---

[13]In fact, having a look at the repository, we counted 2282 books divided into 61 categories; PG decided to join the self-publishing sector only in 2012.
[14] http://self.gutenberg.org/SiteMaintenance.htm?aspxerrorpath=/Blog/default.aspx
[15] http://self.gutenberg.org/SiteMaintenance.htm?aspxerrorpath=/Blog/default.aspx

> With this 40% Narcissus will pay the individual online bookstores (asking for an average of 25-30%), retaining 10% to compensate for the distribution service.
> [Narcissus N/A] It will liquidate to you 60% of sales periodically on the basis of a detailed report of sales made in each library."

The graph directs attention also to other commercial cliques identifiable with famous editors and Italian vendors like Feltrinelli (fig. 8-1, in the upper left) or Mondadori (fig. 8-1, bottom left of the graph) that multiply their distribution channels on the Internet. However, the majority of the nodes in the network adhere to PG's philosophy of free exchange and distribution.

In particular, two cliques that have joined an open access philosophy are interesting because of the numbers of nodes involved and their disposition in a star group (Wasserman and Faust 1995). The first (fig. 8-1, the biggest cluster on the left of PG's star, *see also* fig. 8.4) presents GiardiniBlog.org as the central node while the other (fig. 8-1, the top cluster of the graph) is centred around honkiat.org.

According to our direct observation of the two cliques, Giardini Blog.org appears as the central node for the Italian eBooks' open source network, and it is also the most important gatekeeper for Italian net surfers. Precisely, it suggests where to find websites that do not belong to the mainstream market:

> "GiardiniBlog is a multi-thematic blog at 360° on various topics, focusing mostly on the world of technology and digital entertainment. The idea of sharing our knowledge, was born in the "gardens" of our city, a place that gave birth to the name of the blog." [16]

The other websites which belong to the blue star net obviously support the same scope of the central node, but the most important observation is that a large portion of them focus their activity on writing and not only on reading practices. Indeed, as we can read in the presentation of some websites:

> "If you enjoy writing here you can post free stories, poems, novels collections and much more. If you like to read, read the works of the day or week, or browse the archive."[17]

[16] http://www.giardiniblog.com/info/
[17] http://www.neteditor.it/

> "With the same registration, if you want, you can also upload your own literary works. Immediately and for free, by uploading them directly on this site in the formats you want."[18]

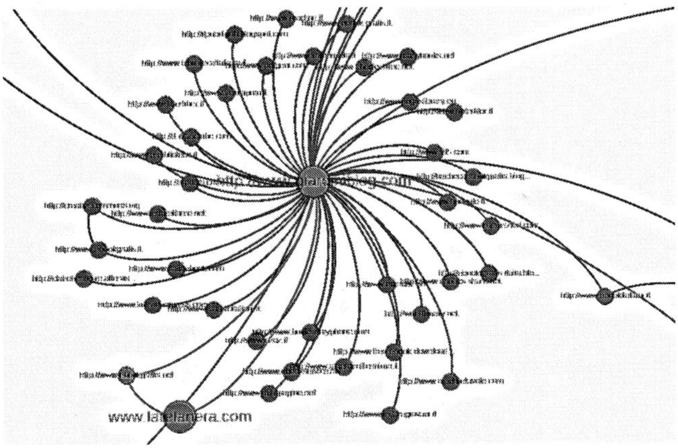

Fig. 8-4: a zoom in on Giardiniblog star group and LaTelaNera.

Among all of these, we have found two websites that share a link to be of much interest: LaTelaNera and EbookGratis (as evidenced in fig. 8-4, with grey hedges). According to a deeper analysis, we observed that the two entities cooperate in the production and distribution of original free e-books outside the mainstream market.

> "LaTelaNera.com in collaboration with our partner site eBookGratis.net, one of the leading portals for the distribution of free e-books in Italy, is pleased to offer to all its readers a wide range of novels, collections, reviews, essays, poems and manuals on the pages of this important site." [19]

LaTelanera gathers fans of horror and thriller literature and fiction, posting articles, reviews and news about these themes. Moreover, LaTelaNera.org regularly promotes competitions in its forum that are forms of collective writing among the members of the community. From one of the pages describing this kind of activity, we can read that:

> "LaTelaNera has produced since 2003, numerous ebooks selecting the most worthy material through its literary competitions in order to ensure a

[18] http://www.Ebookservice.org

[19] http://www.latelanera.com/ebook/ebookslatelanera.asp

good level of quality of the texts and therefore a greater satisfaction for the reader. All its ebooks are free and their purpose is purely informative: we want to promote the work of many authors available to a lot of readers."[20]

This statement reveals that the community, through the interaction between the members, aims to reach a "good level of quality" of the novels. For this reason, this website has created a system based on various contests and prizes that have different norms, goals and awards. The list of competitions is very long and heterogeneous. The purposes of the competitions and the characteristics of the participants underline the huge diversification of the writing practices that are fostered by LaTelaNera.

"Literary contests are very different from each other, by kind, length and timing. Some have been organized in collaboration with other web sites, others are closer to the shape of the "literary workshop." All allow the writer to deal with other authors in the same conditions.
NeroPremio
It is a free contest for horror stories, crime, noir, fantasy. Four issues per year, one per season. The enrollments, therefore, will always be open.
Skannatoio 5 e mezzo
It is a free literary competition that has books as prizes, it takes place on the forum for aspiring writers of our site. The participants exchanged their judgments and opinions on the stories in competition. There is an edition every 40 days or so, entries are open for a week. Minimum 5 members, maximum 15.
Macelleria numero 6
It is a literary free competition without prizes which takes place on the forum of our site. Inside, the participants (max 6) exchange between themselves their judgments and opinions on the stories in competition. At the end of a thorough analysis of the text the winner will be selected. Registration is open until the 6 available places have been filled.
NeroArcano
It's a free contest for reviews (books and movies) and in-depth articles on the topics treated by our site on the pages of "il Pozzo Senza Fondo". Four issues per year, one per season. The enrollments, therefore, will always be open.
(new edition in 2014)
Nella Tela!
It's a contest to charges for horror stories, noir, fantasy. Each winner of the 3 sections will receive the sum of 200 Euros.
(new edition in 2014)
SANguinario VALENTINO

[20] *Ibid.*

It is a contest for stories of all kinds containing a "History of Love gone bad…"
(new edition in 2014)
MortErotica
It's a free contest for horror stories, crime, noir containing a strong erotic/sexual component.
(new edition in 2014)
ConcorsoMorto
It's a free contest for stories that contains the presence of at least one of the undead: vampires, zombies, mummies, ghouls, banshees, etc…
(new edition in 2014)
Una penna per Poe
It's a free contest for stories linked to the theme of Edgar Allan Poe."[21]

The competitions' descriptions show how LaTelaNera configures itself as a proper community where writers and readers are constantly in touch with each other, competing and sharing knowledge and opinions on their stories.

"Macelleria Numero 6 (N/A Butchery #6): what is it? Traditionally, Macelleria Numero 6 is a free literary competition that allows all competitors to have rapid feedback on their works and to improve as writers. It is a writing workshop open to authors willing to work hard, to read other people's material with an open mind and especially to withstand criticism with humility. […]
If you have worked well (if they all worked well), you'll have a beautiful new story to read and use. The stories included in every edition of the contest will be able to participate in NeroPremio then, in Nella Tela! Or in any other literary competition organized by LaTelaNera.com. I repeat: this is not a competition, but a laboratory, no frills, to learn how to improve, a Fight Club that should be taught to all participants as being in the world.[22]
The award ceremony [of Nero Premio N/A] will take place within ten weeks of the closing of entries. There will be no public ceremony. The best stories, with the approval of the authors, will be published in a free e-book that will be distributed on the partner's web pages www.eBookGratis.net. The organization has no obligation of remuneration to authors for this publication, but only the obligation to indicate clearly the name of the ebook author for each story published. The literary work property belongs always to the author."[23]

The excerpt above describes a competition that aims to help the

[21] http://www.latelanera.com/scritturacreativa/concorsilatelanera/
[22] http://www.latelanera.com/macellerianumero6/
[23] http://www.latelanera.com/neropremio/bandoneropremio.asp

participants in improving their writing skills; it is a sort of boot camp for stories whose writers wish to participate in and win the other contests. All the writers' pieces are read and criticised by the other members of the community in an interactive way allowed by the setting of the forum.

The latter, Nero Premio (Black Prize), is based on different premises; while Macelleria Numero 6 is a sort of laboratory where the members aim to be praised for the originality of their plots or the reliability of their characters, this offers a real prize to the winners, the chance to distribute their works in digital text.

The competition spirit demands a serious commitment ("*If you have worked well (if they all worked well)*") that involves both writers that compete and readers that act as judges.

The commitment, the engagement and the seriousness of the tasks are exemplified by the interactions that take place on the discussion forums. The extracted conversations allow us to individuate routine interactions between members within the contests; the forum setting also permits users to establish durable relationships.

> "Hello Rovo, is a bit 'that no one reads each other?' I come straight to the point: I liked your story, is valid and has a nice "colour". You have been able to make very well the Sicilian way, yet there are some things that in my humble opinion, should be improved. Probably it's all due to the very limited number of characters, but the story is perhaps too wordy. With a re-write, and expanding the (almost non-existent) descriptive parts, could come out a great piece. Other defects that I have found are little more than gleanings: in the first sentences you fall into the infodump, when Rossetti explains who is the mother of Caimmé. Necessary from a narrative point of view, but at least redundant, do not you think? Another little problem is the one in the succession of sequences, you are likely to lose the thread of the discourse, perhaps even the use of dialectisms which, although very readable, may cause some confusion. In terms of form, well, I do not think I have to tell you anything-curated and precise as usual-. This tale is different from the usual, with a happy ending, a moral and it lets imagine the ferocity of Mechanical clan. The characters are well defined, although the role of commissioner is interesting and deserves a greater development. So, in summary, the real flaw of this story is not the plot or the characters, rather the style perhaps too dry. In certain passages seem to read a script, and maybe you should venture out to just turn it into a piece of lead foil. Special mention should be made to the metaphor of the paper, taken at the end. That was just a touch of class, well inserted into the structure of the story. I give you the second position: a tale however respectable, interesting and above all with a lot of potential. One last question: but have you got Sicilian origins? I know that you know the environment of the

> Arma (Carabinieri corps), but I notice a familiar knowledge, perhaps first-hand one, of the environment in which you move your characters." [24]

In this excerpt, one member, "Smidolon", before expressing observations and criticism, hails the other member (Rovo), underlying the fact that they used to "read each other" more frequently in the past. We can also observe that Smidolon makes reference to a specific piece of vocabulary that belongs to literary criticism (i.e. "infodump").

> "'Throwing open those beautiful black eyes' as I learned from 'Skannatoio November' (thanks Rovignon), beautiful' is more a personal opinion of the narrator than a fact. Still quoting Brambles, "as the narrator should only highlight the positive aspects, so as to give rise to the idea of looking into the reader without saying."[25]

> "Here is the compulsive! Well! I know you could even have been worse! But why are all convinced that Jonas is beautiful and tall? I've never said that! I think it's my avatar misleading...
> Thanks Gar, I sprinkle ashes on my heads and over the entire line! Concerning punctuation inside and outside of the double quotes, however, come to rescue me. Because from what I remember in the double quotes, the phrase must always end with a punctuation mark and must also start with a punctuation mark outside. And I work my ass off just for putting them, if I do not need them I am three steps ahead! What the hell is the rule?
> For the record, yes! I wrote it in half a day, it was by one Skanna, [Scannatoio 5 and 1/2] I think a 24 hour!
> Thank you dear, as well as useful, it is always a "pleasure" to be criticised by you!"[26]

During this interaction, members, as we said, usually share a common knowledge that is transmitted from member to member. For example, as Nitroneus did with Russel in the forum section called "Stories under the lens", older advice is re-proposed. Also, personal feelings are involved in this practice, as we can see in the second extract, where Polly responds to another member that has strongly criticised her work.

The community's members are deeply involved in writing and in giving suggestions, but also in praising when a novel is well constructed. Members' commitment is linked to the pleasure of reading a good novel that is, at the same time, an enjoyable story to be read and an outcome of

[24] *Ibid.*
[25] http://latelanera.forumfree.it/?t=64233673
[26] http://latelanera.forumfree.it/?t=64233673&st=15

the community itself.

4. Multiple and diverse practices. Multiple and diverse methods

Reading and writing practices have never come to a complete and irreversible closure. Indeed, they are interpreted and performed according to the environment in which they take place, as are other cultural and artistic practices such as music and photography (Magaudda 2013, 2011; Schwartz 2009, 2010). When they are intercepted by the broader digitalisation process, multiple and diverse trajectories are outlined.

The research we conducted on the reading and writing practices' trajectories following the e-book has been an explorative one. The principal aim was to understand how these trajectories have been shaped by the race for digitalisation. We recorded different kinds of data that are useful in stimulating further research both on this topic and on the study of the Internet as informational infrastructure. In concluding this contribution, we would like to remark on two different elements.

First, the approach we used. The naturalised use perspective, combined with practices theory and applied to the Internet, seems particularly appropriate for studying how users and infrastructures are shaping each other, disciplining practices and fostering diverse kinds of processes, for instance, collective and interactive writing or the open access philosophy. Here, the former has been embodied by LaTelaNera in which, through spaces of interaction such as the forum, texts are created and recreated exactly according to the community's own standards and tastes; the infrastructure allows interaction, transforming some websites into laboratories of collective writing. These transformations represent a trajectory with the ability to dodge publishing industry mechanisms.

The latter (open access philosophy) is illustrated by Project Gutenberg that in some way anticipated the open access approach for scientific journals[27]; the openness of the format chosen and its norms are elements that concretely steer clear of standards imposed by the publishing industry. Indeed, it is a two-way trajectory that involves both readers, by choosing the most open format of digitalisation of texts such as the ASCII code for e-books available on Project Gutenberg's repository, and writers. In fact,

[27] As Suber states (2008): "Open Access literature is digital, online, free of charge, and mostly free of copyright and licensing restrictions."
http://dash.harvard.edu/bitstream/handle/1/4322580/suber_oagratis.html?sequence =1

the community of Project Gutenberg has applied the same approach in promoting its services for self-publishing. In this case, it has also chosen a suitable format, even if it was not conceived specifically for the e-book - the PDF.

Taking into account the trajectories of some practices means looking for changes through multiplication and diversification. Practices are stable entities, but they are not static social relations. In fact,

> "the relationship between practices and performances is recursive: practices configure performances, and practices are reproduced, and stabilized, adapted and innovated, through performances." (Southerton *et al.* 2012, 240).

The second goal of this chapter is to define a research method which is able to provide a response to some challenges that the study of the Web is creating (Howard 2002; Schneider and Foot 2004). This research identifies a mixed method (Morgan 2007) that defines a precise overview of the research field, putting in evidence the structure and the dimension of the Net's portion that the researcher intends to observe. Net surfers' activities become visible to the researchers, thanks to the "digital footprints" that the surfers leave on web forums, chat rooms or social network sites (SNS). But the direct observation of the various websites, without any graphical support, runs the risk of proposing a nebulous research field and, worse still, of missing important websites or connections.

Certainly, network analysis provides graphs and measurements that firstly define the research field and secondly describe many of its characteristics: the importance of the websites, how they are connected, whether they form communities and so on. After that, a direct observation can give sense to the result proposed by the graphs and the measurements, visiting websites of interest and collecting data that shed light on the peculiarities of the objects considered.

This is not a linear process divided into two phases, but the two moments of the research move in circles, adding important data, step by step, to the research.

We hope that our contribution may be a stimulus for advancing the development of these two subjects.

References

Borgman, Christine L. 2003. *From Gutenberg to the Global Information Infrastructure: Access to Information in the Networked World.* Cambridge, MA: The MIT Press.

Bourdieu, Pierre. 1977. *Outline of a Theory of Practice*. Cambridge: Cambridge University Press.

—. 1990. *The Logic of Practice*. Stanford: Stanford University Press.

Bowker, Geoffrey C. and Susan Leigh-Star. 1999. *Sorting Things Out: Classification and Its Consequences*. Cambridge, MA: The MIT Press.

Bräuchler, Birgit and John Postill (edited by). 2010. *Theorising Media and Practice* (Vol. 4). Oxford: Berghahn Books.

Callon, Michel. 2005. "Why Virtualism paves the way to Political Impotence: A Reply to Daniel Miller's Critique of the Laws of the Markets." In *Economic Sociology: European Electronic Newsletter*, 6(2): 3-20.

Callon, Michel and Fabien Muniesa. 2005. "Economic markets as calculative collective devices." In *Organization Studies*, 26(8): 1229-1250.

Consalvo, Mia and Charles Ess (edited by). 2011. *The Handbook of Internet Studies*, Oxford: Wiley.

Freeman, Linton C. 1979. "Centrality in social networks conceptual clarification." In *Social Networks*, 1(3): 215-239.

Giddens, Anthony. 1984. *The Constitution of Society: Introduction of the Theory of Structuration*. Berkeley: University of California Press.

Hafner, Katie and Matthew Lyon. 1996. *Where Wizards Stay up Late: The Origins of the Internet*. Chicago: Simon and Schuster.

Howard, Philip N. 2002. "Network ethnography and the hypermedia organization: new media, new organizations, new methods." In *New Media and Society*, 4(4): 550-574.

Law, John. 1994. *Organizing Modernity*. Oxford: Blackwell.

Lieberman, Evan S. 2005. "Nested analysis as a mixed-method strategy for comparative research." In *American Political Science Review*, 99(3): 435-452.

Magaudda, Paolo. 2001. "When materiality 'bites back': Digital music consumption practices in the age of dematerialization." In *Journal of Consumer Culture*, 11(1): 15-36.

—. 2013. "What happens to materiality in digital virtual consumption?" In *Digital Virtual Consumption*, edited by Janice Denegri-Knott and Mike Molesworth, London: Routledge.

Marin, Alexandra and Barry Wellman. 2011. "Social Network Analysis: An introduction." In *The Sage Handbook of Social Network Analysis*, edited by John Scott and Peter J. Carrington, 11-25, London: Sage.

Morgan, David L. 2007. "Paradigms lost and pragmatism regained methodological implications of combining qualitative and quantitative methods." In *Journal of Mixed Methods Research*, 1(1): 48-76.

Oudshoorn, Nelly and Trevor Pinch. 2003. *How Users Matter: the Co-construction of Users and Technology*. Cambridge, MA: The MIT Press.

Rainie, Harrison *et al.* 2012. *The Rise of E-reading*. Pew Research Center's Internet & American Life Project.

Santino, Marco *et al.* 2012. *Rivoluzione e-book: quali novità dal fronte*, A.T. Kearney, http://www.atkearney.it/

Schatzki, Theodore. 1996. *Social Practices: A Wittgensteinian Approach to Human Activity and the Social*. Cambridge: Cambridge University Press.

Schatzki, Theodore R., Karin Knorr-Cetina and Eike von Savigny (edited by). 2001. *The Practice Turn in Contemporary Theory*. London: Routledge.

Schwarz, Ori. 2009. "Good young nostalgia: camera phones and technologies of self among Israeli youth." In *Journal of Consumer Culture*, 9(3): 348-376.

—. 2010. "Going to bed with a camera: On the visualization of sexuality and the production of knowledge." In *International Journal of Cultural Studies*, 13(6): 637-656.

Schneider, Steven M. and Kirsten A. Foot. 2004. "The web as an object of study." In *New Media and Society*, 6: 114-122.

Southerton, Dale, Wendy Olsen, Alan Warde and Shu-Li Cheng. 2012. "Practices and trajectories: A comparative analysis of reading in France, Norway, the Netherlands, the UK and the USA." In *Journal of Consumer Culture*, 12: 237-262.

Star, Susan L. 1999. "The Ethnography of Infrastructure." In *American Behavioural Scientist*, 43(3): 377-391.

Star, Susan L. and Karen Ruhleder. 1996. "Steps Toward an Ecology of Infrastructure: Design and Access for Large Information Spaces." In *Information Systems Research*, 7(1): 111-134.

Warde, Alan. 2005. "Consumption and theories of practice." In *Journal of Consumer Culture*, 5(2): 131-153.

Wasserman, Stanley and Katherine Faust. 1994. *Social Network Analysis: Methods and Applications (Structural Analysis in the Social Sciences)*. Cambridge: Cambridge University Press.

Acknowledgements

The chapter was conceived and discussed by both the authors, but section 1 is attributed to Paolo Giardullo while the second and third section to Gian Paolo Lazzer. The rest of the chapter was jointly written.

CHAPTER NINE

"DISPOSITIVES" OF FLEXIBLE MOBILITY FOR TRAVELLING HUMANS AND TECHNOLOGIES

OANA MITREA

Introduction

Although aiming at unifying and giving meaning to a very diverse world society (Bammé 2010), the current technological civilization built on ICTs depicts some internal tensions.

The publishing of the World Value Survey results (1990-1993) (Inglehart *et al.* 1998) accompanied the beginnings of the extended implementation of the current advanced technologies based on agents, sensors and context awareness. At that time, the survey had testified the marching of post-modern values in Western Europe, USA, and Canada. Post-modernization implies that

> "a new world is gradually replacing the outlook that has dominated industrialized societies since the industrial revolution, reflecting a shift about what people want in life – a change of the value and belief systems." (Inglehart et al. 1998, 9)

According to the authors, the key traits of modernization are: the emphasis on the individual economical achievement, the decrease of communal obligations, and the acceptance of social mobility (with former socialist states rated as a culmination of modernization due to heavy bureaucratization). On the contrary, the postmodern values emphasize the individual choice of life styles and individual self-expression: the pursuit of individual well-being becomes a powerful norm (*Id.*). Individualization has been criticized for its risks and constraints (Beck and Beck-Gernsheim 1994; Hradil 1995; Giddens 1990, 2000; Sennett 1998). Sennett (1998) claims that the culture of the new capitalism, with its short-term perspective and the logic of flexibilization, negatively influences human characters and social relationships.

In the last while, advanced ICTs (mobile internet front-end devices and related back-end infrastructures, pervasive computing devices, intelligent environments) in the areas of mobility, commerce, housing/living, old age, health, etc. are supposed to have made all these areas "smarter" through context-aware and personalized usage. However, "real-time" (mobile) ICTs can also reinforce a type of flexibilization and mobilization of daily life that could become enslaving and unbearable for some humans. Smart phones, tablets, notebooks and dynamic navigation devices that react in real-time to traffic changes, and the positioning and tracking of movements, impose new commandments on modern travelers: Hurry up! Don't lose time! Take risks! Be ready to leave home! Be flexible! Be mobile! Travel farther and faster! Be efficient!

How do individuals, groups, and societies cope with these requirements? What is the nature of the current mobility supported by the advanced system technologies?

1. Problem statement and proposed approach

The active, interactive, decentralized, self-organizing, and pervasive nature of smart technologies seems to find its parallel in the postmodern focus on individual autonomy and self-realization, the depreciation of state planning and control, and the growing respect for market forces (as described by Inglehart *et al.* 1998, 9). It is possible that such traits greatly engrave the visions of current technology constructors. They develop flexible and de-centralized solutions for increasingly mobile humans; respectively flexible and de-centralized institutions. Postmodern societies and their underlying technologies find themselves in a state of mutual interdependence and shaping (see MacKenzie and Wajcman 1985).

This chapter aims to theoretically examine this interdependence and to answer the following questions:

- How do intelligent/smart ICTs and humans co-act and interact in the technological civilization?
- What structuring frames for this can be described?
- What consequences of this interaction can be foreseen?

The present analysis focuses mainly on the spectrum of mobility actions and interactions and its intelligent enabling technologies. Smart technologies, such as mobile social networking, dynamic ride-sharing systems, advanced travel information systems (ATIS), cooperative traffic systems, advanced driver assistance systems (ADAS) and context-aware information systems, receive the status of structure-forming material *dispositives* for flexibilization and mobilization. The key particularity of a material dispositive is the "disposition" or "arrangement" resulting from the interaction of humans with technology as apparatus (devices, software, and infrastructures) on the background of some general social and overall cultural conditions. Such arrangement is supposed to have structuring effects on the design of intelligent system technology and their subsequent usage.

The sociological literature dealing with mobility and mobile communication has helped me to identify and understand the nature of mobility and social interaction, as it is structured in the dispositive. The analysis in this chapter draws on findings of the *new mobilities paradigm* (Sheller and Urry 2006; Urry 2007), *mobility business regimes* (Kesselring 2012), respectively virtual work (Schmiede 1996), the hybridization of the daily life experiences and social relations (Kesselring 2012), the space/time compression (Giddens 1990), the perpetual contact (Rule 2002), and other spatial arrangements such as the movement of portable devices closer to the body (Fortunati 2002).

A model of the intelligent mobility system technology as a dispositive is proposed in the last section of the current chapter. This highlights its key elements and arrangements. The anatomy of the dispositive model benefits from the literature and models on "dispositives" (Foucault 1978; Hickethier 1991; Lenk 1997), the contributions of Actor-Network Theory on the role of technology in the stabilization of social configurations and the programs of action (Latour 1992), the concept of script (Akrich 1992), and the sociology of advanced agent-based technology (Rammert 2007; Weyer 2007, 2009; Schulz-Schaeffer 2009).

In order to better understand the structure and physics of the proposed dispositive model, let's first look at the present situation of mobility in the technological civilization.

2. Chances and tensions of mobility in the technological civilization

At the beginning of *Mobilities*, John Urry observes that "people are travelling farther and faster", even if the time spent on-the-way remains unchanged (Urry 2007, 4). More and more accessible air traffic and faster cars and trains make us see the conquest of distance in a new light. This implies a rapid change of the socio-cultural context, which is differently managed by the various travelers. The accommodation of the rapid context changes is now facilitated more and more by the context-aware information and communication enabled by advanced ICTs.

As "the new mobilities paradigm" (Sheller and Urry 2006; Urry 2007) emphasizes, individuals, goods, ideas, representations, and pieces of communication flow through intermingled Internet and traffic/transportation network systems. Modern mobility is virtual, as well as physical. Next to goods, travelling humans transport with them an invisible luggage of ideas, representations, attitudes, beliefs and stereotypes that are relevant for the succession of contexts they meet on-the-way. Also, these are communicated to others through mobile devices and infrastructures. In this way, travelling humans, contents, and information and communication technologies are continuously challenged by the requirements of the always new contexts and by the imperative character of mobility in our current civilization.

The sociological perspective on modern mobility has emphasized both opportunities and tensions of this situation, as synthesized in Figure 9-1.

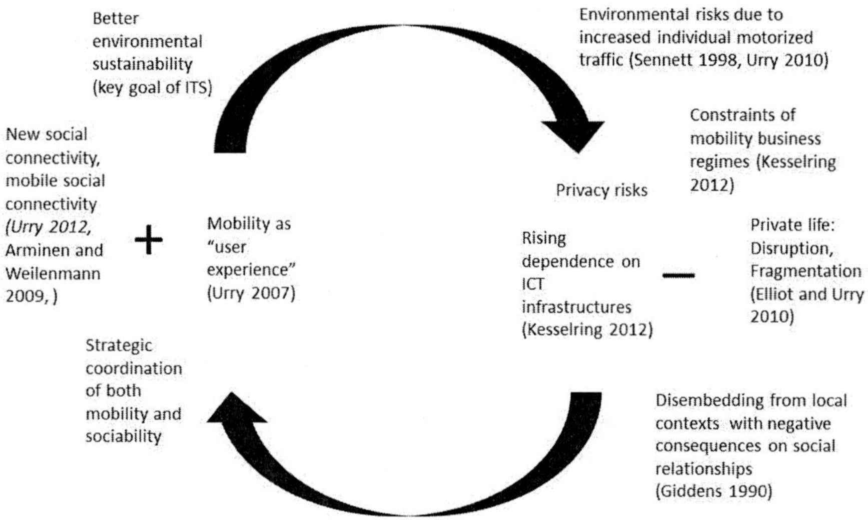

Fig. 9-1: chances and risks for mobility in the technological civilization construction

2.1 Chances

Mobility technologies seem to enable a new type of social connectivity, turning humans into mobile portals who connect parallel worlds on the move (Kesselring 2012, 92). Navigation in the space defined by social and spatio-temporal coordinates involves "the capacity to decide which type of sociality is currently relevant: the space of the city or the *facebook*?" (Kesselring 2012, 93). This simultaneous immersion and switching between multiple offline and online worlds, quite often observed on trains and buses, can also represent a chance for a richer interaction and communication (Mitrea and Kyamakya 2013).

According to Urry (2007), mobility solutions have to integrate the movements of humans and things with the circulation of ideas, action possibilities, imagination, and pieces of communication. The intertwining of information, communication and movement in the current intelligent mobility and mobile communication systems represents indeed a chance to improve the strategic and tactic planning of the daily life of users. Let us consider the needs of urban citizens with their crowded and highly fragmented daily agendas, many places to reach in a short space of time, trapped in the city sprawl and having difficulty integrating distant places in the same daily route. Their situation is even more difficult if they have no car or little opportunity to use their own and so are dependent on an inflexible or unreliable public transportation system. These users can benefit from an intermodality (the successive use of various modes along the same route) supported by reliable and suitable information provided by intermodal travel information systems or ride-sharing systems (Kyamakya and Mitrea 2010).

Moreover, access to travel information on mobile phones, and embedded screens, combined with the possibility of contacting somebody on various channels, has opened new possibilities for the concomitant planning of mobility and sociability, one in function of another: "*when to meet somebody and who*", reformatting of social actions and relationships under mobile conditions. The "mobile contextual configuration" has been studied by Arminen and Weilenmann (2009). They have stressed the high amount of information exchange in mobile phone calls about location, transportation, current activity, etc. In a recent article, John Urry emphasizes the fact that mobile ICTs "make and remake" connections across time and over space, which are essential for the extension and/or consolidation of the social networks (Urry 2012). The same author invites us to reflect also on the user experience value of the time spent on-the-way and the diversity of mobility activities:

> "[...] the time spent travelling is not necessarily unproductive and wasted dead time that people wish to minimize. Movement often involves an embodied experience of the material and sociable modes of dwelling-in-motion, places of and for activities in their own right, to climb a mountain, to do a good walk, to take a nice train journey. There are activities conducted at the destination, activities conducted while travelling including the 'anti-activity' of relaxing, thinking, shifting gears; and the pleasures of travelling itself, including the sensation of speed, of movement through and exposure to the environment, the beauty of the route and so on" (Urry 2007, 11).

The importance of the experiential dimension of mobility in the psychological sense is also confirmed by Canzler and Knie (2002) and Kesselring (2012)[1]. In relation to mobility as social practice, Canzler and Knie (2002) stress that the desire for cozy spaces and safety competes with the need for anonymity, and for open, risky and contradictory worlds of experience. Studies are still necessary on the very different individual dispositions and preferences in different social and cultural contexts (*Id.*).

2.2 Tensions and risks

The environmental risks of mobility are usually among the most perceived and debated issues in many disciplines. Let us consider the environmental pollution, stress, and noise attributed to heavy motorized traffic that should be solved by ITS (Intelligent Transportation Systems). Such types of risks have also been recognized by sociologists (Sennett 1998; Urry 2010).

An interesting problem with mobility in our current world seems to be its newly postulated imperative character in both work and private spheres. The mobilization of work imposes as one of the most important dimensions of society modernization (Kesselring 2012, 84). The fulfillment of work tasks on-the-way and the increasingly distant work-related social interactions are to be supported by smart phones, navigation systems, intelligent travel information systems for intermodal travel planning, mobile Internet infrastructures, etc. Its compulsory nature in business/work mobility regimes (Kesselring 2012, 87)[2] represents,

[1] Kesselring observes that mobile workers use waiting times as phases to think or do phone calls. Traffic congestions are considered suitable to discuss sensitive/problematic issues, also because the intensity can be broken at any time because you have to concentrate on the traffic (Kesselring 2012).

[2] According to Kesselring (2012), the work mobility regime represents a set of principles, norms and rules that control the travel activity within and on behalf of

however, a practice of domination constructed through discourses which pursue the disciplining of individuals' movements, in particular the normalization of work travel, the rationalization of mobility management and the self-optimization of mobility behavior (*Id.*, 88-89). Such domination may take rather indirect and abstract forms. For instance mobile workers control themselves and their navigation, while accepting the necessity to travel and to renounce their free time and self-determination (Kesselring 2012, 89). The consequence (good or bad, depending on the perspective) is to regard successful business travelers as "entrepreneurs of their own mobility skills" (Kesselring 2012, 89):

> "To be successful, one must be mobile and be willing to take risks and to demonstrate the ability to act project-oriented and networked." (Kesselring 2012, 90)

Another aspect is that modern work is now more and more defined by the multidimensional coordinates of time/space/networking. Besides virtuality (Schmiede 2006), this involves physical travel and disruptive mingling with multiple other spheres (the private sphere among them). The negative social implications of the fragmented and short-term perspective of the work life of the "new capitalism" have been highlighted by Sennett (1998). About this point, Kesselring emphasizes the high social costs of work and life as successions of short lived projects. They make long-term life planning and durable relationships extremely difficult (Kesselring 2012, 90).

The resulting fragmentation and flexibilization of time, space, and life experiences also contaminates the private life domain. Increasing distances between communicators impede the coordination with the members of the trusted social space, such as family and close friends. Multiple processes of negotiation and renegotiation over (mobile) ICTs are necessary (Elliot and Urry 2010). The social relations and interactions are more and more dependent on the technological networks operating in the background (Kesselring 2012, 92-93). In this context Kesselring highlights the *ad hoc* nature of the spontaneous network-type mobility based on the connective logic of integration in meaningful contexts and socially-coupled connections and interactions. People no longer meet in a specific place but in the city or city quarter, the exact coordinates of which are communicated by means of the mobile phone (Kesselring 2012, 93).

---

the company. The goals are the disciplining of the mobile subjects by means of a "framework", which specifies who is allowed to move and on what terms. The central principle of the mobility regimes is the mobility imperative: "mobility as guiding principle of action" (Kesselring 2012, 87).

3. The specificity of the intelligent technologies for (communicative) mobility

Current "mobilities" are enabled by a mix of smart information and communication as well as intelligent transportation systems. In what follows an outline of such technologies is provided.

3.1 Smart mobile communication and information

Particularly in recent years, the ensuring of mobile access to communication and to data services – anytime, anywhere – has become the design norm. Smart phones now enjoy rapid acceptance and growth. They are hybrid devices which "are predominantly communication devices, with additional computational power built in" (Beale 2005, 1). The mobile social sensing (Aggarwal and Abdelzaher 2013; Eagle and Pentland 2005) present in smart mobile devices can turn public places into places of mobile connectivity in which contacts between strangers become possible[3]. A user-centered definition of Mobile Social Software (MoSoSo) is given by Lugano[4] as follows:

> "[...] a class of mobile applications whose scope is to support social interaction among interconnected individuals. MoSoSo is an emerging paradigm which exploits the media convergence process and the increasing power of mobile devices to offer a variety of services." (Lugano 2007, 2)

3.2 Intelligent Transportation Systems (ITS)

ITS are defined by the Intelligent Transportation Systems Society as

> "a category of systems and concepts utilizing synergistic technologies and systems engineering concepts to develop and improve transportation systems of all kinds."[5]

In the ITS field, the most recent efforts have been made in the areas of driver assistance (ADAS) and cooperative traffic systems with the purpose of increasing the safety, efficiency, comfort, and the ecological sustainability of driving. In general, the "intelligence" of ITS is mainly related to the way in which the data received from sensors are processed

---

[3] For the recognition of "mutual proximity" at a distance see Licoppe (2009).
[4] For the potential of Mobile Social Software for sustainability see Lugano (2010).
[5] http://www.ewh.ieee.org/tc/its/.

and turned into meaningful information for various "on-demand" situations. According to Wang *et al.* (2006), the intelligence of such system technologies relies on the following components:

1) the active in-and-out vehicle environment sensing and road infrastructure sensing;

2) intelligent spaces defined as "environments that continuously monitor what's happening in them, communicate with their inhabitants and neighborhoods, make related decisions, and act on these decisions";

3) the agent-based control – "distributed operational agents letting networked transportation systems operate on a management-on-demand or service-on-demand basis" (Wang *et al.* 2006, 69).

The most interesting ITS for the purposes of this chapter (emphasizing the intermingling of physical mobility with social interaction and communication) are: dynamic ride-sharing systems, advanced travel information systems, and cooperative traffic systems.

Dynamic ride-sharing systems are defined as

> "systems that facilitate the ability of drivers and passengers to make one-time ride matches close to their departure time. The hallmarks of a dynamic service are flexibility and convenience.[6]"

Examples of ride-sharing systems are: flinc (https://flinc.org/) and zipcar (http://www.zipcar.com/). In recent times, such systems have openly emphasized target goals such as social interaction on-the-way or the promotion of sustainable mobility, as well as allowing faster and more efficient inner-city or between-cities travel.

Advanced Travel Information Systems (ATIS) aim to provide an optimized and efficient access to information as basis for any service supporting urban mobility related processes. An example in Austria is *AnachB* – a traffic information service that aims to offer mobility information for all classes and populations. This information should be based on traffic and environmental policies, dynamic data and the concept of intermodality for the entire Vienna region, and should be free of charge for the users (mobile application + 'www.AnachB.at'). *Context awareness* represents a key feature of the current advanced traveller information systems and intelligent mobility systems in general, in both physical (e.g. a user's location) and social sense. Through the incorporation of the context in intelligent mobility systems, information can be employed for a

---

[6] Definition at http://www.dynamicridesharing.org/ definition.php.

variety of purposes, which may deviate from the initial goal of going from A to B. For instance, systems and services offering participative travel information in urban environments can also provide mobile users with socially-validated orientation and sometimes with unexpected entertaining user experience (jokes related to places, transport means, funny advice which increases the attractiveness of some points of interests or transport means). The impregnation of cities' mobility environments with awareness of social contexts (mobility-oriented social networking platform) has recently been "sold" as a powerful form of user participation.

Cooperative Traffic Systems are defined as

> "systems in which a vehicle communicates wirelessly with another vehicle (V2V – vehicle-to-vehicle communication) or with roadside infrastructure (V2I – vehicle-to-infrastructure communication) or I2V – infrastructure-to-vehicle communication." (CVIS 2010)

The main objectives of their implementation are the increase of traffic safety, efficiency, and environmental sustainability. The majority of cooperative traffic services focus on foresighted driving and early detection of hazards, for instance through ACC (Adaptive Cruise Control)[7], operative platooning, WILLWARN (Wireless Local Danger Warning), INTERSAFE (Cooperative Intersection Safety), etc. Examples of already implemented cooperative traffic services are: incident management, road/weather condition warning, roadwork information, lane utilization information, in-vehicle variable speed limit information, traffic congestion warning. Cooperative traffic systems are very involved with the enhancement of driving/safety based on sensors and telecommunication technologies. However, they can also incorporate in principle a variety of other functionalities for communication and interaction between drivers.

---

[7] Adaptive Cruise Control is defined as "A vehicle cruise control system that automatically slows down the car if it is moving too close to the vehicle in front of it. A radar or laser unit located behind the grille determines the speed and distance of the vehicle in front. When the distance is computed to be safe again, the system accelerates the car back to its last speed setting. Also called 'active cruise control' and 'intelligent cruise control.'" (PC Magazine Encyclopedia, http://www.pcmag.com/encyclopedia/term/37505/adaptive-cruise-control).

4. Steps towards the dispositive model

4.1 What makes current mobility technologies smart and what for?

My common daily life perspective about conventional digging would be that humans plan and anticipate the results, appropriately use the tools, are able to supervise and control the digging process, and finally see the result of their actions: the hole in the soil[8].

The classical view of both engineers and sociologists of conventional technology is that only humans can act with a given suzerainty: setting their goals; choosing means to reach those goals; and evaluating the consequences of actions (Rammert 2006). Conventional tools/technologies are not supposed to act, they can only function (*Id.*). They are instrumental and predictable insofar as they accomplish determined processes, in which the relation cause-effect is quite clear.

This view has been criticized by Actor Network Theory (ANT) scholars (Latour 1987, 1992, 2005). According to ANT, any acting units in an observed context (that also means non-human elements such as tools, keys, hinges, streams, winds, etc.) can obtain the status of actors. The action of non-human elements means in this post-social technology sense (Degele 2002) to have an effect on something/somebody. Agency represents an effect of the co-action of a variety of linked heterogeneous human and various non-human elements and powers (technologies, natural elements and phenomena) – the actor-networks (Kneer 2009).

ANT has the great merit of emphasizing the action possibility of technologies, independent of their form, and their power to give durability and consistence to the heterogeneous associations (Bonz 2007). However, it seems that "advanced technologies" possess some characteristics such as enhanced (autonomous) activity and interactivity (Rammert and Schulz-Schaeffer 2002a, 2002b), which cannot be satisfactorily comprehended within ANT. It is challenging to explain the behavior of technological agents and their interactivity with humans only from the perspective of their effects (low agency concept), interests, and delegation in heterogeneous network building. A theoretical frame for the comprehension of the distributiveness of agency on humans, intelligent acting technologies and social frames has been proposed in the sociology of advanced technology (Rammert 2007; Rammert and Schulz-Schaeffer 2002a, 2002b; Schulz-

[8] For a detailed differentiation between conventional technology and advanced technology, and examples, see Rammert (2006); Rammert and Schulz-Schaeffer (2002a, 2002b).

Schaeffer 2009). This gives a positive answer to the question whether machines can act, on the condition that an external observer is able to produce such attributions about the nature of their co-actions (in cooperation with humans). These attributions are in accordance with a gradual concept of agency (a three-level model of agency) that differentiates among agency as succession of actions inducing changes - the realm of classical machines (level 1); contingent agency as the capacity to act in a different way than intended (level 2); and the intentional and reflexive agency about the induced changes (level 3), reserved so far to human subjects with conscience (Rammert and Schulz-Schaeffer 2002a, 44)[9].

According to this perspective, only advanced or "intelligent" technologies can represent possible subjects of agency in contrast with the "only" technical artifacts (Rammert and Schulz-Schaeffer 2002), and can reach at least the contingency of actions (Rammert 2007, 82). The capacity to act in a contingent manner increases the importance of the interactivity between humans and technical objects: advanced technologies do not only react to previously implemented commands, but they can learn lessons from previous situations, identify a situation, and adapt their actions to it (*situational awareness*, see Schulz-Schaeffer 2009).

In hybrid systems, including humans and intelligent agents acting together (Weyer 2009), this vision is even more intense and obvious due to the higher action potential of the de-centralized and adaptive technologies in which complex tasks are delegated to the agent-systems (for searching, reserving, route-planning, etc.). Rammert (2007) opposes the cooperation of autonomous agents (called "intra-action") to the interactivity between humans and social systems on the one side, and technology on the other. The cooperation of technical units is achieved if their behavior, although based on algorithms, can be described as situated cooperativeness or "artificial" interaction rather than determined operation (Rammert 2007, 83). The notion of interactivity implies a different quality level of the relationship between humans and technology, that goes beyond mere instrumentality: users experience the system as a communicative counterpart, whose reactions may deviate from human expectations and develop in a contingent manner (Rammert 2007, 84). This is characterized by complexity, contingency, and symbolic mediated communication.

---

[9] Nevertheless, Rammert gives examples of BDI (belief, desire, intention) agents that can provide intentional explanations for their actions, based on information and instructions that have the function of beliefs, goals and obligations (Rammert 2007, 82).

In the sociology of advanced technology, the challenge of the future is represented by the "rising 'agency capacity' of artifacts and the 'distributiveness' of activities in socio-technical constellations" (Rammert and Schulz-Schaeffer 2002). Advanced (agent) technologies are meant to observe humans' behaviors and to check them according to specific rules inscribed in the system, with minimal human participation. It is then obvious that they are intelligent insofar as their collective emergent behavior is "noticeable" as such. The decentralized, not hierarchical and adaptive system architectures (self-organizing systems), the ambient intelligence devices and applications, and the delegation of complex tasks to agent-systems (for searching, reserving, route-planning, etc.) usually work beneath human awareness and reach. Usually, humans see only the results of the agents' activities in the form of syntheses or recommendations as results of invisible internal communication between intelligent components. Agent technologies are supposed to know much about the users and usually not the other way around. Maybe it is closer to reality to say that actually neither users nor agents can be sure about the agency range of the other part. Therefore, in hybrid systems the occurrence of unintentional actions and events is quite frequent.

Let's look closely at the following (imaginary) example of decisions and actions in ride-sharing:

(Imaginary scenario):

Lucy is in a hurry. She has to reach the university in 20 minutes and has no car. There is no bus station in the neighborhood. Taxicabs are too expensive for a 23 year-old female student. She reaches for her smart phone and opens the 'urban dynamic ridesharing' application, which automatically recognizes her position. After inputting the destination and desired time, a message notifies her that someone is ready to collect her 100 m ahead in 5 minutes. The profile suits her: also a young female student, even at the same university. She runs to the meeting point, where the car is already waiting for her. A nice person, nice conversation on the way, speedy trip. The small fee is automatically sent to her driver's account by the application. In front of the university she recognizes a girl, as a facebook contact, a friend of her friends, studying at the same faculty. She is a nice person, so much so that Lucy will certainly keep in touch with her. Actually she has just invited her for a coffee (also because she's very good in social statistics). Yesterday it didn't work so well. The matching process had some problems. It took so long to find someone, that Lucy abandoned everything and took a taxi. However, it's nice that there are so many young people subscribed to the system, who are willing to take other passengers spontaneously and trust them.

In this scenario, the social interaction between humans is enabled and initialized by technical agents that hold their characteristics and interests/goals and so enable their matching. In the next step, humans take the lead and interact directly on-the-way. The chain is continued with the post-trip virtual meeting between technical agents that check-out the trip and eventually enable the automatic paying. In the end, humans may meet again face-to-face or virtually.

"Driving together" as the common mission of drivers and passengers represents a chain of contact, communication, and social interaction actions, which are distributed to humans and technologies at different system levels. In this scenario, a ride-sharing application is developed with the intended goal of solving urban mobility problems (many cars with single occupancy, deficits in public transportation). However, unintended consequences may occur: Lucy has reached the destination as wished (supported by the ride-sharing system) and she has made a new friend through direct communication on-the-way, improving thus her chances for participation in social life. The intended usage (mobility) has triggered unintended but welcomed consequences: new social contacts, the information, confirmation, and future support from peers. In dynamic (real-time) ridesharing, successful trip arrangements depend heavily on the perception of the mutual benefits (for drivers/passengers), safety and partners' temporal and spatial flexibility. The construction of trust-building social applications is very important for the successful trip contacts in both instant and previously planned trips.

There are significant differences in how humans and advanced technologies co-act and interact. In dynamic ride-sharing, humans actually meet face-to-face driven by the system's goal: satisfaction of mobility needs. They can also cancel any technical initiative to bring them together. In cooperative traffic systems, unacquainted drivers are usually brought together by the sensor interaction of intelligent cars and don't engage in direct interaction and communication.

Let's take another imaginary example of a technologically enabled interaction with manifest mobility purposes:

(Imaginary scenario):

Marius is entering the hall of the conference, after an interesting presentation. He would like to go home, but he has to wait three hours for the train to his city. Who knows, maybe there is some chance to get a lift. He accesses a dynamic ride-sharing application, hoping that there is a possibility to travel with somebody. Normally the offer should come in a few minutes. However, somebody out of the unknown crowd approaches

> him with a wide smile. No need to look further for distant connections: by means of the short range Bluetooth, another mobile device has detected his query and the owner has accepted it. It is someone from the group nearby, with whom he has already spoken twice. His request, electronically sent, could have also been formulated face-to-face, if the technology weren't so seductive and usage-friendly. Or maybe not … Could the real reason be rather that he preferred to hide his growing social-awkwardness behind technical opportunities?

The scenario above leads to reflections on the grounds of mobile proximity contacts. The "face-to-face" setting does not characterize only human interaction in the same space anymore, but also agents' communication being physically close to one another. For instance, the Bluetooth-enabled smart ICT devices are able to locate and reach the other devices of their kind in the same way as people meet in the same room and start a discussion (*see also* the plug-in properties of the devices in the smart home). What happens further constitutes also a form of face-to-face communication and interaction between neighboring agents that represent users. This machine-machine interaction implies the matching of requests, demands, time-space availability, and human profiles. Up to this point humans might still not know one another. It is possible that they don't even want to be in touch yet. However, their portable proximity-enabled computers have already said "hello" to one another in the same physical space. Taken to the extreme, this would imply that technical agents would meet for a gossip about human profiles and would decide whether their masters have the same interests, hobbies, background, goals, and whether they should also meet. In the end, they would "accidentally" arrange a face-to-face meeting of masters, bursting about their smartness and care for humans.

At the macroscopic level the ICT based intelligent transportation systems have started to gain in relevance for the sociology of technology, mainly because they illustrate how advanced technologies can "act" in cooperation with human and social systems to achieve the global goals of safety, efficiency and sustainability. The view of a global "intelligent" mobility is extended to the level of the heterogeneous constellation of humans, collective actors, technical agents, infrastructures, and signs relevant for mobility (Rammert 2007, 2008; Schulz-Schaeffer 2009). Rammert (2007) stresses that the performance of "intelligent" mobility systems represents a result of a complex interface coordination among the activities of human drivers, car technology, environment, social and political actors. Due to the distributiveness of actions on technological, human, social and political systems in such constellations, it is possible that the performance of the whole system decreases, even if the driving behavior, car control or telecommunications become smarter (Rammert 2007). The performance of the intelligent mobility system is dependent on how the hybrid driver-vehicle-environment system is embedded in the broader socio-technical configuration of the mobility system. This includes a variety of actors: city and traffic planners, producers of alternative mobility technologies, telecom providers, the structure of their products, etc., who decide whether with individual vehicles or not, with how much freedom, in what combination with other means and under what guidance and regulatory system mobility is performed (Rammert 2007, 129). As a consequence, the system engineering of intelligent mobility needs to adopt a holistic perspective including all actors relevant for a given socio-technical constellation.

The examples and comments in this section have mainly emphasized the co-agency power of intelligent mobility technologies and the heterogeneity of the networking in which they are included. In the following section we go further and aim to explain the broader social consequences of this distributed co-action.

5. Towards a "dispositive" model of flexible mobility for travelling humans and technologies

From here on in, the concept of "dispositive" is employed to explain the interdependent structure re-building of communication, information, social interaction, work, and movement due to the integration of intelligent mobile/mobility technologies in the daily life.

Made famous mainly by Foucault's (1978) theses about *power dispositives* (constellations or structures that facilitate and perform the exercise of power), the concept has encouraged modeling attempts mainly in media studies (Hickethier 1991; Lenk 1997). The media and technology *dispositives* and *dispositive models* (Hickethier 1991; Mitrea 2005) include technology as apparatus, its uses or programs, the human/social entities on the background of perceptive/affective and social frames.

The usage of this concept in Science and Technology Studies builds on orientations in the Social Shaping of Technology that highlight the mutual structure-forming relationship between technology and society (MacKenzie and Wajcman 1985). It also draws on important concepts and findings of ANT, mainly concerning the role of technology in the consolidation of the dispositive's arrangements through scripts and programs (Akrich 1992; Latour 1992).

Although Foucault does emphasize the high heterogeneity of the concept as an assembly of discourses, institutions, architectural arrangements, laws, administrative measures, philosophical and moral sentences, etc., I retain for the purposes of this chapter the term "*disponere*" which calls for a mental imagery of a specific arrangement enabled by the usage of a specific technology in a specific context. The arrangements (*les dispositions*) come into being through the co-action of human designers and users, social formations, and intelligent technologies. They are at the same time socially and technically shaped and have structuring powers over the participating elements.

The generic dispositive we propose in relation to advanced system technologies of mobility centrally includes the interaction and co-action of intelligent technology as apparatus (infrastructure, devices, software) with human designers and users. This is influenced by broader societal developments such as globalization (Giddens 1999), individualization (Bauman 2001), societal individualization (Beck and Beck-Gernsheim 2002), virtuality of work (Schmiede 2006), mobilization of work (Kesselring 2012), perpetual contact (Rule 2002), de-traditionalization, disembedding (Giddens 1990) and by some more general frame traits of the technological civilization filled with ICTs such as: space/time compression (Giddens 1990), and the simultaneity of virtual and physical experiences in mobility (Urry 2007; Kesselring 2012; Mitrea and Kyamakya 2013).

Important bricks for the construction of the arrangements of the dispositive are the scripts and programs of action. Scripts are described by Akrich (1992) as the end-product of the designers' hypotheses and visions about the "entities that make up the world into which the object is inserted" (Akrich 1992, 207-208). In the structure-forming dynamics of the dispositive, various designer scripts and enforced programs of action (Latour 1992)[10] combine with user programs (unintentional uses and anti-programs to the enforced programs) and so lead to the emergence of temporary uses (or usage frames) such as: planning of mobility; navigation on maps; physical movement in space; communication and interaction on the move (real and virtual); display of oneself (as absent or present, with certain characteristics given in user profiles), switching (between modes of transportation, between reality and virtuality, between work and private spheres), working on the move, etc. The list above is not exhaustive. Other programs and combinations with resulting usage frames can also be

[10] "The program of action: 'if a car is moving, THEN the driver has a seat belt' is enforced" (Latour 1992, 226).

imagined. It is important to highlight that such temporary arrangements appear and disappear as a result of the accessing of multiple interconnected programs and scripts (as emphasized by the arrows in Figure 9-2).

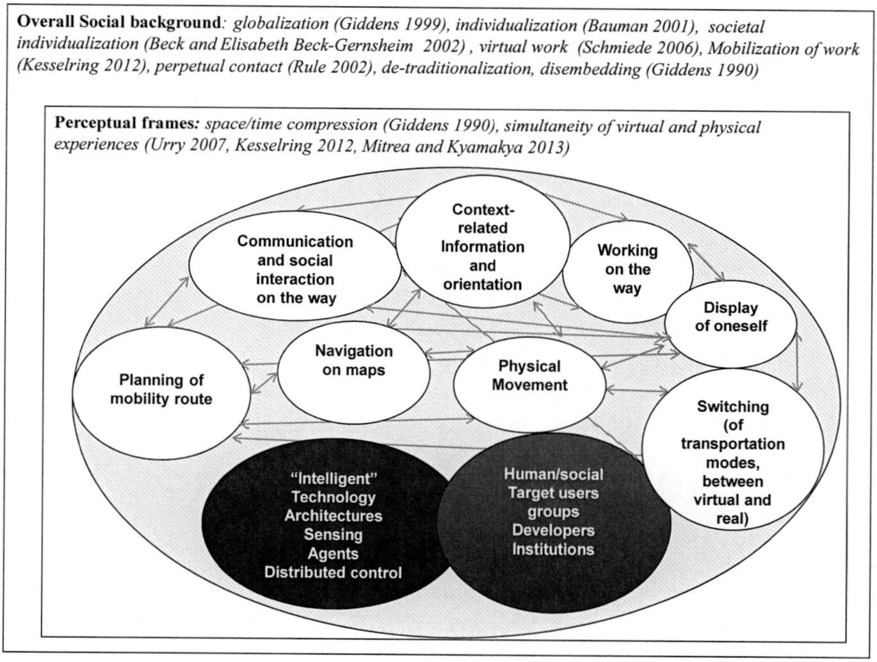

Fig. 9-2: the dispositive model

At their turn, the usage frames can combine among themselves. I formulate below two hypotheses about the possible future emergence of broader arrangements resulting from the aggregation of usage frames:

- **Flexible Mobility** results from the simultaneous mobilization and flexibilization of work/private life activities. This involves almost all highlighted usage frames: working on-the-way, communication, display of presence/absence and user characteristics, social interaction, mobility planning, physical moving, access of context services, switching between modes of transportation and virtuality/reality. The enabling technologies are telecommunication technologies (mobile devices such as phones, laptops, infrastructures), mobile spaces as work spaces integrated with ITS

(Advanced Travel Information Systems, Dynamic Navigation Systems, connected vehicles, Advanced Driver Assistance Systems).

- **Continuity and coherence through disruption**. The disruption and fragmentation of the continuity of mobility routines and work/ private life activities seems to be promoted by a high acceptance and use of switching (between virtual/real worlds and between modes). Interestingly, the intermodal transport – the usage of several transportation modes along the same route – is expected to enhance the continuity and fluidity of mobility, by means of dynamic information provided by the intermodal Advanced Travel Assistance Systems. The paradox of reaching coherence and continuity of social interaction through the apparently disruptive switching between real and virtual worlds can also be encountered in the case of mobile communication on-the-way (humans with mobile phones act as mobile portals linking worlds on-the-way, Kesselring 2012).

To examine the internal physics of the model and visualize one of the structuring frames, Flexible Mobility, see Figure 9-3:

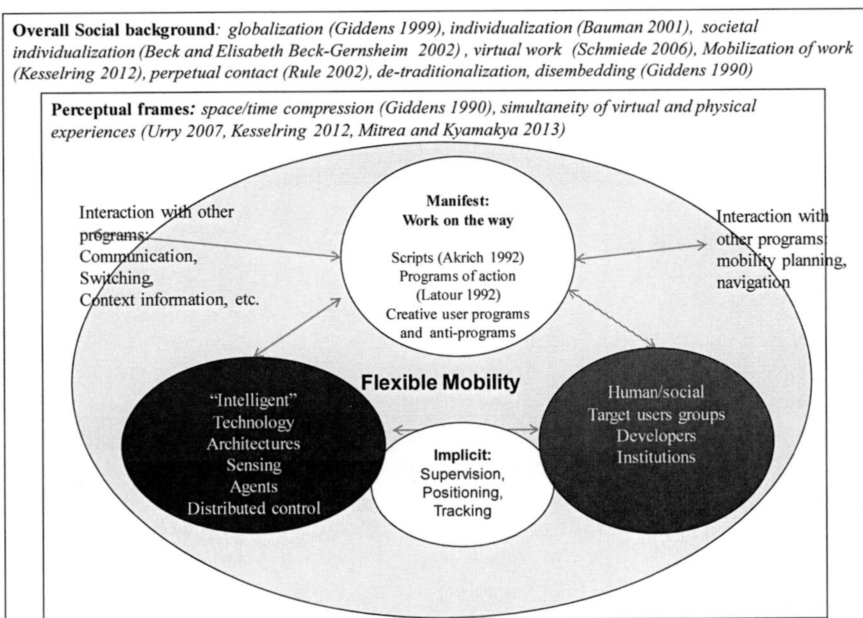

Fig. 9-3: fragment of the dispositive arrangement: Flexible Mobility.

At a broader level, values, representations, ideals and attitudes about sociability, mobility, and work, together with the manner of reception and interaction with technologies that is imprinted in the habits and expectations of humans in a given context, influence the design of technical infrastructures, software, and end-devices. The regulation of technology implementation and usage also plays its role in the general framing of design and usage and the producing of scripts.

Scripts can be manifest or rather hidden/implicit. For instance, tracking and positioning in ITS are not openly disclosed to users as prerequisites for a successful navigation and context-aware information and communication. Nevertheless, many travellers may "read" them as supervision and an "attack on privacy".

The distribution of actions and decisions to humans as users and the technological apparatus in various mobility settings looks like a complicated puzzle. In public transportation humans handle mobile devices, and hold them close to the body (Fortunati 2002); look at various screens and hear audio information on trains, buses, planes; work on laptops; chat with other travellers, etc. While driving, they turn wheels, push gas pedals and brakes while undergoing a multisensory bombardment of in-car information about traffic conditions and/or entertainment.

Advanced technology as material apparatus sets some limits to the capacity of individuals to act and react. For instance, drivers of vehicles equipped with ADAS see or hear the results of agents' actions and can react to recommendations, but are not aware and conscious of their internal interaction processes. Some automated driving processes are beyond humans' awareness and reach. Particularly the pro-active nature of the advanced technology (due to learning capacity based on data mining and anticipating of user behaviors) can lead to the imposing of programs of action such as: "*no vehicle start if alcohol has been consumed*", enabled by smart start ignition interlock systems (http://www.smartstartinc.com/).

The further usage of advanced technologies implies the accessing of several programs or uses in this co-agency modus (distributed agency on humans and advanced technology). For Flexible Mobility we selected the "work on-the-way" as attractor of other designer programs such as: the pre-trip planning of mobility, the spatial navigation, and actual moving (driving on the road, by train, plane, etc.); respectively mobile communication and accessing of relevant information. Users have a certain freedom to access only some of the designer programs of action or to create new user programs: for instance, by integrating private trips for short re-embedding in the family context during the professional trip planning. The actual work on-the-way is derived thus from the combined accessing of linked designer and user programs.

With extended usage, Flexible Mobility can generalize in society and has the potential of turning into a normative requirement for the design of technical systems and the further usage of technologies: how users should plan their next mobile work sequences, interact with colleagues, family and friends on the move, mobilize others, perceive, and exploit space and time. New acquired knowledge and competencies, shifts in values and beliefs may lead to changes of the background frames. These changes put the dispositive in motion again.

6. Discussion: risks and fears in the dispositive arrangements

The possibility for technical failures; users' fears about positioning, tracking, dangers to privacy; users' perception of technology opacity and loss of control represent important investigation matters, when dealing with the dispositive arrangements.

An interesting aspect for sociological analysis is the potential conflict emerging from the relationship between the degree of transparency of technological agency and human adaptability and autonomy (Weyer 2007, 2009). Weyer (2009) has emphasized the decrease in humans' capacity for strategic planning in hybrid systems (humans and technical agents coming together) and the tendency towards adaptive human behavior (in the sense that only short-time reactions to complex situation changes remain possible for humans). While the capacity for strategic agency implies instrumental rationality, the possibility of anticipating the consequences of actions, as well as regularity and predictability, the adaptive agency of humans translates into rapid reactions to situation-adjusted solutions with emergent character generated by IT systems, which cannot be predicted in all the details (*Id.*).

As a consequence of this warning, sociological studies of advanced technology emphasize the fact that the distribution of activities on humans and machines in socio-technical constellations of mobility should be gradually balanced in system design, leaving room for human self-initiative; own responsibility; control of personal data; intervention capacity; and the human decision on the real usefulness of applications (Rammert 2007). Surely the sociological work on advanced technology (Rammert 2007; Rammert and Schulz-Schaeffer 2002a, 2002b; Weyer 2009) has brought valuable contributions to the understanding of complex actions and their consequences in a changed mobility world that is enriched with information and communication. However, there is still room for consideration of the unintended effects of the implemented solutions and the identification of possibilities, as well as for a more strategic involvement of humans interacting with advanced mobility technologies in various arrangements of the presented dispositive.

The preservation and enhancement of the strategic capacity of users can become stringent in the future. Nowadays strong efforts are made for the future implementation of autonomous driving technology. Intelligent environments in general (Weiser 1991; Friedewald and Lindner 2008) should be able, in theory at least, to hide their perceptible presence and conceal their operations much more efficiently and completely than the current advanced technologies. This could represent a true challenge for designers and users coping with them.

References

Aggarwal, Charu and Tarek Abdelzaher. 2013. "Social Sensing." In *Managing and Mining Sensor Data*, edited by Charu Aggarwal, 237-297, New York: Springer.

Akrich, Madeleine. 1992. "The de-scription of technical objects". In *Shaping Technology/Building Society: Studies in Sociotechnical Change*, edited by Wiebe E. Bijker and John Law, 205-224, Cambridge, MA: The MIT Press.

Arminen, Ilkka and Alexandra Weilenmann. 2009. "Mobile presence and intimacy. Reshaping social actions in mobile contextual configuration." In *Journal of Pragmatics*, 41(10): 1905-1923.

Bammé, Arno. *Technologische Zivilisation*. Available at: http://www.iff.ac.at/oe/ifftexte/band3rps.htm

Bauman, Zygmunt. 2001. *Individualized Society.* Cambridge: Polity.

Beale, Russel. 2005. "Supporting Social Interaction with Smart Phones." In *Pervasive Computing*, 4(2): 35-41.

Beck, Ulrich and Elisabeth Beck-Gernsheim. 1994. "Individualisierung in modernen Gesellschaften - Perspektiven und Kontroversen einer subjektorientierten Soziologie." In *Riskante Freiheiten*, edited by Ulrich Beck and Elisabeth Beck-Gernsheim. Frankfurt a.M.: Suhrkamp.

Bonz, Jochen. 2007. *ANT in zehn Minuten Vortrag im Rahmen der Projektpräsentation „techartefact" an der Universität Bremen*, http://www.jochenbonz.de/wp-content/ant-in-zehn-minuten.pdf. Accessed December 11, 2013.

Canzler, Weert and Andreas Knie. 1998. *Möglichkeitsräume: Grundrisse einer modernen Mobilitäts- und Verkehrspolitik*. Wien: Böhlau.

CVIS. 2010. *Kooperative Mobilität in der Stadt. CVIS Handbook*, http://www.polisnetwork.eu/uploads/Modules/PublicDocuments/pol_cvis_handbook_de_03_web.pdf

Degele, Nina. 2002. *Einführung in die Techniksoziologie*. München: Fink/UTB.

Eagle, Nathan and Alex Pentland. 2005. "Social serendipity: mobilizing social software." In *IEEE Pervasive Computing*, 4(2): 28-34.

Elliott, Anthony and John Urry. 2010. *Mobile Lives. Self, Excess and Nature*. New York & London: Routledge.

Fortunati, Leopoldina. 2002. "Italy: stereotypes , true and false." In *Perpetual Contact. Mobile Communication, Private Talk , Public Performance,* edited by James E. Katz and Mark Aakhus, 42-62, Cambridge: Cambridge University Press.

Foucault, Michel. 1978. *Dispositive der Macht. Über Sexualität, Wissen und Wahrheit.* Berlin: Merve Verlag.

Friedewald, Michael and Ralf Lindner. 2008 "Gesellschaftliche Herausforderungen durch "intelligente" Umgebungen." In *Technikfolgenabschätzung – Theorie und Praxis*, 17(1): 78-83.

Giddens, Anthony. 1990. *The Consequences of Modernity*. Cambridge: Polity.

—. 2000. *Runaway World: How Globalization is Reshaping Our Lives.* New York: Routledge.

Hickethier, Knut. 1991. "Apparat - Dispositiv - Programm. Skizze einer Programmtheorie am Beispiel des Fernsehens." In *Medien/Kultur. Schnittstellen zwischen Medienwissenschaft, Medienpraxis und gesellschaftlicher Kommunikation,* edited by Knut Hickethier and Siegfried Zielinski, 421-447, Berlin: Knilli zum Sechzigsten.

Hradil, Stefan. 1955. *Die "Single- Gesellschaft"*. München: Beck´sche Verlagsbuchhandlung.

Inglehart, Ronald F., Miguel Basanez and Alejandro Moreno. 1998. *Human Values* and *Beliefs*: *A Cross-Cultural Sourcebook. Political, Religious, Sexual, and Economic Norms in 43 Societies: Findings from the 1990-1993 World Values Survey*. Ann Arbor: The University of Michigan Press.

Kesselring, Sven. 2012. "Betriebliche Mobilitätsregime. Zur sozio-geografischen Strukturierung mobiler Arbeit." In *ZfS (Zeitschrift für Soziologie)*, 41(2): 83-100.

Kneer, Georg. 2009. "Akteur-Netzwerk-Theorie." In *Handbuch soziologischer Theorien,* edited by Georg Kneer and Markus Schroer, 19-40, Wiesbaden: VS Verlag.

Knie, Andreas. 2010 . "Fahren statt besitzen." In *Der Spiegel*, 9, August.

Kyamakya, Kyandoghere and Oana Mitrea. 2010. "Rethinking Urban Mobility Services and Operations. An Enhanced Individual and User Specific Service for Urban Mobility." In *Data and Mobility - Transforming Information into intelligent Traffic and Transportation Services. Proceedings of the Lakeside Conference 2010,* edited by Julia Düh, Hartwig Hufnagl, Erhard Juritsch, Reinhard Pfliegl, Helmut-Klaus Schimany and Hans Schönegger, 177-186, Berlin, Heidelberg: Springer Verlag.

Latour, Bruno. 1987. *Science in Action: how to follow scientists and engineers through society*. Cambridge, MA: Harvard University Press.

—. 1992. "Where Are the Missing Masses? The Sociology of a Few Mundane Artifacts". In *Shaping Technology/Building Society: Studies in Sociotechnical Change,* edited by Wiebe E. Bijker and John Law, 225-258, Cambridge, MA: The MIT Press.

—. 2005. *Reassembling the Social. An Introduction to Actor-Network-Theory*. Oxford, New York: Oxford University Press.

Lenk, Carsten. 1997. *Die Erscheinung des Rundfunks. Einführung und Nutzung eines Mediums 1923–1932*. Opladen: Westdeutscher Verlag.

Licoppe, Christian. 2009. "Recognizing mutual 'proximity' at a distance: Weaving together mobility, sociality and technology." In *Journal of Pragmatics*, 41: 1924–1937.

Lugano, Giuseppe. 2007. "Mobile Social Software: Definition, Scope and Applications."
http://mobspace.cosix.it/docs/Lugano_eChallenges2007.pdf.

—. 2010. *Digital community design. Exploring the role of mobile social software in the process of digital convergence*. Jyväskylä: University of Jyväskylä.

MacKenzie, Donald and Judy Wajcman. 1985. "Introduction: the social shaping of technology." In *The Social Shaping of Technology,* edited by Donald MacKenzie and Judy Wajcman, 1-25, Milton Keynes: Open University Press.

Mitrea, Oana. 2005. "Wireless telephony and the rebuilding of interpersonal communication. Formative effects on communication structures." In *Soziale Technik. Zeitschrift für sozial- und umweltverträgliche Technikgestaltung*, 2: 12-14.

Mitrea, Oana and Kyandoghere Kyamakya. 2013. "'The Journey is the Purpose': A Concept for Public Transportation as Social Transient

Space." In *Proceedings of the IEEE-ITSC2013, IEEE Intelligent Transportation Systems Society Conference*, The Hague, October 6-9.

Rammert, Werner. 2006. "Technik, Handeln und Sozialstruktur: Eine Einführung in die Soziologie der Technik." In *Technical University Technology Studies Working Papers TUTS-WP-3-2006.* http://www.ts.tuberlin.de/fileadmin/fg226/TUTS/TUTS_WP_3_2006.pdf. Accessed December 11, 2013.

—. 2007. *Technik – Handeln – Wissen: zu einer pragmatischer Technik- und Sozialtheorie*. Wiesbaden: VS Verlag für Sozialwissenschaften.

Rammert, Werner and Ingo Schulz-Schaeffer (edited by). 2002a. *Können Maschinen handeln? Soziologische Beiträge zum Verhältnis von Mensch und Technik*. Frankfurt/New York: Campus.

Rammert, Werner and Ingo Schulz-Schaeffer 2002b. "Technik und Handeln - Wenn soziales Handeln sich auf menschliches Verhalten und technische Artefakte verteilt." In *Technical University Technology Studies Working Papers TUTS-WP-4*-2002, http://nbn-resolving.de/urn:nbn:de:0168-ssoar-11076. Accessed May 7, 2011.

Rule, James. B. 2002. "From mass society to perpetual contacts: models of communication technologies in social context." In *Perpetual Contact. Mobile Communication, Private Talk , Public Performance.* Edited by James E. Katz and Mark Aakhus, 242-254, Cambridge: Cambridge University Press.

Schmiede, Rudi (edited by). 1996. *Virtuelle Arbeitswelten, Arbeit, Produktion und Subjekt in der "Informationsgesellschaft".* Berlin: Ed. Sigma.

Schulz-Schaeffer, Ingo. 2009. "Die Frage nach der Handlungsträgerschaft von Technik und das Konzept gradualisierten Handelns." In *Das Tätigsein der Dinge. Beiträge zur Handlungsträgerschaft von Technik,* edited by Wilhelm Berger and Günter Getzinger, 37-59, München: Profil Verlag.

Sennett, Richard. 1998. *Corrosion of Character: The Personal Consequences of Work in the New Capitalism*. London: W.W. Norton.

Sheller, Mimi and John Urry. 2006 "The new mobilities paradigm." In *Environment and Planning A*, 38(2): 207–226.

Thom-Santelli, Jennifer. 2007. "Mobile Social Software: Facilitating Serendipity or Encouraging Homogeneity?" In *IEEE Pervasive Computing*, 6(3): 46- 51.

Urry, John. 2007. *Mobilities.* Cambridge: Polity Press.

—. 2010. "Consuming the Planet to Excess." In *Theory, Culture & Society*, 27: 1–22.

—. 2012. "Social networks, mobile lives and social inequalities." In *Journal of Transport Geography*, 21: 24–30.

Wang, Fei-Yue, Daniel Zeng, and Liuqing Yang. 2006. "Smart Cars on Smart Roads: An IEEE Intelligent Transportation Systems Society Update." In *IEEE Pervasive Computing*, October-December: 68-69.

Weiser, Mark. 1991. "The Computer for the 21st Century." In *Scientific American,* 265(9): 94-104.

Weyer, Johannes. 2007. "Autonomie und Kontrolle. Arbeit in hybriden Systemen am Beispiel der Luftfahrt." In *Technikfolgenabschätzung - Theorie und Praxis*, 16(2): 35-42.

—. 2009. "Die Kooperation menschlicher Akteure und nicht-menschlicher Agenten. Ansatzpunkte einer Soziologie hybrider Systeme." In *Das Tätigsein der Dinge. Beiträge zur Handlungsträgerschaft von Technik*, edited by Wilhelm Berger und Günter Getzinger, 61-92, Munich/Vienna: Profil.

CHAPTER TEN

BETWEEN STANDARD AND SITUATED USES: CONFIGURING THE USER IN A TECHNICAL CALL CENTRE

SIMONA ISABELLA

Introduction: configuring and re-configuring users

The relationship between technology, users and designers is a frequently debated one: within the field of Science and Technology Studies, scholars have provided different angles to approach these three concepts/objects and the relationships between them.

In the field of the Social Construction of Technology (as in Bijker, Hughes and Pinch 1987; or Bijker 1995), first attempts to translate some of the concerns of social constructivism are offered: technological innovation results in fact from the struggle and interaction of different social groups with different interests and interpretations. Within this approach, Steve Woolgar (1991) proposes his idea of a "configured user". According to Woolgar, the act of configuring is not only a way to define the putative identity of users and to set constraints upon their future actions, but also a process that shapes a technology. In his study on the design of a personal computer, Woolgar demonstrates that the process of configuration always has to do with boundaries: he uses the metaphor of "technology as a text" in order to analyze the process of construction (writing) and use (reading) of a machine. The relationship between *readers* and *writers* is conceived as a relationship mediated by the machine in the sense that the designers always try to define and delimitate users' actions through the production of the machine. Woolgar claims that, whereas experts know the machine, users could only have a configured relationship with it, which means that only certain forms of access and use are encouraged.

By reference to Woolgar's research outcomes, since the machine transforms itself in relation to the users and vice versa, the technological artefact becomes the metaphor of the company so the boundaries of the machine are the boundaries of the company. What Woolgar stresses in his study is that boundaries between technical and non technical, technological artefacts and users, and between the company and the "outside world" are blurred.

Another theoretical framework is that of Actor Network Theory, within which Madeleine Akrich (1992) introduces the idea of a "projected user". According to the French scholar, the designers of a technology inscribe a vision or a *script* about the world and about users into the technical content of the object in order to predetermine the setting and the type of use users are asked to imagine for a particular technology. The projected user is defined with specific tastes, competences, aspirations and prejudices that do not always correspond to real users. The prediction about the user is thus scripted into the technology. Madeleine Akrich recognizes that the projected user is an imaginary user and asks what happens when the projected user does not correspond to the actual one, but does not offer a solution.

The ideas of "configured" and "projected" user give a representation of the user as a one-time static subject: both authors stop their analysis at the point in which designers/innovators construct the technology and imagine a virtual user who will use it. Even if Woolgar draws attention to the importance of interaction between users and technology, he doesn't study the actual uses of a technology by people. Even recognizing Woolgar's work as an important basis for studying how users are imagined in the development of a new technology, further research (see Mackay *et al.* 2000) argues for an extension of the notion of configuring users. Drawing on an ethnographic study on RAD (Rapid Application Development – a methodology for information systems development) the scholars elaborate four arguments and focus particularly on the idea that: "whilst it is clear that designers *do* configure users, configuration is not a one-way process: designers, in turn, are configured-by both users and their organizations" (Mackay *et al.* 2000, 739) and on the fact in the process of configuration "the boundary between designers and users, far from given or fixed, is fluid, negotiated, constructed, managed and, indeed, configured" (*Ibid.*).

As Christina Lindsay (2003) suggests in her research on the design and development of a new computer (the TRS-80 in the late Seventies), the process of "configuration" can be well analysed only if the researcher takes into account both the process of the designing of a technology and the real uses users make with the artefact. She proposes indeed

> "that 'user representations' encompass many other imagined users, and that these user constructions are not built, and do not exist, in isolation. Each of the social groups involved with a technology throughout its life history, even those that are not directly involved, will have its own ideas about who and what the user is." (Lindsay 2003, 32)

Lindsay suggests studying the interplay of these various user representations and the ways in which social groups use these constructions to reinforce or challenge their own ideas of the user and, at the same time, how users themselves co-construct their identity and those of the designers.

Drawing on these theoretical perspectives, the main purpose of this chapter is to analyze some trajectories of stabilization of a mobile phone device that allows people to use Wi-Fi network in order to make VOIP calls and to surf the Internet. In particular, the categories of "user" and "uses" are challenged and redefined by the study of an informative infrastructure as working tool in a customer service office.

1. Method and context of the empirical research

Arguments and reflections presented in this chapter emerge from the analysis of data gathered during an empirical research[1] carried out in a period of about a year[2]. The focal point of this empirical study was a software for smartphones developed by an Italian telecommunication company (*Tiscali*), which allows people to make calls and to surf the Internet by using Wi-Fi or 3G technologies. While retracing the different trajectories of stabilization of the device (here called *Idevice*) the research focuses on users, their practices and the reciprocal configuration between them and the device.

Strictly from a methodological point of view, data were collected in different sessions of participant observation in a call centre for technical support where I had the possibility to listen to phone call interactions between clients and operators, to record the technical requests users send by e-mail, to participate in some meetings of the different work groups and to observe operators in their day-to-day work routines. During the same period I collected some non-structured interviews with managers, team

---

[1] The whole research received financial support from RAS L.R.7/2007, Promotion of Scientific Research and Technological Innovation in Sardinia.

[2] The empirical research started in April 2010 and lasted until October 2011. Data were collected at three different points: the first two between May and July 2010 and the last one in October 2011.

leaders and representatives, and some others with *Idevice* users, some of which privileged or 'power' users since they were also *Tiscali* employees.

The possibility of accessing the research field at different intervals over a period of more than one year allowed me to study the *Idevice* from its first launch as an experimental project to its consolidation in the market of mobile phone applications and its inclusion as a stable product in the *Tiscali* offers.

Moreover, although the *Idevice* is very similar to some other applications[3], *Tiscali* advertised it as a great novelty in the application market and among its services. Actually, *Idevice* was launched in a particular period of corporate restructuration, which included the reintegration of the technical support call centre. From this point of view, it really represented the symbol of the new image of innovation *Tiscali* wanted to project of itself in Italy and abroad.

With regards to the organizational context in which the research took place, *Tiscali* is one of the most important telecommunication companies in Italy, both for land line services and for mobile ones. Like all the other companies in the same branch, and following a general tendency already initiated in the '90s (Whalen *et al.* 2002), it entrusted the management of customer services activities to outsourced companies. In January 2010, four months before the beginning of the empirical research, the whole process of reintegration of the technical call centre was formally completed. This event had a crucial impact on the research for different reasons. On the one hand, the real interest the management showed in knowing "how the integration process was going" (a manager)[4] through an empirical research, eased access to the research field: the fact I was introduced in the technical inbound office by one of the managers of *Tiscali* assured me and my research a sort of interest and curiosity from representatives, and their complete availability and formal collaboration. As a counterpart, my daily presence in the call centre was often perceived as an intrusion, a form of control on behalf of the management[5] and I had

---

[3] The *Idevice* software is very similar to some other software like Viber (www.viber.com) or Skype (www.skype.com). It was first developed for Symbian, Windows Phone Operating Systems and, in a second phase, also for Android, iPhone, Blackberry, Windows and Mac OS. Launched by *Tiscali* during the early months of 2010 with a name, and only for some mobile phone models of certain brands, it reached a commercial success only in 2011 with a new name and when it entered the application market for both Android and iPhone.

[4] All interviews and fieldnotes are in Italian and translated by the author.

[5] During my stay at the company, I was once explicitly asked by a representative, what I was taking notes for. Even if I explained that my research focus was not

some difficulties in winning representatives' trust. In the beginning, people interacted with me by using the very formal and codified language they utilized during phone calls, placing me at a certain distance and excluding me from their everyday informal routines. Indeed, it took a certain time in order to gain the community's trust and also to be involved in informal conversation among representatives during work time or breaks.

On the other hand, the organizational ferment and a general positive atmosphere due to the reinstatement as a part of the company[6], allowed me to observe a "reality" in its doing: in the first part of my ethnographical study the technical inbound call centre was just reorganized in work teams with a team leader who monitored the group. A lot of people were just hired or relocated from other offices (i.e. the commercial call centre for example).

As Kunda (1992) explains in the methodological appendix of his research about the organizational culture in a technological company, the way the researcher gets into the community he/she wants to study, the period and the time he/she spends on the observation, and the role people attribute to him/her during the stay in the organization are only some of the variables that can influence the observation. The ethnographer always has to take into account the fact he/she is contributing in the construction of the world or culture he/she is observing. His/her presence and participation in the activities of the studied community influence the community itself.

2. Working in a technical inbound service, managing users

According to some studies on work and organization (see, among others, Suchman 2007 and 1996; Kunda 1992) studying the real or virtual spaces where work practices occur and analyzing how people organize them is useful in order to understand the meanings they give to their work practices and the culture behind that world.

---

observing single workers but the way the technical inbound office worked, he gave me no other chance to talk with him for the rest of the time I was there.

[6] The process of reintegration was not a complete success since there were some strikes which affected the timing of the research: in those periods I was not allowed to visit the company. The interesting thing was that when I tried to get information about the strikes everybody, senior and junior representatives, played down the events as something accidental, as did the management. The particular relationship the company has with the geographical area in which it is situated could explain the deep "brand loyalty" showed by the employees also in the case of the strikes.

With regard to the technical inbound call centre at *Tiscali*, some considerations on how spaces and people inside it are organized and distributed, could help focus on the work of representatives and their idea of users.

The technical call centre of *Tiscali* is divided into two different levels of customer support: at the first level representatives work directly with clients both in front end and back office activities. At the second level agents rarely have to deal directly with clients on the phone but they solve problems by directly accessing customers' landlines, mobile lines or circuits. Technical troubles, which cannot be solved by the second level of customer support, are managed by other levels of technical support that are not part of the call centre and that I did not explore in my research.

Since the reinstatement in *Tiscali,* the technical call centre has been located in two big bright rooms organized in open space workstations: each agent sits at a big desk shared with other colleagues and has a personal work place with a computer desktop and a headset. Even if agents always occupy the same place they are not allowed to personalize it: no sticky notes on the desktop, no paper sheets or pencil box on the desk, no pictures or photos as screensaver or on the desk. This sort of aseptic work place gives a sense of precariousness, but also reveals the great mobility across groups and offices which characterizes this kind of job and the organization of the company.

In the open space the formal division in teams is "visible" only by some big signs, hanging from the roof, which indicate the specific work field (i.e. land line, mobile, business etc.) of support of each team, which is led by a team leader -a senior agent who has the responsibility of the team in terms of productivity and quality of work. This intentional non-division of spaces suggests that the call centre is conceived as a collaborative work place. However, this idea is contradicted by the fact that representatives are not allowed to move freely from their workstation[7] or to talk to each other during phone calls. The open space organization seems only to foster control by the management: two big displays, which hang at two opposite walls of the room, show in real time incoming calls, answered and lost ones. It seems like a sort of warning for representatives and team leaders who know well that the productivity of the inbound call centre is calculated and evaluated according to the percentage of answered calls and solved problems.

---

[7] Drawing on EU directives, in Italy the 626/94 law introduced important rules for health and safety in workplaces. Specific attention is given to the regulation of alternation between work time and breaks for inbound and outbound call centre workers.

The second level of technical support occupies another big open space room separated from the first one by an informal area with coffee machines, toilets and some meeting rooms and an entrance with lifts. This area is often the place where informal meetings among team leaders, managers and workers of the second level of technical support happen. In these kinds of meetings, from which junior representatives are always excluded since they do not have the same mobility during work hours as other employees, differences and hierarchies are evident[8].

Technicians who work in this specialized customer support office, have no direct contact with clients: they solve technical problems that call centre agents can not. The two levels of technical support "communicate" through a tool of trouble ticketing: it is a software, connected to a Customer Relationship Management (CRM), that allows call centre agents to recognize clients during a call and to make a diagnosis of the technical problem. If the representatives cannot solve the problem on the phone with the client, they have to open a trouble ticket that will be managed by technicians of the second level support who can, eventually, call the customer back.

Apart from a short period of observation and some interviews, my research focused principally on the relationship between representatives of the inbound call centre and users of *Idevice*.

At the moment I started my research, *Idevice* had been launched on the application market for a few months and was also still in an experimental phase from a technical point of view[9]. What was the role of technical customer service in the reciprocal configuration between the users and the device?

Even if the management considered the technical contact centre just as "the face" of the company, because through it customers communicate with *Tiscali*, it actually seems more as a "liminal place": a space *in between* the inside and the outside of the company, whose role is that of mediating between management's requests of quality and efficiency, and

---

[8] There are many other organizational rules that underline hierarchical position in *Tiscali*. First of all, even if there is a common cafeteria where all the employees can eat, call centre agents have the priority since they have a more rigid time schedule while other employees can manage their lunch break as they want. Inside the fist level of technical support hierarchical division can be retraced in the language: team leaders and managers call junior representatives as "the guys", underlining in this way the different kind of experience they have in this job.

[9] During the first part of the research (April-July 2010) designers were still implementing the device both considering the application market's responses and the technical troubles users complained about through the customer service.

customers asking for assistance. It is in this "frontier land" that managing/solving customers' technical problems becomes a way of configuring the user. Apart from the idea of a model user/projected user which designers have during the projecting and developing process of the *Idevice*[10], the real user "appears" during the phone calls with representatives.

As Mackay and colleagues underline in their research on RAD (Mackay *et al.* 2000, 738) "users are not a monolithic or straightforward group, but are complex and fragmented in nature, and are attributed with varying significance". Definitions and classifications of users could also depend on the position of the observer in the company: in *Tiscali* I found three different ideas of users and different layers of abstraction of these figures according to the role people have in the company. At the management level users are defined simply as "flows of calls, volumes of calls" (manager) that have to be monitored: a user is just an incoming call that has to be managed as best as possible in terms of length and efficiency of the call.

As regards the specialized technical support, the definition of users emerges from technicians' work: that is solving technical problems. Since they do not have direct contact with clients on the phone, their idea of users is similar to that of the management: users are not real people, they are just the univocal identification each client has in the technical database of *Tiscali*.

> "a telephone number (Calling Line Identifier), an ID in the telephone Exchange (a PBX), a caller ID." (a technician of the special support)

Technicians do not have the same abstract vision of users as managers, but this is strictly connected with work tools they use to solve technical troubles: all the information they have about customers is what representatives write on the trouble tickets created on the shared tool as well as the information they find on the technical and commercial CRM of the company. Since they "see" only the technical problems users complain about, they sustain and consolidate the stereotyped idea among *Tiscali* technical employees that users are "technologically illiterate" people[11].

This culturally connoted image of users, typically produced by

---

[10] On the role and work practices of designers of *Idevice* see Mongili *in this volume*. For some interesting examples of research on designer's practices *see also* Parolin 2010; Storni 2012.

[11] For a further reading on designers' works and their idea of users see Mongili *in this volume*.

technical workers as designers, engineers and technicians, can have a strong influence on the work practices of the weakest link of this technical chain: inbound representatives. They indeed are not supposed to have real technical competences, since their function is to

> "manage the client and to understand, in the shortest time possible, what his/her request is, try to solve it or create a trouble ticket." (an inbound manager).

At the lowest level of the technical support, the user is a real person with whom agents talk on the phone. Representatives are really impatient of general customers' attitude of complaining and attribute the cause of problems to their "technological illiteracy":

> "they don't really have problems with *Idevice*, which works well; they simply cannot use their mobile phone!" (a representative)

However they are also proud to play the role of "interface" of the company. Although representatives' job seems very repetitive and standardized, they are required to have a personal attitude in dealing with users as people needing to be "reassured", "cuddled" and "managed". They are so used to this stereotyped image of ignorant, angry and insolent users to the extent of being really puzzled and disoriented when they have to deal with polite and competent customers.

In the next paragraph I will try to show how the vision of the user is also strongly influenced and configured by the complex information infrastructure of the inbound call centre, composed by interrelated software with different functions and work tools.

3. The improvisational choreography of informative infrastructure: between standards and situated uses

Inbound representatives' work is supported by a series of tools: a diagnostic software (an Issue Management System), a commercial and a technical CRM (Customer Relationship Management) as well as a Workforce Management System. All this software allows agents to access in a simplified way, on one side a database with all the information about customers and, on the other side, the telephone network.

According to *Tiscali* management

> "a good inbound technical support must do more things with the same number of people or with a fewer amount." (an inbound manager)

This is an idea that carries an oversimplification of representatives' work tools. During a call, inbound workers gain access to an Issue Management System (or a Trouble Tracking System) through a guided path by asking questions to the customer, in order to understand what the problem is and how to solve it[12]. At a first glance, a representative's work is a very standardized one and could be schematized in a few passages: answer the call, identify the client, take note of the problem, give a solution or create a trouble ticket, close the call. All these passages are recorded in the system creating a sort of history of users' complaints and configuring them.

The refinement of the work procedures that representatives have to follow is the way in which managers think they obtain a quality inbound service: the more representatives' work practices are embedded in the technology, the less time a call lasts, the better the customer service is. Of course the result is not always a good technical support lowering margins of errors, but just a way to answer as many incoming calls as possible.

As Whalen, Whalen and Henderson (2002) well describe in their research, the "routine" actions of sales representatives during their calls with customers constitute an "improvisational choreography":

> "We recognize that by describing this organization as 'improvisational choreography', we are joining together two concepts that appear to be incongruent. To improvise means to compose extemporaneously or fabricate out of what is conveniently on hand, while choreographed action is commonly understood to mean a carefully arranged or directed sequence of steps and movements, and thus actions that follow a pre-determined structure. However, we will show that this apparently incongruity can be solved through an appreciation of how workspaces, technologies, and other sources can be carefully arranged to afford what must necessarily be a somewhat extemporaneous composition." (Whalen *et al.* 2002, 241)

In *Tiscali* inbound call centre, the pre-determined structure of the agents' choreography is given by protocols and standards they have to follow in their work. The 'routine' actions they perform are so deeply inscribed in the technologies they utilize that they cannot "see" the limits of the infrastructure (Bowker and Star 1999; Edwards *et al.* 2007) which is

---

[12] Actually, the greatest numbers of solved troubles have to do with information about *Idevice* (costs, compatibility with a specific smartphone etc.), while for real technical problems (configurations, network management) representatives use to create a trouble ticket that would be solved by the specialized support.

culturally connoted[13]. For junior representatives work tools are completely "naturalized" (Bowker and Star 1999, 299); they use these software systems taking them for granted, without questioning their real usefulness. On the contrary, senior representatives (or team leaders), who have been working in *Tiscali* since the early years of its foundation, know very well how the process of managing technical troubles has changed as well as the constraints of technology:

> "In the beginning we solved technical problems with customers here, talking with him/her face to face... now everything is quite different!" (a team leader)

> "Look at this [showing me the software she is using], today it doesn't work... no way! And I cannot work with this old computer…" (from my fieldnotes-a technician of the specialized support)

The history of the inbound informative infrastructure is strictly connected with the history of the company and, in particular, of the technical call centre. Senior representatives could, in fact, retrace the major steps of its changes. According to their accounts of differences between working as an outsourcer and working inside *Tiscali*, it seems that the process of call centre reintegration inside the company goes hand in hand with the launch of the *Idevice* and with the implementation of the new informative infrastructure, that is the principal work tool of the technical support as well as an instrument of communication across different groups in *Tiscali*.

> "Since we have been back everything is different… we can talk more with the other sectors, we have a lot of meetings to talk about… everything! It is more difficult to meet representatives because of their shifting... but we work [he refers to the implementation of a new phone bar] in order to simplify their job!" (a team leader)

---

[13] In *Tiscali*, the development of the complex informative infrastructure which allows different offices (inbound technical support, specialized support, management) to communicate to each other and share information, is an ongoing process whose contributors are several subjects: managers with their request of proficiency, developers with their technical competence and users that, in this case, are junior and senior representatives. As Alessandro Mongili (chapter one *in this volume*) well explains, the roles of developers and users are often blurred: here representatives are the users of the technology, considered as common users and described as *utonti* (the mix of two Italian words: 'utente' that means user and 'tonto' that means dumb). At the same time they often help designers in implementing the technology, becoming themselves developers.

This process of refinement and simplification of the Issue Management System seems to be a way of imbricating representatives' work practices into the technology they utilize and it is mainly pushed by the administration. Since management principal interest is to maintain the relationship between the number of incoming calls and the percentage of solved problems at an acceptable level, any minimal gap between them is "filled" with new procedures and protocols that seem more focused on "managing the user" than on really solving problems. In this process a key role is played by a group of senior representatives-called "the staff" group-who mediate between managers' requests and representatives' needs.

> "The staff is a group of 5 people, including me… Everybody has a task: [name of a colleague] organizes turns,[name of a colleague] is in charge of representatives' training, [name of a colleague] deals with the monitoring of data and writes procedures…I manage and organize team leaders because I have the data and I can say 'let's do it this way, let's do it the other way!... and then I manage that side of... knocking the rough edges off... at a personal level I mean… different people, different characters…" (a member of "the staff")

The work of articulation (Corbin and Strauss 1993 in Bruni and Gherardi 2007, 28) of "the Staff" group is mostly a one-way process: even if they assert to mediate among different groups, they actually put into practice managers' choices.

> "I work in a group called "product design" and I have worked on the *Idevice* project from the beginning. It's a ghost group now: it exists and doesn't exist at the same time… we should coordinate different people who work on the *Idevice* project but, now, as a matter of fact, it is coordinated by 'the top' of the company… so we are useless…" (a designer of *Idevice*)

The previous excerpt expresses well how strongly the management influenced the development of the *Idevice* project and the way *Tiscali* decided to manage *Idevice* users through the work of the inbound call centre. During the empirical research I shadowed[14] some representatives of a team "dedicated to the technical support for *Idevice* users". It was a new

---

[14] Because of the shift rotation I could not observe any representative in particular. The shadowing consisted principally of standing beside them in their work routines, listening to the incoming calls with a double headset, following them in the pauses and in the few informal conversations they had with colleagues. Most of the time I sat at the team's desk and participated in their activities and conversations.

group composed of representatives of 3/5 years work experience in the company, some of them coming from other positions. Observing this new group in two different periods (May-July 2010 and October 2011) allowed me to record and explore the trajectories of stabilization of the *Idevice* and of the infrastructure that sustained their work. In the first phase of observation, agents worked with the same Issue Management System as other teams, which provided technical support for other *Tiscali* services as landline or mobile phone services, adsl services, etc. Even if much software is in common (as for example the CRM and the work force management system), each team has a different Issue Tracking Software. In 2010, when I arrived at *Tiscali*, the *Idevice* technical support team was an experimental group with a certain autonomy in managing users: since they did not have rigid protocols and a specific issue tracking system like the other teams, representatives could work with relative freedom:

> "…it is a document I create in order to be quick! Customers call to ask about *Idevice* configuration problems and I have to be quick in answering… giving a solution…" (a representative, M1)

> "Look at this! [proudly showing me a word file with text] I wrote down some standard answers to common problems that I use when I work in the back office and I have to answer to customers' emails…" (from my fieldnotes, a representative, F2)

In both examples, agents showed the twofold meaning of this "invisible work" (Suchman 1995; Star and Strauss 1999): creating files with draft emails or technical information about mobile phones is a specific support that single representatives activate in order to facilitate their activities of solving technical problems. It is not a required job, rather something more they have to do in order to cover a lack in the system. At the same time "being quick" means being competent in their job: from this point of view representatives know well they are "the face of the company" and they have to present a good image of it. In this sort of representation of the Self in public (Goffman 1969), agents perform their improvisational choreography both for customers on the phone but also for their co-workers[15] and superiors who will evaluate them based on their

[15] Working in a call centre is not an individual activity, on the contrary, there is a strong complicity among co-workers. During my stay at *Tiscali* I noticed how agents "worked together": they have a series of shared gestures to let colleagues understand what type of costumer they were managing on the phone at that moment, or to let a colleague know they needed help, etc.

performances (see, among others, Pentimalli 2008).

Lucy Suchman explains how some technologies are created with the aim of coordinating and controlling complex, distributed activities, and describes the example of the reporting system in airline operations: while it was a manual activity, workers

> "afford a kind of autonomous space or manoeuvring room that can be used by workers to maintain a reasonable relation between prescriptive representations like schedules and the actual contingencies of getting airplanes off the ground. New tracking devices (such as sensor plane wheels that automatically record the time of 'push back' of a plane from a gate) at least reconfigure, if not further close down, that space." (Suchman 1995, 59)

In the same way, one year from the *Idevice* launch and from call centre reintegration in the company, agents' work practices had been completely embedded in the new Issue Management System the *Idevice* inbound team utilize: representatives are now "quick" and can perform a competent or an "intelligible call" (Whalen *et al.* 2002) because they have a new trouble tracking guided path which, according to the management, allows them to identify and solve a technical problem in a few minutes[16].

At this point in the history of *Idevice* and of the technical call centre, some of the representatives' invisible work has emerged: certain draft emails a representative (see fieldnotes' excerpt on representative F2) had written in order to quicken her back office work, became the standard answers all the representatives started to use in replying to users' emails. But, as Susan Leigh Star and Anselm Strauss underline

> "the relation between invisible and visible work is a complex matrix, with an ecology of its own. It is relational, that is, there is no absolute visibility, and illuminating one corner may throw another into darkness." (Star and Strauss 1999, 24)

While always being under the constant pressure of the management, call centre agents still carve out small "spaces of autonomy" in managing technical problems, often in order to really face users and their requests and not only "manage them", that is, keeping them outside the company, which is the main aim of the management. In this arena of mediation, a particular role is played by team leaders whose principal task should be that of managing their team and controlling that workers/agents apply the

---

16 Of course, how long a "good" call should be is a standard decided by the management.

right procedures published weekly in the company Workforce Management System. On the contrary, they also infringe protocols and standard procedures when, for example, trying to solve users' issues utilising old procedures and software which require a certain technical competence. This kind of technical "virtuoso performance" seems to have a double meaning: on one side team leaders try to emphasize the distance from junior representatives' competence in order to consolidate their position as leaders; on the other side it is a risky way to be noticed by their superiors and try to move up to the specialized support office. At the same time, senior representatives encourage junior representatives who have an attitude to find "their way to solve problems": in these cases they allow the violation of procedures, especially when they have to manage a competent user and they know they cannot just imbricate him/her in the system.

All these little torquings (Mackenzie 2003) to the system, consequence of situated uses of small groups, seem to modify the whole infrastructure and, as a consequence, to contribute in shaping the figure of user.

4. Some closing remarks

According to Paul Edwards, Steven Jackson, Geoffrey Bowker and Cory Knobel

> "the growth of complex infrastructure and the form it takes are the result of converging histories, path dependencies, serendipity, innovation, and "bricolage" (tinkering). Speaking of cyberinfrastructure as a machine to be built or a technical system to be designed tends to downplay the importance of social, institutional, organizational, legal, cultural, and other non-technical problems developers always face" (Edwards *et al.* 2007, 6).

The ethnographical research in the inbound technical service of *Tiscali* allowed me to observe and analyze the different factors, and trajectories of naturalization which contribute to give a certain shape to the informative infrastructure that sustains the whole work process of the technical customer care.

When analyzing daily work routines, co-workers' cooperation and practices, rules and structures that sustain the call centre organization, two converging stories emerge: on the one hand, the management decision to reintegrate the inbound call centre in a phase when most enterprises still outsource these services, creates a euphoric atmosphere among workers who better welcomed all the restrictions and limitations imposed by the new Issue Management System. On the other hand, the launch of *Idevice* and the good economic result it seems to have had, encouraged representatives to perform intelligible calls in order to sustain the *Tiscali* image outside the company, often bypassing their role of mediators between the company and the world outside.

In these converging stories, users play an interesting role: even if configured by designers and technicians as a "technically illiterate people" and treated consequently as such by representatives, they are able, in some cases, to impose changes in the *Idevice* developing process. *Idevice* was initially designed for mobile phones with Symbian OS (in particular for Nokia phones) but the drastic changes in the market of smartphones, forced *Tiscali* to implement *Idevice* for Android OS smartphones and iPhones. At the same time, because smartphone users are "competent users", representatives' performances on the phone have to be more specialized and more technical. Representatives' frustration in managing a "competent user" through a system, which is culturally connoted, leads to a sort of "bricolage" (Edwards *et al.* 2007): they try to enact new ways of using technologies in order to get around infrastructure's constraints. These situated uses of the Issue Management System or of the Workforce Management System, produce small fractures in the integrity of the infrastructure that appears less and less transparent.

This case study shows how the history of such complex infrastructure, which manages work routines, data, and technical devices, is an ongoing process which aims for a sort of stabilisation without reaching it definitively. As Bowker and Star underline, "classifications are powerful technologies. Embedded in working infrastructures they become relatively invisible without losing any of that power" (Bowker and Star 1999, 319). As the two scholars demonstrate, classifications and standards, "which disappear into infrastructures, into habit, into the taken for granted" (*Ibid.*), could be revealed as a consequence of political and moral choices/works and could be reclassified. In the *Tiscali* case study, the ongoing process of implementing the *Idevice*, managing users, and solving technical problems tells how the organizational, technical and managerial requests, enacted by different actors involved in the process, could determine standards and classifications which sustain the informative infrastructure that technicians use in their work.

As a consequence, any attempt to classify/configure users by managers or technicians is the result of a process that moves between managerial efforts to standardize work procedures creating a constrictive informative infrastructure, and a series of situated uses technicians perform in order to affirm a sort of autonomy in very repetitive work activities.

References

Akrich, Madeleine. 1992. "The de-description of technical objects." In *Shaping Technology/Building Society. Studies in Sociotechnical Change*, edited by Wiebe E. Bijker. and John Law, 205-224, Cambridge, MA: The MIT Press.

Bijker, Wiebe E. 1995. *Of Bicycle, Bakelite and Bulbs*. Cambridge, MA: The MIT Press.

Bijker, Wiebe E. and John Law. 1992. *Shaping Technology/Building Society. Studies in Sociotechnical Change*. Cambridge, MA: The MIT Press.

Bijker, Wiebe E., Thomas P. Hughes and Trevor Pinch. 1987. *The Social Construction of Technological Systems: New Directions in Sociology and History of Technology*. Cambridge, MA: The MIT Press.

Bowker, Geoffrey. 2000. "Biodiversity Datadiversity". In *Social Studies of Science*, 30: 643-683.

Bowker, Geoffrey C., and Susan L. Star. 1999. *Sorting Things Out: Classification and Its Consequences*. Cambridge, MA: The MIT Press.

Bruni, Attila. 2003. *Lo studio etnografico delle organizzazioni*. Roma: Carocci.

Bruni, Attila and Silvia Gherardi. 2007. *Studiare le pratiche lavorative*. Bologna: il Mulino.

Edwards, Paul, N., Steven J. Jackson, Geoffrey C. Bowker and Cory P. Knobel. 2007. *Understanding Infrastructure. Dynamics, Tensions, and Design*. Report of a Workshop on "History & Theory of Infrastructure: Lessons for New Scientific Cyberinfrastructures."

Goffman, Erving. 1969. *The Presentation of Self in Everyday Life*. London: Allen Lane.

Kunda, Gideon. 1992. *Engineering Culture. Control and Commitment in a High-Tech Corporation*. Philadelphia: Temple University Press.

Latour, Bruno. 1996. *Aramis or the Love of Technology*. Cambridge, MA: Harvard University Press.

Lindsay, Christina. 2003. "From the Shadows: Users as Designers, Producers, Markets, Distributors, and Technical Support." In *How Users Matter. The Co-construction of Users and Technology,* edited by Nelly Oudshoorn and Trevor Pinch, 29-50. Cambridge, MA: The MIT Press.

Mackenzie, Adrian. 2005. "Untangling the Unwired. Wi-Fi and the Cultural Inversion of Infrastructure". In *Space and Culture*, 8: 269-285.

—. 2003. "These Things Called Systems: Collective Imaginings and Infrastructural Software". In *Social Studies of Science*, 33: 365–387.
Mackay Hugh, Chris Carne, Paul Beynon-Davies, and Doug Tudhope. 2000. "Reconfiguring the User: Using Rapid Application Development". In *Social Studies of Science*, 30: 737-757.
Mattozzi, Alvise (edited by). 2006. *Il senso degli oggetti tecnici*. Roma: Meltemi.
Mongili, Alessandro. 2008. "Oggetti e allineamenti: qual è l'oggetto della sociologia della tecnoscienza?" In *Rassegna Italiana di Sociologia*, 3: 404-425.
Mongili, Alessandro. 2007. *Tecnologia e società.* Roma: Carocci.
Oudshoorn, Nelly and Trevor Pinch (edited by). 2003. *How Users Matter: The Co-Construction of Users and Technology*. Cambridge, MA: The MIT Press.
Parolin, Laura. 2010. "Sulla produzione materiale. Qualità sensibili e sapere pratico nel processo di stabilizzazione degli artefatti". In *Tecnoscienza. Italian Journal of Science & Technology Studies*, 1: 39-56.
Pellegrino, Giuseppina. 2007. "Rhetoric, Practice, and Context-Sensitivity in Sociotechnical Action." In *Issues and Trends in Technology and Human interaction* edited by Bernd C. Stahl, 172-193, London: IRM Press.
Pentimalli, Barbara. 2008. "Pratiche di cooperazione in un call center: uso congiunto e situato delle tecnologie e messa in scena di 'performance' intelligibili", paper presented at the II National Conference of STS Italia "*Catturare Proteo: Tecnoscienza e Pratiche di conoscenza in Europa"*, Genova, June 19-21, 2008. [available at http://www.stsitalia.org/conferences/papers2008/1_4_Pentimalli.pdf]
Ruhleder, Karen. 1995. "Reconstructing Artifacts, Reconstructing Work: From Textual Edition to On-Line Databank". In *Science, Technology, & Human Values* , 20: 39-64.
Stahl, Bernd C. 2007. *Issues and Trends in Technology and Human Interaction*. London: IRM Press.
Star, Susan L. 2010. "This is Not a Boundary Object: Reflections on the Origin of a Concept". In *Science, Technology, & Human Values* , 35: 601-617.
Star, Susan. L. and Geoffrey Bowker. 2002. "How to Infrastructure." In *Handbook of the New Media,* edited by Leah A. Lievrouw and Sonia Livingstone, 230-245, London: Sage Publications.

Star, Susan L. and Karen Ruhleder. 1996. "Steps Toward an Ecology of Infrastructure: Design and Access for Large Information Spaces." In *Information Systems Research,* 7: 111-34.

Star, Susan L. and Anselm Strauss. 1999. "Layers of Silence, Arenas of Voice: The Ecology of Visible and Invisible Work." In *Computer Supported Cooperative Work*, 8: 9-30.

Storni, Cristiano. 2012. "Unpacking Design Practices: The Notion of Thing in the Making of Artifacts." In *Science, Technology, & Human Values* , 37: 88-123.

Suchman, Lucy. 2007. *Human-Machine Reconfigurations. Plans and Situated Actions,* 2nd Edition. Cambridge: Cambridge University Press.

Suchman, Lucy. 1995. "Making Work Visible". *Communications of the ACM*, 38: 56-68.

Wahlen, Jack, Whalen Marilyn and Henderson Kathryn. 2002. "Improvisational Choreography in Teleservice Work". In *British Journal of Sociology*, 53: 239–258.

Woolgar, Steve. 1991. "Configuring the User: The Case of Usability Trials." In *A Sociology of Monsters*, edited by John Law, 57-102, London: Routledge.

IV. On Boundary Objects, and On Multiplicity

CHAPTER ELEVEN

ORGANIZATIONS AS BOUNDARY OBJECTS: ORGANIZING BUSINESS BETWEEN RESEARCH AND THE MARKET

FRANCESCO MIELE

Introduction

For a good many years now, a significant part of organizational research has been turning its attention away from organizations – in the sense of solid structures with well-defined borders – to *organizing*, in the sense of a fluid, ever-changing activity process that involves heterogeneous groups of actors (single individuals, large and small organizations, pieces of larger organizations, objects and technological infrastructures, and so on). Over the past twenty years, a number of different concepts have been used to analyse organizational processes that are increasingly spread across time and space, and which are increasingly difficult to trace back to actions taken by a pre-defined complex of formal organizations. Theoretical constructs such as community of practice (Lave and Wenger 1991; Nicolini *et al.* 2003; Gherardi 2006), actor-networks (Czarniawska and Hernes 2005), networks of practice (Brown and Duguid 2001) or action-nets (Czarniawska 2004, 2008) have all dedicated their attention to practical and situated acts by a wide, varying collection of actors who work together and in conflict with each other in the pursuit of their objectives.

Within the ambit of this broad debate, an increasing amount of attention has been paid to the role of objects, which have often been seen to be stabilizing elements that make it possible for various communities of actors to cooperate with each other (see, for example, Suchman 2005; Lanzara and Morner 2005; Gherardi 2010). The intention of this work is to offer a contribution towards an understanding of the role of objects in today's organizational processes by means of an analysis of the case of Snow Tech[1], a spin-off from a university in the North-East of Italy.

By looking at the case of Snow Tech, our intention is to contribute to the wider current organizational debate in two ways. First, we will show how organizations, which can be interpreted as being *ad hoc* tools built to realize certain collective objectives (Perrow 1986), can turn out to be obstacles to organizing by obstructing the links among various actions, and thereby the attainment of common ends (Czarniawska 2013). Second, organizations will be proposed as *boundary objects*: that is, as objects constructed to facilitate cooperation among groups of actors from different social realms (Star 1989, 2010). The organizational literature has often stressed the stabilizing value of boundary objects (Lindberg and Czarniawska 2006; Lindberg and Walter 2013), but in this case we shall show how they can become a source of instability, and how they can jeopardize collaboration among different groups of actors.

In the pages that follow, we will dedicate our attention above all to the concepts of action-nets, which are used to analyse current organizational processes, and boundary objects, which are a useful device for interpreting the role played by spin-offs. Subsequently, following a brief presentation of the case study and the methodology selected for analysing it, space will be dedicated to the empirical material we collected and to our data analysis. We will conclude with some final reflections.

1. Action-nets, organizations and boundary objects

Action-nets (Czarniawska 2004, 2008) were born as a concept in an area of the organizational debate that has for some time now been focusing its analytical attention away from organizations and towards organizing (Clegg *et al.* 1996), in the sense of an open, continuous process that often involves actors who are located in different organizations. If we apply the action-net approach, we find that contemporary organizational processes can be conceived as groups of actions that are sometimes connected in profoundly stable ways, and sometimes unpredictably. The concept of action-nets was developed to demolish certain ideas that had been taken for granted, and that had dominated the organizational debate for years, above all the assumption that there must be formal organizational structures-such as individual enterprises, joint ventures, or inter-

[1] For privacy and confidentiality reasons organizations, people and places are hereinafter referred to with fictitious names. All names and sensible information were also removed from the empirical material used in this chapter.

organizational partnerships-at the heart of organizational studies. Through the concept of action-nets, researchers are invited to place certain organizational processes (for example, publishing a book, developing and marketing a product, or placing technology on the market) at the centre of their work, to focus on the actions that go towards making them up, and to consider those actors who are important for the phenomenon under consideration (individuals, organizations, pieces of organizations, professional communities, technology, and so on) only at a later stage.

From this perspective, rather than being points of departure to be focused on from the outset, formal organizational structures can be interpreted as emerging products that are the fruit of complex organizational processes which in most cases involve actors located in different spaces and places (Czarniawska 2004). If organizational processes can be conceived as flows of actions, then organizations would seem to be an attempt to give stability to these flows by establishing collective roles, procedures, and objectives to ensure that the processes are continuously reproduced. Let us take the case of the London School of Economics as an example. This is a formal structure that can undoubtedly be viewed as an attempt to provide stability to a network of pre-existing actions consisting of teaching and research activities in the field of human and social sciences (Czarniawska 2009). In this case, a movement of young intellectuals gives life to a formal structure that will be able to provide their projects with continuity, but in doing so, they have to involve new actors, thereby creating a wide action-net within which their interests, values, beliefs, and interpretations of reality will be partially modified.

In a recent work (2013), Barbara Czarniawska makes the claim that organizations can be defined as objects designed for the realization of collective purposes, and that like all objects, they may turn out to be inadequate or obsolete, or can be used in unforeseen ways. The author notes that despite the fact that organizations are frequently planned in great detail, they frequently turn against their creators, become obstacles to the organizational processes in which they are involved, and hinder the attainment of the objectives on which they are founded. This work will take this controversial claim and attempt to dig more deeply into the relationship between organizational processes in the sense of complex, intricate action-nets, and formal organizations, by which we mean objects that have been specifically designed to ensure the attainment of certain collective goals. We especially wish to focus on the case of spin-offs, which are organizations that have been created across differing social realms, behind which lie complex organizational processes the purpose of which is to market products and services originating with university research. In these processes, how does the emergence of a formal organization contribute towards stabilizing – or destabilizing – wider organizational processes?

In order to answer this question, in addition to the concept of action-nets, we will also consider that of boundary objects (Star and Griesemer 1989), which have their origin in Science and Technology Studies (STS), and which we will use to interpret the role of spin-offs.

As Star has recently confirmed (2010), this concept was developed with a view to analysing the nature of cooperative work in the absence of a consensus: that is, the ways in which actors from different social realms succeed in cooperating while maintaining differing interests, values, beliefs, and-more generally-interpretations of reality. If objects are conceived as "something people act towards and with" (*Id.*, 603), the characteristic of boundary objects is that they have a certain degree of interpretative flexibility, since they are placed at the centre of very varying interpretations, and put various communities of actors in a position where they can work side by side. It is precisely within these flexible, ambiguous elements that boundary objects find themselves at the centre of dense interactions involving groups of actors who would probably not meet under other circumstances; they may take the form of objects that differ greatly from one another, such as repositories (Star and Griesmer 1989), artefacts and engineering prototypes (Henderson 1991; Carlile 2002), patient records (Berg and Bowker 1997), project management tools (Sapsed and Salted 2004), and software specifications (Barrett and Oborn 2010).

Following its wide-scale diffusion, the concept of boundary object has broken out of the confines of STS and is also widely used in organizational studies. Certain recent studies in particular have focused on the ways in which boundary objects can give stability to an action-net, and especially to the connections of which it is composed (Lindberg and Czarniawska 2006; Lindberg and Walter 2013). We could consider for example, the study carried out by Kajsa Lindberg and Barbara Czarniawska (2006) in which the establishment of a "chain of care" project in an area of Gothenburg is analysed. The main purpose of this chain was to coordinate various organizations involved in providing care and assistance to elderly patients. The main feature of the project was that it involved various different organizations (such as the University Hospital, primary care units, and municipality care centres), all of which were involved in the care of the same elderly patients, who in this case became true boundary objects. What the various organizations involved in the project had in common was that they acted *with* and *towards* the patients by providing them with care and assistance, and through periodic meetings they became aware of their differences and the importance of cooperation for the wellbeing of the patients. A complex action-net therefore developed, in which actions taken by organizations that were very different from one another were linked in a stable manner, because the element they all shared was that they were oriented towards the patients, to whom very heterogeneous meanings were attached, which pushed the actors to act jointly for different reasons.

Through this work, therefore, one finds oneself at the heart of that area of the organizational debate that has focused in recent years on the role of objects in contemporary organizational processes, following the stimuli and suggestions offered by the STS. We will show in particular how formal organizations, in the sense of boundary objects designed to enable groups of actors from different social realms, can end up being an obstacle to the reproduction of wider, more complex organizational processes.

2. Context and method

Spin-off companies from university research programmes are a particular kind of entity, which aim to market certain results of academic research. There has been an exponential increase in the number of spin-offs in Italy in recent years: in 2009 there were 802 spin-offs here, 75% of which had been formed since 2002 (Balderi *et al.* 2011). At the same time, there was a significant spread of incentives to encourage the creation of these enterprises: university regulations and government tenders aimed at funding start-ups, financial subsidies to facilitate access to premises in incubators and university centres, meetings and exhibitions dedicated to this kind of enterprise, and so on.

The case presented in this chapter is part of a larger study on the dynamics of stabilization and institutionalization under way in certain spin-off processes in Italy. The focus in this work will be on the case of Snow Tech, a very young company created less than a year ago by a Professor at the Department of Environmental Engineering and three researchers from the University of Città Bianca, and was the university's very first spin-off. At the time of its formation, the company's principle objective was to continue to market the Earth Soft software programme, which had been developed in the university's laboratories to predict hydrogeological instability in mountain areas and to provide consultancy to companies and public bodies.

During the period of research in the field (April 2010-March 2011), in line with the theoretical approach of action-nets (Czarniawska 2004,

2008), the focus was less on the formal organization of the spin-off than it was on the organizational processes of which the spin-off was a part. Our attention was therefore directed to the sets of actions occurring prior to and after the creation of the enterprise that contributed to the development and marketing of the Earth Soft software. For this reason, a multi-sited logic was adopted (Marcus 1995; Bruni 2008), as well as a set of techniques that permitted us to reconstruct the manner in which actions undertaken by groups of actors located in heterogeneous spaces and organizations interconnect and intermingle, and give rise to a wider action-net.

Participant observation was used principally to reconstruct the everyday organizational life of the spin-off and the relationships its members have with other groups of actors (clients, partner enterprises, suppliers, certification bodies, and so on). Subsequently, we used the ethnographic interview technique (Spradley 1979) so as to be able to obtain a detailed understanding of the ways in which actors who are formally outside the spin-off cooperate with it. Finally, by analysing the documents relating to the various actors involved in the development and marketing of the Earth Soft software, we focused on reconstructing the action-nets that emerged in the most complete and exhaustive manner possible.

3. Act I. The creation of Snow Tech, a boundary object between research and the market

Our story begins at the start of the new century, when Professor Rossi of the Department of Environmental Engineering of the University of Città Bianca, assisted by a group of doctoral students and young research assistants, began thinking about a software programme that would effectively monitor changes taking place in mountain areas. Professor Rossi's intention was to create cutting-edge technology that incorporated the latest and most innovative results of research at the university, to provide a detailed, in-depth reading of what was occurring in areas where there was a high risk of natural disasters. These were years of intense activity, in which various young researchers worked with Professor Rossi to combine heterogeneous knowledge and know-how to jointly develop a special software programme-Earth Soft-and to make it increasingly complete and up-to-date. The development of Earth Soft was profoundly linked to the development of theoretical models created by Professor Rossi and his co-researchers with the aim of creating a "cutting-edge" programme able to carry out analyses and data processing that the scientific community engaged in taking effective technologies in the area of prevention of natural disasters to market and presenting them as innovative. Around this technological object, therefore, a complex action-net was born in which academic research activity was linked in a continuous manner to sets of actions aimed at the development of the software and an analysis of mountain areas.

A few years later, an important change took place within the action-net under review. Three of Professor Rossi's young co-workers, Sara, Matteo, and Fausto, founded Snow Tech, a spin-off which had Earth Soft as its core business. A number of different interpretations immediately arose around the creation of this organization. As Professor Rossi explains:

> "To produce academic research, it is absolutely essential to have well-qualified co-workers (...). One of the ways of achieving this, at least for a part of my activity, was to make it possible to apply the models and theories I had developed in my research, and this could be done through people undertaking a business project (…). In order to have people who would work with me, I thought the only way they wouldn't be misled would be to seek to build some kind of professionalism around their positions, for example through Snow Tech (...). This is a problem we have to face these days because partly as a result of the recent laws, it is hard to imagine that we will be able to go in the direction of increasing staff who are the direct responsibility of the university".

For Professor Rossi, therefore, Snow Tech seems above all to have been a model for continuing his research and development activities on Earth Soft while confronting the dizzying reductions in resources available for public research. Here, the interviewee was referring especially to the reduction in funding for public research as a result of Law 133 of 2008, which is a part of the budget 2009 manoeuvre, in which funding to universities and the number of staff they could take on were drastically reduced. According to Professor Rossi, by founding an enterprise, he can avoid "deceiving" his co-workers by offering them the prospect of continuous work outside the university, while continuing to have the availability of a group of trained, trusted professionals with whom he can develop his lines of research.

The three working partners in the enterprise, who have worked both in companies and as self-employed professionals in the course of their careers and university research experience, seem to view Snow Tech differently. Sara, the young partner and founder of the company, explains:

> "I have always had an entrepreneurial bent, the desire to do things first hand has always fascinated me (...) I liked the idea of putting what I had learnt in all my branches of research into engineering practice (…). [After

the spin-off project had been approved] we got going, and finally we had what we had been dreaming of in our own hands a little – a company on the border between research and the market, between entrepreneurship and engineering".

In this case, Snow Tech is not simply a way of giving continuity to existing research activities: it is an ambitious and innovative project, a company with a plan for transferring knowledge acquired from research on to the market, and for creating a business from this knowledge. This interpretation is quite different from Professor Rossi's; while he believes that Snow Tech is merely a tool for "continuing to do what we were doing before", for his partners and founders, the spin-off is a way of marketing research results and therefore attracting financial resources from the market.

The spin-off therefore seems to have been a centre of different interpretations between them from the time of its formation, and its redesign irremediably reflects this diversity. In the first place, the company's mission revolves around two activities: to provide consultancy services through Earth Soft to public entities and companies interested in various types of construction projects (motorways, rail tunnels, ski runs and the like) in mountain areas, and to understand the potential impact on the environment; and to continue to develop the software with Professor Rossi's research group to keep it up to date with the most innovative results from the university research. Secondly, the organization has structural characteristics that are intended to ensure the continuation of the mission:

- as required by the regulations covering spin-offs from the University of Città Bianca, 10% of the company's share capital is held by the University of Città Bianca;
- after illustrating the continuity between its mission and the research areas developed within the Department of Environmental Engineering to a special committee, the company received the "Spin-off Company from the University of Città Bianca" mark. This mark certifies the continuity between the company's activities and the results from the academic research, thereby guaranteeing the innovative nature of the technology and services produced;
- in addition to the three working partners and directors of the company (Marco, Fausto and Sara), the Board of Directors includes Professor Rossi, who was a non-working director;
- the persons chosen to act as executive directors of the company seem to possess a level of knowledge and know-how that will

> ensure that the company's marketing and administrative activities are carried out (Marco and Fausto have a good level of experience both in companies and as self-employed professionals) and to guarantee the consultancy and research activities with Earth Soft (which have been entrusted above all to Marco and Sara, both PhD with wide experience in university research groups).

Our story therefore begins in a university research laboratory, and for now ends with the creation of a spin-off company. If we take Star's (2010) definition of *object* as a reference point, Snow Tech can undoubtedly be classified as such: that is, something with which and towards which people find themselves acting. Snow Tech was born after the emergence of a broad organizational process, and developed and presents itself as something with the potential for being used by various groups of actors for the purpose of carrying out various activities: keeping a software programme up to date with university research, and producing services using the software. It appears, in fact, as though the company has been designed to be able to carry out these activities in a stable and continuous manner: the qualifications required of those who fill the role of working partner and executive director of the company, the presence of Professor Rossi in the Board of Directors, the granting of the spin-off company mark by the University of Città Bianca, and the ownership of a part of the share capital by the university would all seem to guarantee that both the research and development objectives and the marketing aims will be attained.

Snow Tech therefore immediately establishes a place between the world of research and the market, since an appropriate boundary object has been designed for it to interact with groups of actors who are quite different from one another, who have an interest in carrying out activities of various kinds with Snow Tech, and who therefore interpret it in very differing ways: on the one hand, we have Professor Rossi, who heads the research group that gave life to the company's software core business, and who describes the business as a "broadening of his research group" which will help to stabilize the existing connections between the theoretical elaborations carried out in the university laboratories and development of the Earth Soft software; while on the other hand, the partner-founders interpret Snow Tech as a "transformer of research results into business", and in line with this interpretation, propose that both research and development and consultancy *ex novo* with the company be carried out, thereby widening the existing action-net. The creation of this object appears on the one hand to assist the stabilization of certain connections that make up the existing action-net, and on the other, to represent an opportunity for expanding it by including new activities.

The actors involved in the planning of the Snow Tech object seem to be aware of its confines. Although Professor Rossi interprets Snow Tech principally as a way of continuing his research, he knows that for certain co-workers of his it might be viewed as a source of income and professional fulfilment; for their part, although the partner-founders are mostly interested in doing business through the new company, they interpret the contact with university research as a competitive advantage in the market. As has been seen in other cases (Lindberg and Czarniawska 2006), the various actors involved in action-nets know that they are acting on the border between different social realms, and especially that they are dealing with an object that is subject to multiple interpretations. Up to this stage in the story, however, this lack of interpretative unity seems to be viewed as a resource for stability more than as a problem: it is precisely as a result of the different ways of interpreting the enterprise that it is possible to give life to a complex organizational process within which research and development activities and consultancy go hand in hand.

Finally, we should note that Snow Tech does not appear out of nowhere, but has close ties with government policies, which over the past ten years, as a result of reductions in funding for public research and the incentives offered by contractual flexibility (Sylos Labini and Zapperi 2010), have made it increasingly difficult for young researchers to plan a long-term career at a university, and have pushed them towards other professional paths. At the same time, initiatives promoted by groups of influential actors (such as ministries, universities, technological centres, and networks of companies) have become increasingly frequent, the aim here being to stimulate the creation of companies established to market scientific research (see, for example, the university's shareholding in Snow Tech or the granting of the "spin-off company of the University of Città Bianca" mark). The idea of Professor Rossi and his team to create a company that would be able to carry out both business activities and research and development is closely connected with the actions of a group of *institutional entrepreneurs* (Rao *et al.*, 2000; Suddaby and Greenwood 2005; Tracey *et al.* 2011) which for some time now has been placing a substantial quantity of resources in order to give legitimacy to, and generally facilitate, the diffusion of a way of creating a company that is still innovative in Italy.

4. Act II. When the borders catch fire

In the first few months of the company's life, a complex action-net was formed of various sets of actions that had continuous contact with each other: on the one hand, taking part in public tenders for research and development projects, creating networks of organizations to present them, and pursuing the objectives of the company plans; while on the other finding clients for the consultancy services to be carried out using Earth Soft, publicizing Snow Tech and its products, performing consultancy services with Earth Soft, and so on. What all these sets of actions have in common is the participation of Snow Tech as a project partner, as a group of professionals able to carry out research and development, as a potential supplier with which to negotiate a business service, and as a company producing services using a sophisticated software programme. Most of the actors involved in the action-net under review are acting both *with* and *towards* the object that is Snow Tech (Star 2010): they represent it through narration and graphic images, they abide by the rules written for it, they supply it with the financial resources to permit it to continue to exist, they cross the physical space where it is located, and they interact with the machinery that makes it up.

As we have seen, the emergence of the Snow Tech object leads to the expansion of the action-net and therefore to the involvement of various groups of actors, public bodies, and private enterprises interested in receiving services, and the professionals and companies that wish to collaborate with the company. From the time of its formation, the three partner-founders of the company have interacted with these actors on many occasions, and with increasing frequency, such as at trade shows for companies working in mountain areas, meetings to present the company and the services it offers, informal conversations to update clients on the services being performed for them through Earth Soft, and so on. Alongside this unexpected immersion in the market, two types of problems emerge in our action-net.

On the one hand, Professor Rossi's request that the working partners of Snow Tech dedicate time to his projects begins to be excessively invasive, according to Marco:

> "Rossi looks at us a little as though we were his doctoral students; he always says we need to make an effort and suffer during this period [in which university funds are scarce], but we are a company, and we have the costs of a company."

On the other hand, within Professor Rossi's research group, the work done by Snow Tech on the common research projects is sometimes imprecise, as explained by a research assistant working closely with Professor Rossi on the Prometeo Project, a large applied research project in which the founding partners of Snow Tech are also actively involved:

> "Working on Prometeo, I have realized that it isn't a research project, but a more applied project with companies (...) [and then], although I am not a computer expert, I didn't like the way in which inputs and outputs were managed – in my opinion, it was something to... that work should have been done on immediately, because it didn't seem to me that it was being managed professionally (...) the product has been botched (...). I don't know if [the companies involved] are working on other things ... but then I see that the performance level between the deadlines they give in the meetings and the facts, well … it could be 50% at the most, but possibly less, in the sense that things are not complied with."

These various problems that occur periodically in the action-net can be traced back to a common path. With the creation of the spin-off, a new way of interpreting Snow Tech seems to have come into the picture, in which the ties to the university are very much in second place: the company begins to be seen by its working partners not so much as a "transformer of research results into in business", but more as "a tool for satisfying the needs of the client". As this way of looking at the Snow Tech object takes hold, the continuity between the services supplied through the Earth Soft software programme and Professor Rossi's research programme becomes far less important. As Carlo, a self-employed professional who frequently assists that founding partners of Snow Tech with their research and consultancy puts it:

> "[Snow Tech works in] a bit of a niche sector that I think operates more at a political contact level than one based on ability [...]. If you have contacts, what they are doing they also have them do – let's say – at the firm of the President of the Engineers' Association [...]. But it doesn't matter that they are a spin-off with the mark of the University of Città Bianca, that's not what will get them more work than someone else."

The limited significance that is starting to be attached by the working partners of Snow Tech to their proximity to the research group from which they came can be traced back to the type of market they are in. In the field of protection of the environment and the prevention of natural disasters, it would appear that the ability to create satisfactory relationships with the managers and functionaries of public bodies is more important than the innovative nature of the services and products offered.

After having been located on the border between research and the market, therefore, the Snow Tech object becomes the subject of a new interpretation, this time one that is in conflict with the original interpretations that had accompanied the corporate planning phase. The various groups of actors involved in the action-net now begin to act *with* and *towards* the Snow Tech object on the one hand by interpreting it partly as an extension of a university research group which must be capable of carrying out high-quality research and development, and on the other by using it as a means of embodying the needs of the organizations present on the territory in technologies and marketable services. Snow Tech's working partners apply this new interpretation, and begin to invest less and less time in Professor Rossi's research and development project, while on the other hand the academic research group created around Professor Rossi begins to be increasingly impatient about the way the members of the spin-off work, which they have judged to be not especially appropriate for conducting high-quality research. As a result, Snow Tech's position on the border becomes a source of destabilization, and puts the existing ties between research and development and consultancy in serious difficulty.

At the same time as the interpretations given to the company change, so the working partners also begin to redefine its design. A perfect example of this change is what was happening at the time of the research among the three working partners: Marco and Fausto on the one hand, and Sara on the other. Alongside her activity as executive director of the company, Sara was doing post-doctoral work at a foreign university on a subject extremely close to that of the research and development project on which Snow Tech and Professor Rossi's research team had been working up to that time. This decision begins to be viewed with some impatience, as the following episode at Snow Tech's offices illustrates:

> [One *or* Early] afternoon, a call arrived on Marco's mobile phone from Sara. Sara spoke for a few minutes about the fact that Marco and Fausto had written an e-mail the day before in which they had asked her to give up a part of the salary that she would have been paid, normally, by dividing the turnover by three. As a result of her post-doc research post, in fact, she had worked far fewer than the required 200 hours a month. Marco answered "*for 2010, we should have been at 100%; if you would be so kind as to take another look at what we took from 2009...*". After Sara had talked for a few minutes about how her research activities should be viewed as useful training for the company, Marco said "*this is what I said initially, but we never got away from the problem, here we are running the real risk of obstruction from Fausto – I'm saying this because I know him*

> *(...) if you also use the argument I used last year of saying "I have invested in the doctoral studies", we'll be stuck, because they're not physical things, there we got stuck (...): that is, I realized that in the end his investments have a more practical value compared with our investments in culture, I realized this".* When the call ended, I went to lunch with Marco and tried to understand better what had happened during the morning, and especially if it was a problem to coordinate workloads with Sara. Marco explained that *"the problem is that Sara doesn't like this work, she wants to do something else (...). For example, you can't say you don't have a talent for business; if not, then don't be an executive director (...) all three of us are executive directors because this means having certain responsibilities, duties and tasks (...). If not, you take a step back and say 'I'm leaving the Board and becoming a self-employed professional person»' ..."*

There appears to be a fairly delicate topic at the heart of this episode: what does one need to do to be able to be an executive director of Snow Tech? Initially, the Snow Tech object had been designed in line with the interpretations surrounding it: it was partly an "extension" of a university research group, and partly a tool for marketing certain outcomes of academic research. For this reason, all the executive directors of the company had, or had had, experience of research at a university, thereby demonstrating their ability to interact *with* the Snow Tech object in university research and development. The later interpretation of Snow Tech as a tool for satisfying the needs of clients, regardless of what was discovered and communicated by the research group from which it had emerged, also seems to have altered the characteristics required of those who wished to act as the company's executive director. The proximity to the knowledge and know-how from the university appears to be increasingly of secondary importance, and it is the "practical abilities" that really count, such as looking for potential new clients, and negotiating and concluding agreements with them.

At this point, it would seem that some reflection is required on what occurred in the action-net we are studying.

In the first case, the arrival of a new interpretation of Snow Tech, which begins to be understood as a tool with the primary purpose of satisfying needs expressed by the market. The arrival of this interpretation appears to be closely linked to the interactions among the three working partners of the company and the organizations that wish to receive services from it. The market into which the action-net has made its entry is undoubtedly not known for its long tradition of innovation and ties between companies and university research. As has been noted on a number of occasions (Fligstein *et al.* 1993, 1996), markets may be interpreted as organizational forms characterized by their own rules, values, and interpretations of reality in general. There is a wide range of actors around Città Bianca who have an interest in construction projects in mountain areas and accordingly in seeking out consultancy services that will help them understand the potential consequences of their projects. The business done in these markets is noteworthy for a certain lack of interest in the innovative nature of the services received and the technologies used to provide them, with great importance being attached to the capacity of the service providers to understand and satisfy the needs of the client and to create lasting relationships with it. Snow Tech, in the sense of a tool for providing consultancy services relating to the possibility of building in mountain areas, therefore begins to be surrounded by interpretations that differ from those that stimulated its creation. As a result, even the three main creators of the company, and certainly those who interact the most with Snow Tech on an everyday basis, begin to look at it differently, and no longer use it so much with the intention of transforming research results into innovative services as for satisfying requests from the market as rapidly as possible.

In the second place, it should be noted that once it was put to use, the Snow Tech object displayed a certain level of *fluidity* (de Laet and Mol 2000), altering its characteristics in proportion to the ways in which it was used. In this case, the object was gradually changed as it was used, and at the same time the knowledge, know-how, and interests deemed necessary to "use" it were redefined. One might use the example of a car which is used for a kind of trip that was not foreseen at the time it was designed, and so the seats, the type of steering wheel, the type of gears and so on are changed. This range of changes also leads to an alteration in the characteristics of the people who can drive it: either the original drivers need to show that they know how to adapt to the new characteristics of the object, or they must stop driving it. Similarly, the Snow Tech object is frequently used in ways that were not foreseen at the time it was first used (it is increasingly becoming a tool for satisfying clients totally and without delay), while at the same time, certain of its characteristics undergo change (in the example we have given above, the meaning given to the role of executive director has been redefined). As a consequence, any of the three working partners who do not show that they have the knowledge and know-how to "drive" Snow Tech must either stop interacting with the object or begin to interact with it in a different way (for example, by no longer being executive director, but an occasional collaborator).

The Snow Tech object is therefore interpreted and put to use in ways that had not been foreseen at the time of its creation, and at the same time, its design is altered in the course of its use. The object that emerges from this assortment of changes is not at all fit to remain within a heterogeneous action-net in which it would continue to be interpreted partly as a tool for proceeding with university research and development. On the one hand, the founding partners of the company begin to lose their desire to use the Snow Tech object for university research and development: while at the time the company was formed, these activities were seen as being useful for taking possession of research results to be marketed by producing innovative services, they are now seen as being not especially interesting, because Snow Tech is now mainly presented as a means of satisfying the demands of the market. On the other hand, the members of Professor Rossi's research team are somewhat unenthusiastic about the form the Snow Tech object has taken, given that when it is used in research and development, it produces results of little significance.

5. The moral of the tale ...

The story we have recounted began with an action-net for the development of the Earth Soft programme, followed by the entry on to the scene of Snow Tech, a spin-off company formed to carry out research and consultancy activities using the software. In our story, the formal organization makes an appearance only later on, in the form of a tool used by a number of actors to provide stability and continuity to existing activities (research and development carried out with Earth Soft) on the one hand, and to undertake a new type of activity (the consultancy services provided by interacting with the software) on the other. At the time of its formation, Snow Tech was a small, sophisticated object designed by actors with quite different interests, priorities, and interpretations of reality, and for this reason, the company in turn is subject to varying interpretations. The design of the Snow Tech object reflects this diversity, and it is able both to carry out university-level research and development and to carry out consultancy services of a type the market appreciates.

Anyone who imagined that our story would have an unexpected happy ending in which the arrival of a boundary object causes actors from different social realms to cooperate within a wide and complex organizational process is going to be disappointed, however. A few months after Snow Tech became operational, one section of its creators started viewing it through different eyes, and started to interpret it as a means of satisfying clients who had little interest in the world of research and innovation. As they frequented this "bad company", the working partners soon came to attach a declining level of importance to research collaborations with the university, and looked on them with indifference. Conflicts between the creators of the company – with Professor Rossi on one side and his former co-workers on the other – therefore became a frequent phenomenon which placed the connections between research activities and consultancy seriously at risk. In the meantime, the Snow Tech object seemed to change shape, and became ever more adapted to consultancy purposes and at the same time less desirable for the refined research and development activities performed by Professor Rossi and his research group. Having reached this stage, we now need to ask what this story has taught us.

There are at least three lessons to be learned here. First, we have seen that an organization that had been designed to provide stability to a pre-existing action-net may unexpectedly become an obstacle to its reproduction (Czarniawska 2013). In this case, we have seen that although Snow Tech was carefully designed from the outset to satisfy the expectations of its various creators, its actual use brought about a series of unforeseen events: some of its creators began to see it and use it in different ways, and at the same time, the company began to change shape. As a result of this, Snow Tech became a threat to the reproduction of the action-net, and placed a considerable part of the connections that made it up at risk.

There is a second lesson to be learned from the story we have related: boundary objects do not always lead to a happy ending in which heterogeneous groups of actors cooperate within complex, well-developed action-nets. Unlike other cases (Lindberg and Czarniawska, 2006; Lindberg and Walter, 2013), we have seen how an object on the borderline can quickly become a source of conflict and instability when it is placed at the centre of contrasting interpretations. In the story of the Snow Tech object, market pressures played a key role in this change by bringing about the emergence of a new way of interpreting the company's work. It is precisely the rules, values, and beliefs of a certain type of market that in this case led to conflict arising around the Snow Tech object, and therefore to the destabilization of the action-net.

Finally, this story has been recounted in such a way as to emphasize the influence of an element that has often been ignored in studies that have used the concept of boundary objects up to now: the institutional scenario. As we have said before, the birth of the Snow Tech object does not come out of nowhere; it has close ties to the work of powerful groups of institutional entrepreneurs (Rao *et al.* 2000; Suddaby and Greenwood 2005; Tracey *et al.* 2011) who are providing both direct/indirect and local/national incentives for the creation of companies which have the purpose of placing the results of scientific research on the market. As has been seen in other cases (Miele 2013), the efforts these actors have made have been partly in vain; where the Snow Tech object is put to use in its market it becomes subject to interpretations that make it increasingly less suitable for marketing the products of academic research.

References

Balderi, Chiara, Alessandra Patrono and Andrea Piccaluga. 2011. "La ricerca pubblica e le sue perle: le imprese spin-off in Italia." In *Quaderni dell'Istituto di Management*, 1.

Barrett, Michael and Eivor Oborn. 2010. "Boundary object use in cross-cultural software development teams." In *Human Relations*, 63: 1199-1221.

Berg, Marc and Geoffrey C. Bowker. 1997. "The multiple bodies of the medical record." In *Sociological Quarterly*, 38: 513-537.

Brown, John S. and Paul Duguid. 2001. "Knowledge and Organization: A Social-Practice Perspective." In *Organization Science*, 12: 200-213.

Bruni, Attila. 2008. "Attivare e stabilizzare un network nel settore delle biotecnologie: il caso Tek-Tek." In *Apprendimento tecnologico e tecnologie di apprendimento*, edited by Gherardi, Silvia, 147-178. Il Mulino: Bologna.

Carlile, Paul R. 2004. "Transferring, translating, and transforming: An integrative framework for managing knowledge across boundaries." In *Organization Science*, 15: 555-568.

Clegg, Stewart, Chyntia Hardy and Walter Nord (edited by). 1996. *Handbook of organization studies.* London: Sage.

Czarniawska, Barbara. 2013. "Organizations as obstacles to organizing." In *Organizations and Organizing. Materiality, Agency, and Discourse,* edited by Daniel Robichaud and François Cooren, 3-22, New York: Routledge.

—. 2004. "On Time, Space and Action-nets." In *Organization*, 11: 777-795.

—. 2008. *A Theory of Organizing.* Cheltenham: Edward Elgar Publishing.

—. 2009. "Emerging Institutions: Pyramids or Anthills?" In *Organization Studies*, 30: 423-441.

Czarniawska, Barbara and Tor Hernes (edited by). 2005. *Actor-Network Theory and Organizing.* Copenhagen: Liber & Copenhagen Business School Press.

de Laet, Marianne and Annemarie Mol. 2000. “The Zimbabwe Bush Pump. Mechanics of a Fluid Technology.” In *Social Studies of Science*, 30: 225-263.

Fligstein, Neil and Linda Markowitz. 1993. "The Finance Conception of the Corporation and the Causes of the Financial Reorganization of Large Corporations, 1979-1989.” In *Sociology and the Public Agenda*, edited by William Julius Wilson, 185-206, Beverly Hills, CA: Sage.

Fligstein, Neil and Iona Mara-Drita. 1996. "How to make a market: reflections on the European Union's Single Market Program.” In *American Journal of Sociology*, 102: 1-33.

Gherardi, Silvia. 2006. *Organizational Knowledge: The Texture of Workplace Learning* Oxford: Blackwell Publishing.

—. 2010. "Telemedicine: A Practice-based Approach to Technology." In *Human Relations*, 63: 501-524.

Henderson, Kathryn. 1991. “Flexible sketches and inflexible data bases: Visual communication, conscription devices, and boundary objects in design engineering.” In *Technology & Human Values*, 16: 290-307.

Lanzara, Giovan Francesco and Michèle Mornèr. 2005. “Artifacts rule! How organizing happens in open source software projects.” In *Actor-Network Theory and Organizing*, edited by Barbara Czarniawska and Tor Hernes. Copenhagen: Liber & Copenhagen Business School Press.

Lave, Jean and Etienne Wenger. 1991. *Situated Learning: Legitimate Peripheral Participation*. Cambridge: Cambridge University Press.

Lindberg, Kajsa and Barbara Czarniawska. 2006. “Knotting the action net, or organizing between organizations.” In *Scandinavian Journal of Management*, 22: 292-306.

Lindberg, Kajsa and Lars Walter. 2013. “Objects-in-Use and Organizing in Action Nets: A Case of an Infusion Pump.” In *Journal of Management Inquiry*, 22: 212-227.

Marcus, George. 1995. “Ethnography in/of the World System: The Emergence of Multi-Sited Ethnography.” In *Annual Review of Anthropology*, 24: 95-117.

Miele, Francesco. 2013. "Banalizzare l’innovazione: uno studio sugli spin-off universitari". In *Lavoro e ricerca sociologica: un confronto fra giovani ricercatori italiani,* edited by Michele La Rosa and Umberto Pallareti, Milano: Franco Angeli.

Nicolini, Davide, Silvia Gherardi and Dvora Yanow (edited by). 2003. *Knowing in Organizations: a Practice-based Approach.* Armonk, NY: Sharpe.

Perrow, Charles. 1986. *Complex Organizations: A Critical Essay.* New York: Mcgraw-Hill.

Rao, Hayagreeva, Calvin Morrill and Mayer N. Zald. 2000. "Power plays: How social movements and collective action create new organizational forms." In *Research in Organizational Behavior*, 22: 239-282.

Sapsed, Jonathan and Ammon Salter. 2004. "Postcards from the edge: Local communities, global programs and boundary objects." In *Organization Studies*, 25: 1515-1534.

Spradley, James P. 1979. *Ethnographic Interview.* Orlando: Harcourt Brace Jonanovich College Publishers.

Star, Susan L. 2010. "This is not a boundary object: Reflections on the origin of a concept." In *Science, Technology, & Human Values*, 35: 601-617.

Star, Susan L. and James R. Griesemer. 1989. "Institutional Ecology, "Translation" and Boundary Objects: Amateurs and Professionals in Berkeley's Museum of Vertebrate Zoology." In *Social Studies of Science*, 19: 387-420.

Suchman, Lucy. 2005. "Affiliative Objects." In *Organization*, 12: 379–399.

Suddaby, Roy and Royston Greenwood. 2005. "Rhetorical strategies of legitimacy." In *Administrative Science Quarterly*, 50: 35-67.

Sylos Labini, Francesco and Stefano Zapperi. 2010. *I ricercatori non crescono sugli alberi.* Bari: Laterza.

Tracey, Paul, Nelson Phillips and Owen Jarvis. 2011. "Bridging Institutional Entrepreneurship and the Creation of New Organizational Forms: A Multilevel Model." In *Organization Science*, 22: 60-80.

CHAPTER TWELVE

INFLUENCE AND MULTIPLICITY OF RISK THRESHOLDS IN PREVENTATIVE MEDICINE: THE CASE OF ADVANCED MATERNAL AGE

MAURO TURRINI

Introduction: the multiplicity of risk and risk thresholds as boundary objects

This chapter proposes an approach to the socio-analysis of risk focused on risk thresholds. Drawing on Science and Technology Studies of classification, risk thresholds are analysed as "boundary objects" that are able to mediate, articulate, connect, disconnect, and shape knowledge, technologies, interests, experiences, and subjectivities (Star and Giesemer 1989; Bowker and Star 1999). In spite of their seemingly objective nature due to their numerical form, these composite entities provide an original perspective on the socio-cultural analysis of risk, which is one of the most important and still disputed research subjects of the last three decades (Beck 1992; Douglas and Wildawski 1983; Giddens 1990; Fressoz 2011). The influence of risk thresholds lies in their role as backbone of the new regimes of prediction of adverse events. Without them it would be impossible to articulate the ever-increasing amount of data about risk indicators with activities aimed at monitoring, assessing and managing the occurrence of negatively valued outcome categories like disasters, epidemics, calamities, and, in our case, congenital anomalies. Their classificatory influence may not be estranged by their multiplicity. Risk cut-offs may also be deemed as ubiquitous objects able to move through heterogeneous sets of practices, disciplines and actors. Although mundane and apparently unremarkable, risk cut-offs play an agential role in the processes of negotiation, assessment, and (re-)signification of risk, and thus participate actively to extend, resist and mould risk.

These numerical entities, thus, seem to be worthy of being explored by means of social sciences, especially in a field like medicine, where questions of scientific validity, clinical utility, technological innovation, economic interests, social legitimacy and moral considerations are at stake. The purpose of this article is therefore to demonstrate both the heterogeneous dimensions of cut-offs, and their influence as an active, multiple vector where technological innovations and cultural changes related to risk are negotiated and mediated. A brief digression on a recent news event provides a good starting point to illustrate the argument.

In May 2013, in a *New York Times* op-ed, Angelina Jolie disclosed that she had undergone a preventative double mastectomy to reduce the likely eventuality of developing breast cancer[1]. After having described the sad, early death of her mother at the age of 56, the actress wrote about her "faulty gene", BRCA1, which increases her personal risk of developing breast and ovarian cancer. She continues by describing in minute medical detail the articulated procedures involved in mastectomy. She discusses family values, emphasizing how important it was, as a mother, to assure their children that she was not at high risk anymore, and, as a wife, to be supported by her husband. There is even a final remark about the cost of genetic screening, which is presented as an obstacle for cancer prevention in the fight against cancer[2]. Although directed to a non-expert reader, technical language is not left out. On the contrary, her story is articulated around risk percentages (mentioned five times). The reduction of breast cancer risk from 87% to 5% is depicted as a sort of liberation from danger and a return to a safe life. Being either at high risk or low risk is a condition of being that involves intimate emotions, personal memories, new medical practices of screening, body modifications, and responsibilities towards children and relatives. On a broader level, it also addresses the social fight against cancer.

Moreover, this article is of interest not only for the way in which it combines medical details and lived experiences, social implications and familial feelings, but also for the media-storm it provoked in international public opinion. Opinion-makers, politicians (including, among others, British Foreign Minister William Hague), medical experts, cancer activists,

---

[1] A. Jolie, "My Medical Choice", *New York Times*, May 14, 2013.

[2] This final remark, although brief and without a critical tone, draws on a question which has been long debated in the United States, where a multinational biotech company, Myriad Genetics, patented the screening procedure of the two genes (BRCA1 and BRCA2) connected with a predisposition to breast and ovarian cancer. Cancer patients' associations have condemned this company for years for maintaining the price of these tests well beyond the market price and for hindering the research in this sector; they have even brought the case to trail.

and even laypeople[3] commented on Jolie's "medical choice" in the press, on news sites and in cable-news conversations all over the world. The majority of comments were in awe of what the famous Hollywood star did. They considered this disclosure as supportive for women who suffer from the same condition, informative for those who have a family predisposition to cancer, bravely regardless for her iconic image of female sexuality, and, finally, meticulous in providing medical details about risk assessment and the cost of screening. Moreover, to rephrase the *Times'* front-cover title, "the Angelina effect" was not limited to the expression of opinions. Many women went public about their genetic susceptibility to breast and ovarian cancer and their decisions about it, and many others went to cancer centres to get the same screening.

From our perspective, Jolie's editorial may be deemed a concise, bold compendium of the multiplicity and relevance of risk thresholds in contemporary societies. Usually these risk factor numbers are naturalized and presented as results of scientific considerations. On the contrary, we would like to reconsider these cut-offs as "boundary objects" (Star and Greisemer 1989), namely as symbolic-material entities ubiquitously intertwined with a heterogeneous assemblage of elements which, in the medical domain, call into question statistical calculations, material needs, professional practices, patients' experiences, civic responsibilities, and health conceptions. Drawing on Geoffrey Bowker and Susan Leigh Star, risk thresholds will be analysed here as "means of translation" (Bowker and Star 1999, 297) that are aimed at moving and regulating information across different practices, and, in so doing, shaping them.

The multiplicity of risk thresholds is not taken into serious consideration even by those who have pointed out their influence. In a recent book on *Risk, Safety and Clinical Practice*, Bob Heyman and colleagues (2010) brilliantly deconstruct the apparent scientific nature of risk thresholds, and interpret them as a crucial passage of "risk consolidation"–a complex process whose trajectory consists of risk individualization, risk-prevention linkage, institutional embedding, commercial entrenchment, iconography accumulation, and risk moralization. In this framework, the demarcation of a binary cut-off is seen as a necessary operation for the virtual presence of a risk to be registered in the realm of preventative medicine.

> Thresholds provide triggers for preventive options which in practice must either be, or not be, adopted. [...] Lower and higher probabilities of an

[3] Just to give an example, the NYT online version of the article received 1712 comments in only two days.

unwanted outcome, distinguished in terms of an arbitrary dividing line, become transformed into its virtual absence or presence (Heyman *et al.* 2010, 50-51).

This chapter intends to build on this argument by showing that risk thresholds are not only complex entities embedded in the process of risk consolidation in healthcare, but that they are also dynamic, non-stabilized entities. Rather than a mere step in a linear process, as assumed by Heyman and colleagues, cut-offs are a crucial terrain where technical, political, cultural, and economic aspects of risk interact with and configure one another. Their same materiality is a condition that varies across time, and across the different contexts in which they are embedded.

This argument is articulated through a specific case study, *prenatal diagnosis*, in the form of genetic tests that are provided to "asymptomatic patients", who are in this case pregnant women. By prenatal diagnosis we mean diverse forms of invasive genetic tests used to identify foetal genetic anomalies (most notably, but not only, Down's syndrome); principally in the form of amniocentesis and chorionic villus sampling (CVS), but also, very rarely, percutaneous umbilical blood sampling. Since they are long, expensive, and overall risky tests, they are only recommended and covered by the majority of national health services or private health insurance plans to pregnant women for whom the risk of having an affected foetus is deemed "high". Because the risk of these anomalies increases with advancing maternal age, policies to offer invasive prenatal diagnosis are generally focused on women who are at least 35 years old at the time of pregnancy.

From our point of view, it is particularly interesting to trace the processes through which "advanced maternal age" (AMA), a cut-off age usually corresponding to 35 years, has been developed, practiced, discussed, re-signified and questioned in situations involving doctor-patient relationships, motherhood (and parenthood), parental responsibility, political or administrative bodies of healthcare, technological innovations, reproduction, and life. In order to grasp the multiplicity of AMA definition and practice, different sources have been combined, including medical and scientific literature on this topic (academic articles, medical societies bulletins, official guidelines and position papers, informed consent forms), in-depth interviews with gynecologists, geneticists, and pregnant women, discourse analysis of the popular guides to pregnancy, and the vast corpus of sociological and anthropological literature on prenatal diagnosis. The empirical materials have been organized in four different parts: the first is devoted to the socio-historical account of AMA definition in the 1970s; the second sketches its most important revisions within biomedicine; the third ethnographically explores *in practice the work it does* in the complex process of care; and, finally, the fourth investigates, by analysing its representation in popular culture, the emerging subjectivities with which it is associated.

1. An arbitrary choice? Economy and politics of advanced maternal age

An historical reconstruction of the scientific debate on this topic reveals that all research converged, in an astonishingly spontaneous way, on the decision that 35 years should be the cut-off age for recommending prenatal diagnosis, starting from the introduction of amniocentesis as a medical routine in the early 1970s. The role that AMA played in the process of legitimation of such a controversial medical practice like prenatal diagnosis should be considered. Without doubt, prenatal diagnosis could not even have been imagined without the major legal and cultural changes that occurred between the 1960s and the 1970s–notably the feminist movement's critique of reproduction control and the legalization of abortion in most Western countries (with a few exceptions like Ireland). However, risk cut-offs was also a very useful tool to introduce a disputed medical practice in a rather soft way, by addressing it only to a small minority of pregnant women, around 5%. In the debate that we found in scientific literature, no political concerns are mentioned.

According to an authoritative article on this topic (Kuppermann *et al.* 1999), there were at least four reasons why this cut-off age was selected: *1)* the escalation of Down syndrome incidence with advancing maternal age; *2)* the shifting balance between the risk of delivering a child with chromosomal disease and that of experiencing a miscarriage caused by the invasive procedure of amniocentesis or villocentesis; *3)* cost-effectiveness calculations of prenatal diagnosis; *4)* and, finally, a logistical lack of adequate cytogenetic laboratories and providers. Although the need to provide amniocentesis only to a small minority of women was never called into question, the criteria that justified this choice were presented and discussed separately, and, despite sharing the same conclusions, the debates were not devoid of friction.

Ironically, what apparently seems to be the most scientific criterion, the association of increasing risk of Down syndrome with advancing maternal age, played an ambivalent role. Described in the 1930s by Penrose (1933; 1934), this relationship was actually at the base of the choice to use an age-related individual model of risk for selecting the women to whom amniocentesis was recommended. However, when in the 1970s new research refined and revised the previous findings, establishing "a linear increase in rates between 20 and 30 (or just over 30) and a logarithmic increase in rates from 33 to 49 with a transitional region between these intervals" (Hook *et al.* 1983), the resulting suggestion to have 33 years as the cut-off was discarded in favour of other clinical and economical criteria.

According to the clinical argument, AMA resulted from a balance between two different kinds of risk: the risk of having a miscarriage; and that of having a child with a genetic anomaly (for a historical reconstruction, see Berkowitz *et al.* 2006). In other words, the age-related ascending risk-line of genetic disorders crossed the flat risk-line of the invasive diagnostic procedure exactly at the age of 35 (see fig. 12-1). As early data on amniocentesis showed, at 35 years, a woman had the same chance of delivering a child with chromosomal disease as experiencing a miscarriage from the procedure (Lubs 1972).

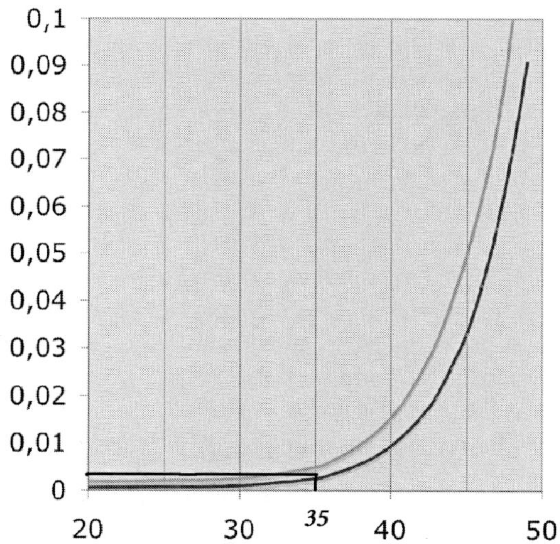

Fig. 12-1: the horizontal line shows the logical passages behind the establishment of 35 years old as the AMA cut-off. The graph illustrates the relation between the probability of having a foetus either with the Down's Syndrome (the darker curve on the bottom) or another congenital anomaly (the lighter curve on the top) and the age of the pregnant woman. Original elaboration based on the data presented by D'Alton and DeCherney (1993).

Even accepting the rationale behind this argument, it was impossible to reach a straightforward agreement due to divergence regarding the degree of amniocentesis safety. The North American collaborative studies indicated an increase of miscarriage rate among amniocentesis recipients as equal to 0.5% or even 0% as compared to non-recipients, while on the other side of the Atlantic, British studies reported doubled figures. Not coincidentally, the first consensus conference on "antenatal" diagnosis, organized in March 1979, considered cost-effectiveness analysis a more valid indicator for AMA cut-off figures. The report of this important conference extensively renders the concerns and hopes of those years regarding the emergent, flourishing field of what was then called "foetal medicine". If moral concerns were taken into consideration, they were left, in any case, to the moral private sphere of the "patients". Rather, the desirability and feasibility of amniocentesis was discussed first and foremost as a public health issue to be evaluated in terms of "the economic impact of hereditary diseases and congenital defects" (Kaback *et al.* 1979, vol. I, 135). The economic burden of offering prenatal diagnosis to women over the AMA was balanced against the savings accrued in averting the birth of babies affected with Down syndrome and other congenital births, which included costs of medical care, institutionalization, special education and training, as well as a loss of productivity. The level of complexity was so high that the conclusions of the report itself admit that "the age limit is *arbitrarily* decided by logistical concerns and is not the consequence of a sudden biological difference between women below and above any given age" (*Id.*, 201, *my emphasis*).

The agreement on risk thresholds, although arbitrary, was an important step in support of the process of establishment and legitimization of such a controversial practice as antenatal diagnosis. This mediation allowed the smooth introduction of a morally controversial procedure like antenatal diagnosis, which also represents the first clinical genetic test to be practiced outside the small group of people affected by the rare genetic disease.

In this sense, to diminish moral concerns raised by pre-natal diagnosis, and render it acceptable, was remarkable. As pointed out by Carine Vassy (2006) in the French case, biomedical practitioners carried out a crucial role of social and political mediation by promoting cautiously these innovations to a small percentage of pregnant women. As Bowker and Star (1999) have shown regarding the roles that disease classifications have played in the establishment of the modern welfare state, we can consider these processes as part of a mode of governance, where the pursuit of general interests, resource considerations, innovative prevention tools and emerging at-risk groups conflate. Since then, amniocentesis changed from an experimental procedure limited to a small percentage of mature pregnant women (in the 1970s in many Western countries only about 5% of pregnant women were 35 years old or older), into a routine service for many women and healthcare services. This passage can be explained by the increase in the number of women who are either pregnant after 35 or decide to undergo prenatal diagnosis regardless of their age and risk. As a result, risk thresholds have remained a topic of debate, but the concerns have shifted from economic collective considerations to more individual clinical questions of safety.

2. Erasing, refusing or strengthening risk thresholds. Epistemology of technological innovations

Far from being discarded with the canonization of amniocentesis and CVS, debate and research on risk cut-offs have continued intensely. First, advanced knowledge regarding cost-effectiveness ratios made economic concerns obsolete, and accordingly, collective resource considerations were replaced by clinical considerations. Second, technological advancements and biomedical research were conducted in order to either revise the level of safety of prenatal diagnosis or to assess more precisely which pregnancies were at high risk.

The first argument was aimed at demonstrating that prenatal diagnosis was a perfectly safe procedure and, as such, should be proposed to all pregnant women[4]. Some prenatal medical gurus have conducted research to revise concern on the implied dangers involved in amniocentesis. Documenting about 35,000 cases collected in over 15 prestigious U.S. medical centres, the research consortium First and Second Trimester Evaluation of Risk (FASTER) reported that second-trimester amniocentesis carries a risk of foetal loss equal to 0.06%, a statistically unremarkable figure (Eddleman *et al.* 2006). Despite the large amount of data presented, a 2007 systematic review of this topic cautiously warned: "the lack of adequate controls tends to underestimate the true added risk of prenatal invasive procedures" (Mujezinovic and Alfirevic 2007, 687). The controversy continued in 2008, when a U.S. clinic team published an original research

[4] It is also important to point out that, while some prominent obstetricians tried to demonstrate the safety of amniocentesis, others specialised in CVS were concerned with demonstrating that foetal loss related to this procedure is not higher than that for amniocentesis, entailing an equivalent foetal loss (see, e.g., Brambati and Tului 2005; Caughey *et al.* 2006).

based on an even larger number of cases (50,000), which indicated the very low amniocentesis-related pregnancy loss of 0.13% (Odibo *et al.* 2008).

Far from being settled, the dispute has important practical implications. As we mentioned, according to the most important argument in favour of AMA, this age cut-off is the result of balancing two kinds of risk: the risk of having a miscarriage and that of having a child with a genetic anomaly. Therefore, lowering or even eliminating the risk of foetal loss associated with prenatal diagnosis also eliminates any clinical reasons that hamper the spread of this service. The moral questions regarding the ethicality of aborting a foetus affected by a congenital anomaly, as well as the restrictions of this test for economic reasons, still remain. However, this research seemed to address women who want this procedure, independent of age or personal risk, and who would probably still have it in a private clinic. Indeed, all of the aforementioned research associate the dramatic decrease of procedure related risk of foetal loss with features, like provider experience and technologically advanced equipment, which may be found only in a few excellent, and usually private, clinical centres (see Turrini 2011). In this regard, the proposal to gauge institution-specific risk is interesting (see, e.g., Odibo *et al.* 2008).

A second shift regarding risk thresholds emerged as a result of the development of new non-invasive screening procedures based on foetus ultrasound indicators and biochemical markers that are easily detectable by blood tests. These so-called screening techniques are in fact *risk assessment tests* designed to provide a personalized risk evaluation, and therefore improve the selection of at risk pregnant women, who are qualified for second-tier, invasive and costly examination, like amniocentesis and CVS. In other words, these tests are a refined alternative option to the risk assessment process based exclusively on AMA. Compared with AMA, screening tests increase the percentage of affected foetuses in pregnancies that are deemed high-risk (higher detection rate), and decrease the percentage of unaffected foetuses found to be at increased risk (lower false-positive rate). Experimented first in the late 1970s, these tests became quite accurate and popular only at the beginning of the 1990s. The increased sensibility radically changed the approach to risk, which, according to the advocates of this innovation, should reject any notion of cut-off ages (Haddow *et al.* 1994; Pauker and Pauker 1994; Kupperman *et al.* 1999; Berkowitz *et al.* 2006). According to this perspective, the concept itself of risk thresholds is an arbitrary decision grounded on disputable assumptions. As pointed out in a *New England Journal of Medicine* editorial – tellingly entitled *Why is 35 a magic number?* – new information provided by prenatal screening procedures to identify

> "cases in which the risk exceeds 1 in 270 (that of a 35-year-old woman) further institutionalizes the arbitrary choice of a threshold age made two decades ago. If a chronologic age of 35 years is arbitrary, is a risk threshold based on equating the burdens of a miscarriage induced by amniocentesis and of the birth of a child with Down's syndrome any less arbitrary?" (Pauker and Pauker 1994)

AMA is criticized for the clinical criteria, which compare and equate "the two very different outcomes of giving birth to a child affected by a chromosomal abnormality and having a procedure-related miscarriage" (Kupperman *et al.* 1999, 163). With the establishment of screening technologies, some physicians have started to consider specific cut-off ages as by-products of a paternalistic approach to risk that should be replaced by pregnant women's private reflections on risk probability. According to this perspective, all pregnant women and couples should be told about the risks and benefits of genetic testing, and should have the chance to decide freely about prenatal diagnosis, beyond any cut-offs. After being proposed in several editorials published in important scientific reviews, this model was recommended by the American College of Obstetricians and Gynecologists guidelines issued in 2007 (ACOG 2007).

On the other side of the Atlantic Ocean, however, things are very different. Screening technologies are not used to turn precise cut-off ages into fluid, personalized risk thresholds that are negotiated in each case with the help of a genetic counsellor. On the contrary, they are used to limit and restrict the offer of these services to the steadily increasing percentage of over 35-year-old women who are offered prenatal diagnosis. In practice, screening tests are offered to women with an AMA as a second-tier analysis, and only those of them with a personal risk of 1 in 275 or higher are given the chance to have further tests. In this case, where neat cut-off ages are still required, risk thresholds are identified with that of 35-year-old women, 1/275 (see fig. 12-2)[5]. In this case, articulating risk cut-offs through two passages – screening tests and prenatal diagnosis – is a strategy to reduce the cost of publicly provided medical services.

[5] The procedures described here are those that are suggested and supported by national health services.

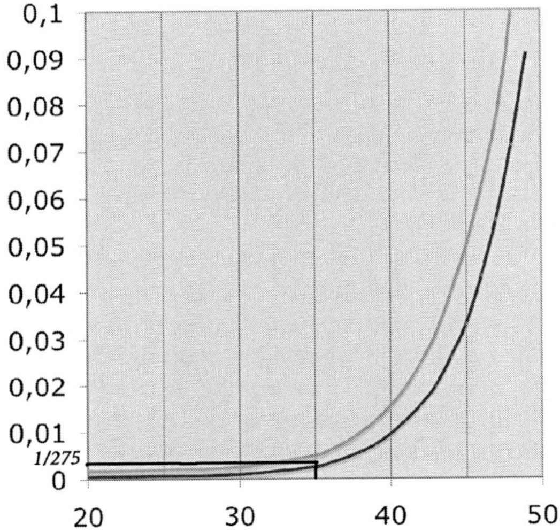

Fig. 12-2: we use the same graph of fig. 12-1 to show the reverse logical process which lies behind the so-called "consolidation" of risk thresholds. Even if AMA is to be replaced by personal risk assessment, risk thresholds are still based on the possibility that the pregnancy of a 35-year-old woman results in a baby with Down syndrome.

Risk thresholds are a strategic terrain of biomedical innovations, and they may be lowered by, become fluid as a result of, or be resistant to them. In any case, they collectively point toward an increasingly technological and privatized medical scenario, where the body may be less deeply invaded, and patient-specific risk assessment may become a crucial part of healthcare.

3. Risk thresholds as communicative devices

Currently, AMA is still used as a cut-off by the majority of health insurance plans and health services of many countries, including many European countries such as Austria, France, Germany, Italy, and the United Kingdom. Those pregnant women who want prenatal diagnosis regardless of their age or risk go directly to a private clinic. For the others no genetic counselling is provided before prenatal diagnosis, and the decision to have these tests is presented to the patients by professional figures, like gynecologists or midwives, who do not have a specific education in genetics. For these professionals, risk thresholds are very useful tools for solving the morally controversial issues surrounding prenatal diagnosis by limiting communication only to those pregnant women who are deemed high-risk.

> "You try to individualize the treatment. When a 25-year-old, healthy woman without any familiar condition comes, you do tell her that she will undergo an ultrasound scan at the 12th [*gestational*] week, but you do not insist on diagnostics. Only once the ultrasound scan is done, if some suspects emerge, then [*you mention prenatal diagnosis*]. Otherwise, if a woman is at low risk, she remains at low risk until some ultrasound findings raise a suspicion. Otherwise, you just take time and you do not send her into the other category [*that of high risk pregnancy*]. When a 43-year-old pregnant woman comes, we can consider the fact that – if she agrees, because any woman can decide to do nothing – it is advisable to have her chorionic villi tested, and I explain what it is, when it should be done, and so forth. Then you try to figure out the woman who is in front of you and you listen to what she says." (gynecologist working for the Italian National Health Service, *my interview and transl.*).

This quotation is very typical of the common approach of Italian gynaecologists in the public health care service. This doctor states that he "individualizes treatments", while what he is actually doing is using risk thresholds as an easy means to answer the tricky question of prenatal diagnosis. In particular, he reduces the technical aspects of risk to a risk threshold in at least three respects: excluding the large majority of pregnant women who are younger than 35 (and do not have other problems with pregnancy); presenting risk thresholds as objective entities grounded on medico-scientific reasons and not as arbitrary decisions prescribed by the National Health Service; and presenting the possibility of a certain condition in terms of absence or presence (for the last point see Heyman *et al.* 2001).

In the same settings risk cut-offs may be associated with more complex communicative strategies that do not exclude the moral reflection on prenatal diagnosis, as in the following case:

> "You try to communicate… communication is like dancing, you lead and you are led by the other. You try to figure out where the woman is leading you. It would make no sense to say abruptly: "With this test, we want to know whether the baby is Down's or not". Do you see what I mean? Of course, there is this issue and the abortion, but I want to introduce the topic gradually […]
>
> Finally, they always ask me 'What would you do if you were in my shoes?' That is very typical! I say everything that is possible to do, but after all everyone has to make her own decision according to her

> experience and beliefs. I can even say what I did with my children but that's just me, and I do not see why one should follow what I did." (gynecologist working for the Italian National Health Service, *my interview and transl.*)

In this case, the doctor tries in depth to understand the woman's moral and cultural attitude to Down syndrome and elective abortion before presenting the possibility of amniocentesis. Nonetheless, risk threshold still frames the doctor/patient interaction, in that it frames the women's decision-making process as an acceptance or refusal of a percentage.

Finally, there are private clinical settings that promote totally free reflection on the possibility of having amniocentesis after assessment through screening tests regarding the risk of having a child with a congenital condition. Compared to this method, AMA seems like an approximate, arbitrary and paternalistic means of communication. Nonetheless, the role of cut-offs in risk discourse should be considered again. According to Robert Crawford (2004), the importance of risk factors depends on their capacity to end the escalating spiral of control and anxiety that characterizes contemporary medicine. In a situation where the more we know about a certain condition, treatment, or test, the more we are aware of the potential dangers it produces, risk serves as a way to domesticate this pervasive and relentless process. In this regard, risk cut-offs provide neat boundaries through which it is possible to organize practices, decisions, and even moral values in a way that limits the expansion of risk discourse. On the contrary, the borderless fluctuation of risk, instead of empowering pregnant women, may reveal itself as an instrument of the intensification of the feeling of uncertainty and the willingness to subject oneself to biomedical control. The attempt to free risk assessment from cut-offs definitely responds to "the logic of choice", namely a way to understand and practice healthcare as an exercise of the liberty of choice promoted by biomedical innovations (Mol 2008). Different from the ideal-typical figure of the consumer, who freely chooses commodities, or that of the citizen, who freely chooses political programmes, the patient has to face a bodily condition from which he or she cannot escape. According to the "logic of care", the point is not to foster and build upon the biomedical promise of a limitless capacity to conquer disease, but rather to work out the problems that arise in any condition with the patient and to tinker them with the other professional, familial, and daily needs. In the case of prenatal diagnosis, the situation is different, because it concerns a decision-making process, but requires nonetheless a careful tinkering with many aspects of pregnancy, family, and disability (Mol *et al.* 2011).

4. From the side of pregnant women

If medical doctors rely on risk thresholds to deliver information about prenatal diagnosis, the same is not always true for pregnant women and their close relatives or friends. Risk thresholds come out of technological and scientific innovations on pregnancy – their vocabulary and grammar are intrinsically medical and technical. In this regard, they are subject to the same criticisms that some feminist scholars have had for "new reproductive technologies": they are medical attempts to take control of pregnancy by checking the blossoming female body and to negate women's self-awareness (Duden 1993; Kats Rothman 1994).

Even if these critiques have effectively questioned the linear development of biomedical progress, they share a similar, although symmetrically reversed, reductionist approach: medical doctors are villains, instead of heroes, and pregnant women are victims, instead of the beneficiaries of technological advancements. We can find a subtler analysis in qualitative investigations of the experiences, understandings and feelings of women who are facing the decision of undergoing or not undergoing prenatal diagnosis (Rapp 1999; Heyman and Herikson 2001; Heyman *et al.* 2006; Williams *et al.* 2005). A common aspect underlined by this literature is the agency that women have in negotiating, appropriating and re-signifying the meaning of risk itself. For this reason, responses to prenatal diagnosis are extremely complex and variable. To some, AMA is a reasonable cut-off to adopt; others refuse to have amniocentesis or CVS; others want to assess their own risk in a more detailed manner; others opt for screening tests just to be reassured about their own healthy condition; others choose prenatal diagnosis independent of their age; and so on.

Some women reject their ascribed high-risk status by questioning the interpretation of probabilities presented to them. The following quotation is taken from a British woman interviewed by Heyman and colleagues (2006, 236-7).

> "One in 61 was what my risk level was, or whatever. And my dad came back to me and said, 'That's only 2%. It's a very, very small number'. If someone had said to me 2%, I think I would have understood that more than 1 in 61. (2.14, aged 34, screened at higher risk but declined diagnostic test, standard site, interview post-screening)."

In this case, the woman comprehends the meaning of the risk ratio only thanks to her father, who is able to shift the understanding of the 1 in 61 risk and, thus, to consider his daughter's pregnancy low-risk. The same re-appropriation of risk thresholds also happens in opposite cases, in which a mother follows doctor recommendations.

> "According to her [the doctor] I was at risk. My orientation was to do it [amniocentesis], because when you are 36 years old that's the way it is. Then, she talked to me about the opportunity to have a bi-test, and to me it was sensible. But at the public service you have to wait for 4 or 5 weeks, so you have to book it at about the second or third gestational week, otherwise it is too late. I was very disappointed by that and I decided not to do it. That day I visited my mother, and she said to me: 'You have to do it [bi-test] in my town!' So I phoned at the hospital of ****** and there I found a place the same day. (…) Luckily my risk was not higher than the national average. In that case, I would have done amniocentesis, even if I think that to wait for two weeks for a response about the possibility that your baby has congenital anomalies is a very stressful experience." (Italian woman with two children, *my interview and transl.*)

Sometimes even if the pregnant woman trusts doctor and follows their recommendations, she may decide not to have the test after she encounters the waiting list at the clinic. She then may decide to have it only after having talked to her mother. This extract clearly shows the importance that contingencies, both material (e.g. the waiting list of the clinic) and relational (the visit to her mother), play in the decision-making process. Risk thresholds configure the increasingly intertwined relationship with technologies, but when this information is presented to pregnant women it is necessarily mediated by experiences, relationships, values and aspirations. As shown in both cases presented, the risk is translated from the realm of medically abstract calculations to that of real life through a process of cultural re-appropriation that is never trivial. What is found here is the same process that Rayna Rapp has depicted in her notable work on the socio-cultural impact of amniocentesis.

> "When viewed collectively, all the women using or refusing the test can be seen as moral pioneers. […] Whether as willing conscripts or draft resisters to biomedicine' technicist promise of more control over pregnancy outcomes, all made conscious set of values, ethics, and choices which were located in the realm of the private but were shaped and in turn helped to shape a more social terrain." (Rapp 1999, 309)

According to Rapp, prenatal diagnosis technologies are not neutral technologies that help women in their decision about pregnancy, but are rather instruments with a deep cultural and social impact on the ways in which pregnancy is monitored, managed and experienced. In the new regime of reproductive technologies, risk thresholds are a crucial point of passage. If risk thresholds are, on the one hand, paternalistic means of communication that are used to select not only a small percentage of pregnant women to whom prenatal diagnosis is addressed, but also the information provided regarding this procedure, on the other hand erasing common cut-offs means extending the probabilistic logic of risk to all pregnancies. Moreover, while private clinics promise a medical future in which the material body may be less deeply invaded, and patient-specific risk may be assessed for each single pregnant woman, public health services are unable to keep up with the fast-moving pace of biomedical innovations. To put it in Adele Clarke's and colleagues' terms, risk thresholds are a crucial terrain of the process of "biomedicalization" (Clarke *et al.* 2010), whose social impact is not limited to the privatization, personalization, and preventative aspects of health, but also to the formation of emerging "techno-scientific identities" (*Id.*), which will be the subject of the next section.

5. Mature maternity between stigmatization and pride

The implications of being categorized high-risk should not be restricted to a purely medical domain. Several qualitative studies on the experiences of women who have been diagnosed with cancer have shown the deep impact on the lives of at-risk patients (Gifford 1986; Robertson 2001). Only at an early stage is risk perceived as a purely medical state and associated with a special clinical regime. Afterwards, it is internalized as part and parcel of the identity of "at-risk" people, who increasingly shape their behaviour accordingly. "Being at-risk" thus becomes a new medically disadvantaged condition, to which negative values, new responsibilities and, sometimes, stigma are ascribed. Social blaming regarding overeating, smoking or having un-protected sexual intercourse are just some examples that indicate the magnitude of the phenomenon. However, as we have already seen in the section about re-appropriation of risk thresholds, risk goes beyond its mere "medical nature".

The four decades of experience with AMA are instructive in that they present the opposite movements of stigmatization and pride. The unexplored and yet interesting corpus of popular pregnancy guides gives a clear representation of both aspects. Here we report some findings from a selection of books written in English, French and Italian of the relatively recent literary genre that lies in between medical and scientific popularization, and true-life experiences. Including life sketches, curiosities, recipes, many pictures of pregnant women, foetuses and usually nuclear heterosexual families, the major parts of these books are devoted to providing updated information on the ideal pregnant lifestyle. They include nutrition regimes, physical exercises, medical tests, the development of the foetus, as well as treatments and tricks for resolving the most common pregnancy conditions.

AMA is usually evoked as an unusual and non-optimal condition for getting pregnant, and it is usually associated with stricter medical controls. A clear example is provided by the last edition of a classic book of this genre, *J'attends un enfant (I'm expecting a child)* by Laurence Pernoud (2013), where the topic of maternal age is treated in the chapter on the *Medical surveillance of pregnancy* in the following manner:

> "A favourable period in which to be pregnant exists. This age can be defined as being approximately between 20 and 35 years. […] Happily, nowadays very young women, as well as older ones, may benefit from a high quality surveillance." (Id., 208, *my transl.*)

A similar yet sterner description of mature pregnancy is found in *Le grand livre de ma grossesse (The big book of my pregnancy)*, the official book of the French gynecological and obstetric medical association (le Collège national des gynécologues et obstétriciens français).

> "ATTENTION! After 40 years, a rigorous follow-up is necessary.
> Do a health check. (…) Gather information on the potential anomalies of the foetus: good information on the benefits of prenatal screening is required for an early diagnosis of chromosomal anomalies and foetal malformations. You have to both assess the advantages and drawbacks of this screening, and decide in accordance with your personal believes; this information will help you to make a decision in respect to whether to carry your pregnancy to term or not."(Lansac and Evrard 2013, 32, *my transl.*)

In both passages, however, AMA designates a specific condition, where congenital anomalies are only one of the potential dangers that women incur during pregnancy. The so-called "elderly prima gravida" are also associated with a range of other clinical complications. Particularly for this reason, mature pregnancy is presented as an unfavourable condition that needs stricter medical control and, especially, prenatal diagnosis technologies and/or screening. The second example makes the responsibility of choosing to terminate a pregnancy clearer.

Although depicted as socially unusual, medically dangerous and morally controversial, mature pregnancy is the starting point of a new way of conceiving and experiencing maternity in which medical technologies play a crucial role. In this sense, the rise in the second half of the 1990s of a new subgenre of self-help books devoted to mature pregnancy is particularly interesting. Especially telling is the experience of Lindsey Harris, a mature mother who created the website www.mothers35 plus.co.uk during her pregnancy in order to exchange experiences. What started as a hobby became her main employment within a few years. The website provides older mothers, mothers to be, and would-be mothers, with a large source of rich information as well as commodities, including special books on pregnancy. A common feature of these books is that they define mature pregnancy by mentioning a cut-off age, which is usually 35, sometimes 40[6], and play with the ambivalence implied by older motherhood. The trepidation of having a baby so "late" is balanced with the idea of a rationally pondered choice that is supported by recent medical advancements. In this sense, *Conception and Pregnancy Over 35: Expert Advice for Every Stage, from Planning a Baby to Giving Birth* (Goetzl 2005) and *Everything Guide to Pregnancy Over 35: From Conquering Your Fears to Assessing Health Risks* (Sember 2007) are particularly evocative of the importance of medical expert knowledge and current medical options. Paradoxically, therefore, the biomedicalization of pregnancy is used here to reverse traditional medical concerns about mature pregnancy.

> "Overall the age-related risks have been exaggerated, particularly in the popular press, although women over 35 are more likely than their younger counterparts to already suffer from common chronic diseases such as arthritis, hypertension, and diabetes prior to pregnancy. They are also at greater risk from developing high blood pressure and diabetes during pregnancy. However, modern techniques mean that these can be successfully detected, treated and monitored." (www.mothers35plus.co.uk)

Even clearer are Laura Goetzl's words: "I want you to leave guilt at the door as you enter the experience of your pregnancy" (Goetzl 2005, 6). On another website, after having discussed the potential pregnancy/birth dangers of an older motherhood, the new screening techniques are mentioned such as non-invasive options that assess one's personal risk of having a baby with a congenital anomaly.

---

[6] *First-Time Mothers, Last-Chance Babies: Parenting at 35 +* (Cain 1994); *Older Mothers: Conception, Pregnancy and Birth After 35* (Berryman *et al.*, 1995); *Choosing Older Motherhood: Essential Guide to Becoming a Mother Over 35* (Jones 1996); *Motherhood After 35: Choices, Decisions, Options* (Jones 1998); *Not too late: Having Baby after 35* (Thorn 1998); *So Glad We Waited! A Hand-Holding Guide for Over-35 Parents* (Nachamie 2000) and others (Curtis e Schuler 2001; Goetzl 2005; Tournaire 2005; Moore and de Costa 2006; Sember 2007; Schmid-Altringer 2009).

> "Although it is no longer unusual to have your first child in your 30's, having a baby after the age of 35 can require special considerations. Fortunately, there are new tests available to detect possible complications. There are also interesting, positive statistics about babies born to older mothers." (www.tinyprints.com/pregnancy-after-35-thirty-five-years-old.htm)

Biomedical expertise is increasingly placed in the position of a resource to be drawn on in life planning, rather than as a master discourse in arbitrating decisions on procreation in light of risk. Pregnant women's various engagements with regimes of risk and surveillance imply an extended responsibility over the promotion of health beyond a purely medical discourse. This implies the formation of biomedically inflected identities founded on a certain pride in opting for a different type of motherhood. Postponing having children until later in their late thirties or even their forties is not only represented as a medically unfavorable situation, but also seen through its potential benefits in terms of self-realization, financial arrangements, and possibly better life conditions for the children regarding breastfeeding or school performance.

> "Another positive aspect is that the offspring of older parents do better at school than those born to the very young and studies have shown that children of older mothers tend to do better in ability tests. Research shows that older mothers are more likely to breastfeed and that they may be less prone to postnatal depression than younger women."
> (http://www.mothers35plus.co.uk/pregnancyover35.htm.)

A choice that fits in with other demands on one's own life and professional career indicate a mature personality, an open-minded attitude, and even a good level of education.

> "The trend towards later maternity is strongest among women with better educational qualifications. (…) On the whole, babies are more likely to be planned and wanted by women in their thirties. There is evidence that older women express greater satisfaction and feel they are ready to have a child because they have been fulfilled in their lives before that time. The notion of sacrifice is more often talked about in younger mothers. Older mothers may want to spend more time with their children." (Berryman *et al.* 1995)

In general, older mothers are described as responsible and aware, and their babies as planned and wanted. To put it in Carlos Novas and Nikolas Rose's terms, knowledge about predisposition of genetic disease creates "an obligation to act in the present in relation to the potential futures that come into being" (Novas and Rose 2000). In this regard, AMA is at the centre of new forms of "biopower", through which subjects work on themselves *qua* living beings (Rabinow and Rose 2006). These subjectivities go beyond a purely medical discourse, in that they also affect how women understand motherhood and reproduction, and how they evaluate options according to currently available knowledge and technologies.

6. Final remarks: The influence and multiplicity of boundary numbers

The development of preventative medicine poses new questions about the viability and the desirability of a new kind of medicine, and is an important chapter in the current medical expansion and imbrication with technology and the market – the so-called "biomedicalization". A recent editorial of *British Medical Journal* concludes by raising questions regarding the changing borders between what constitutes disease and normality in preventative medicine. Tellingly entitled "Are we at risk of being at risk?" (Godle 2010), it discusses the case of "prehypertension", i.e. of the risk of developing hypertension, a condition which, in turn, increases dramatically the possibility of more serious pathologies. The problem of risk thresholds – which can be defined as the decision of what should be screened and tested and what is not – becomes more than a medical question, it is an attempt to translate into practices the concerns at stake in the direction and development of preventative medicine. The impact of Jolie's disclosure on global public opinion, as well as our brief examination of AMA in a wide range of situations, from expert assessment debates to medical risk communication and guides to pregnancy, has shown not only the magnitude of the risk thresholds' importance, but also their complex intertwinement with a heterogeneous set of elements. Risk thresholds are boundary objects whose assessment process crosses economic resource considerations, clinical concerns, statistical calculations, political projects, processes of innovation and research, and private company strategies. At the same time, they inform medical practices, and particularly doctor/patient communication. They are numerical entities, whose simplicity and precision are easily translatable between experts and lay people, as well as between the disciplines of statistics, medicine, economics and politics. Even if they are necessary classifications for dealing with the large amount of information produced by medical research, they should not be understood as merely technical aspects of risk management. These *boundary numbers* instead constitute the terrain of negotiation around which new medical practices, sets of civic rights and duties, conceptions of health, and other phenomena like motherhood are configured and reconfigured.

As Theodore Porter (1995) argues, the language of mathematics is a communication strategy well suited to embodying the development of modern objectivity, and, as such, lends authority to decisions which otherwise would be criticized for being subjective and personal[7]. As part and parcel of a modern epistemology that responds to the moral demand for impartiality and fairness, numbers have the power to "create new things and transform the meaning of old ones" (*Id.*, 17). In the case of preventative medicine, the use of numerical risk cut-offs provides a privileged point of view on the processes of technological, political and socio-cultural change. Risk thresholds are the target of biomedical innovations as well as a means used to oppose resistance to them; they are communicative devices of simplification, but also a barrier against the spread of risk discourse; they are associated with a paternalistic model of healthcare, but also with the emergence of new subjectivities. In any case, the definition, negotiation, and use of risk thresholds are crucial processes which play a crucial role in defining health policies, technological innovations, patient/doctor relationships, styles of healthcare, new modes of governance and emerging subjectivities.

References

ACOG. 2007. - Association College of Obstetricians and Gynaecologists. "Invasive prenatal testing for aneuploidy." In *ACOG Practice Bulletin*, 88.

Beck, Ulrich. 1992. *Risk Society: Towards a New Modernity*. London-Thousand Oaks -New Delhi: Sage.

Berkowitz, Richard L., Roberts Jaclyn and Howard Minkoff. 2006. "Challenging the Strategy of Maternal Age–Based Prenatal Genetic Counseling." In *Journal of American Medical Association*, 295(12): 1446-1448.

Berryman Julia, Karen Thorpe and Kate Windridge. 1995. *Older Mothers: Conception, Pregnancy and Birth After 35*. London: Pandora.

Bowker, Geoffrey C. and Susan L. Star. 1999. *Sorting Things Out: Classification and Its Consequences*. MIT press.

[7] I thank one of the anonymous referees for this inspiring suggestion.

Brambati, Bruno and Lucia Tului. 2005 . "Chorionic villus sampling and amniocentesis." In *Current Opinion in Obstetrics and Gynecology*, 17: 117–20.

Cain, Madelyn. 1994. *First-Time Mothers, Last-Chance Babies: Parenting at 35+*. Far Hills, NJ: New Horizon Press.

Canick, Jacob A., Devereux N. Saller and Geralyn M. Lambert-Messerlian. 2003. "Prenatal screening for Down syndrome. Current and future methods." In *Clinics in Laboratory Medicine*, 23: 395-411.

Caughey, Aaron B., Linda M. Hopkins and Mary E. Norton. 2006. "Chorionic Villus Sampling Compared With Amniocentesis and the Difference in the Rate of Pregnancy Loss." In *Obstetrics &* Gynecology, 108: 612-616.

Clarke, Adele E., Laura Mamo, Jennifer R. Fosket, Jennifer R. Fishmanand Janet K. Shim (edited by). 2010. *Biomedicalization: Technoscience, Health, and Illness in the U.S.* Durham, NC and London: Duke University Press.

Crawford, Robert . 2004. "Risk Ritual and the Management of Control and Anxiety in Medical Culture." In *Journal for the Social Study of Health, Illness and Medicine*, 8(4): 505-524.

Curtis, Glade B. and Judith Schuler. 2001. *Your Pregnancy After 35*. Cambridge, MA: Perseus.

D'Alton, Mary E. and Alan H. DeCherney. 1993. "Prenatal Diagnosis." In *New England Journal of Medicine*, 328: 114-120.

Doubilet, Peter, Milton C. Weinstein and Barbara J. McNeil. 1986. "Use and Misuse of the Term 'Cost Effective' in Medicine." In *New England Journal of Medicine*, 314: 253-256.

Douglas, Mary and Aaron Wildawsky. 1983. *Risk and Culture: An Essay on the Selection of Technological and Environmental Disasters*. Berkeley and Los Angeles: University of California Press.

Duden, Barbara. 1993. *Disembodying Women: Perspectives on Pregnancy and the Unborn*. Harvard, MA: Harvard University Press.

Eddleman, Keith A., Fergal D. Malone, Lisa Sullivan, Kim Dukes, Richard L. Berkowitz, Yara Kharbutli *et al.* 2006. "Pregnancy Loss Rates After Midtrimester Amniocentesis." In *Obstetrics and Gynecology*, 108 (5): 1067–72.

Fressoz, Jean-Baptiste. 2011. *L'apocalypse joyeuse. Une histoire du risque technologique*. Paris: La découverte.

Giddens, Anthony. 1990. *The Consequences of Modernity*. Stanford: Stanford University Press.

Gifford, Sandra M. 1986. "The Meaning of Lumps: A Case Study of the Ambiguities of Risk." In *Anthropology and Epidemiology*, edited by

Craig Janes, Ron Stall and Sandra M. Gifford, 213-246, Dordrecht: Reidel.

Godle, Fiona. 2010. "Are We at Risk of Being at Risk?" In *British Journal of Medicine*, 341: 4766.

Goetzl, Laura. 2005. *Conception and Pregnancy Over 35: Expert Advice for Every Stage, from Planning a Baby to Giving Birth*. London: Dorling Kindersley.

Gregg, Robin. 1995. *Pregnancy in a High-Tech Age: Paradoxes of Choice*. New York and London: New York University Press.

Haddow, James E., Glenn E. Palomaki, George J. Knight, George C. Cunningham, Linda S. Lustig and Patricia Boyd. 1994. "Reducing the need for amniocentesis in women 35 years of age or older with serum markers for screening." In *New England Journal of Medicine*, 330: 1114-1118.

Heyman, Bob and Mette Henriksen. 2001. *Risk, Age and Pregnancy: A Case Study of Prenatal Genetic Screening and Testing*. London: Palgrave.

Heyman, Bob, Gillian Hundt, Jane Sandall, Kevin Spencer, Clare Williams, Rachel Grellier and Laura Pitson. 2006 . "On Being at Higher Risk: A Qualitative Study of Prenatal Screening for Chromosomal Anomalies." In *Social Science & Medicine*, 62: 2360-72.

Heyman, Bob, Monica Shaw, Andy Alaszewski and Mike Titterton. 2010. *Risk, Safety, and Clinical Practice*. Oxford: Oxford University Press.

Hook, Ernest B., Philip K. Cross and Dina M. Schreinemachers. 1983. "Chromosomal Abnormality Rates at Amniocentesis and in Live-Born Infants." In *The American Journal of Medical Association*, 249(15): 2034-2038.

Jones, Maggie. 1996. *Choosing Older Motherhood: Essential Guide to Becoming a Mother Over 35*. London: Vermillon.

Jones, Maggie. 1998. *Motherhood After 35: Choices, Decisions, Options*. Tucson: Fisher Books.

Kaback, Michael M. *et al.* 1979. *Antenatal Diagnosis: National Institute of Health Consensus Statement*. Bethesda: National Institutes of Health.

Katz Rothman, Barbara. 1994. *The Tentative Pregnancy: How Amniocentesis Changes the Experience of Motherhood*. Rivers Oram Press/Pandora List.

Kuppermann, Miriam, James D. Goldberg, Robert F. Nease and A. Eugene Washington. 1999. "Who Should Be Offered Prenatal

Diagnosis? The 35 Year-Old Question." In *American Journal of Public Health*, 89: 160-163.
Lansac, Jacques and Nicolas Evrard. 2011. *Le grand livre de la grossesse*. Paris: Eyrolles.
Lejeune, Jerôme, Marthe Gautier and Raymond Turpin. 1959 "Etude des chromosomes somatiques de neuf enfants mongoliens." In *Compte Rendue de l'Académie des Sciences de Paris*, 248: 1721-1722.
Lubs, Herbert A. 1972. "Indications for amniocentesis." In *Antenatal Diagnosis*, edited by Albert Dorfman, 17-27, Chicago: The University of Chicago Press.
Mol, Annemarie. 2008. *The Logic of Care: Health and the Problem of Patient Choice*. London: Routledge.
Mol, Annemarie, Ingunn Moser, Enrico Maria Piras, Jannette Pols, Mauro Turrini and Alberto Zanutto. 2011 "Care in Practice. On Normativity, Concepts, and Boundaries." In *Tecnoscienza. Italian Journal of Science & Technology Studies,* 2(1): 73-86.
Mooney, Gavin and Mette Lange. 1993. "Antenatal Screening: What Constitutes 'Benefit'?" In *Social Science & Medicine*, 37(7): 873–878.
Moore, Michele C. and Caroline M. de Costa. 2006. *Pregnancy and Parenting after Thirty-Five: Mid Life, New Life*. Baltimore: Johns Hopkins University Press.
Mujezinovic, Faris and Zarko Alfirevic. 2007. "Procedure-Related Complications of Amniocentesis and Chorionic Villus Sampling: A Systematic Review." In *Obstetrics & Gynecology*, 110 (3): 687-694.
Nachamie, Lois. 2000. *So Glad We Waited! A Hand-Holding Guide for Over-35 Parents*. New York: Three Rivers Press.
Novas, Carlos and Nicolas Rose. 2000, "Genetic Risk and the Birth of the Somatic Individual." In *Economy and Society*, 29 (4): 485-513.
Odibo, Anthony O., Diane L. Gray, Jeffrey M. Dicke, David M. Stamilio, George A. Macones and James P. Crane. 2008. "Revisiting the Fetal Loss Rate After Second-Trimester Genetic Amniocentesis A Single Center's 16-Year Experience." In *Obstetrics & Gynecology*, 111: 589–95.
Pauker, Susan P. and Stephen G. Pauker. 1994. "Prenatal Diagnosis: Why is 35 a magic number?" In *New England Journal of Medicine*, 330: 1151-1152.
Penrose, Lionel S. 1933. "The relative effects of paternal and maternal age in mongolism." In *Journal of Genetics*, 27: 219-224.
—. 1934. "The relative aetiological importance of birth order and maternal age in mongolism." In *Proceedings of the Royal Society of London.*

Series B, Containing Papers of a Biological Character, 115(795): 431-450.

Pernoud, Laurence. 2013. *J'attends un enfant.* New edition. Horay.

Porter, Theodore. 1995. *Trust in Numbers: The Pursuit of Objectivity in Science and Public Life*. Princeton (NJ): Princeton University Press.

Rabinow, Paul and Nikolas Rose. 2006. "Biopower Today." In *BioSocieties*, 1: 195-217.

Rapp, Rayna. 1999. *Testing Women, Testing the Fetus. The Social Impact of Amniocentesis in America*. London: Sage.

Robertson, Ann. 2001. "Biotechnology, Political Rationality and Discourses on Health Risk." In *Health*, 5(3): 293–309.

Schmid-Altringer, Stéfanie. 2009. *Ma grossesse après 35 ans*. Paris: Vigot.

Sember, Brette M. 2007. *Everything Guide to Pregnancy Over 35: From Conquering Your Fears to Assessing Health Risks: All You Need to Have a Happy, Healthy Nine Months*, Avon, MA: Adams Media.

Star, Susan L. and James R. Griesemer. 1989. "Institutional Ecology, "Translation" and Boundary Objects: Amateurs and Professionals in Berkeley's Museum of Vertebrate Zoology." In *Social Studies of Science*, 19: 387-420.

Thorn, Gill. 1998. *Not too late: Having a baby after 35*. London: Bantam.

Tournaire, Michel. 2005. *Le bonheur d'être mère. La grossesse après 35 ans*. Paris: Jacob.

Turrini, Mauro. 2011. " "Anch'io se dovessi infilarmi un ago nella pancia sarei preoccupato". La dimensione simbolica del rischio sanitario tra pratiche rituali e valutazione scientifica." In *I rischi in sanità. Un nuovo fenomeno sociale*, edited by Tatiana Pipan, 76-101, Milano: Angeli.

Vassy, Carine. 2006. "From a Genetic Innovation to Mass Health Programmes: The Diffusion of Down's Syndrome Prenatal Screening and Diagnostic Techniques in France." In *Social Science & Medicine*, 63: 2041-2051.

Williams, Clare, Jane Sandall, Gillian Lewandohundt, Bob Heyman, Kevin Spencer and Rachel Grellier. 2005. "Women as Moral Pioneers? Experiences of First Trimester Antenatal Screening." In *Social Science & Medicine*, 61(9): 1983-1992.

Acknowledgements

I would like to thank my doctoral supervisor Federico Neresini, Tatiana Pipan, Marina Maestrutti and Nadav Even Chorev, as well as the two editors of this book for their encouragement and intellectual generosity, and Daniela Crocetti for having revised the article.

CHAPTER THIRTEEN

FROM SCIENCE PARKS TO INFRASTRUCTURES

MICHELA COZZA

Introduction

Starting from the 1980s, science parks (henceforth SPs) have become an international phenomenon. Usually, they are defined as intermediate organizations that provide social environment, technological and organizational resources, and managerial expertise for the transformation of a technology based business idea into an efficient economic organization. Such an environment is deeply embedded in the political, social and economic systems and it could also be influenced by its geographic location. For this reason, no general theory is possible. However, to make such a conclusion at this time would probably represent a rush to judgement on very little evidence, which is why I believe the following analysis is timely.

In this chapter, I will examine the work of Italian science parks focusing on their national network: the Italian Association of Science and Technology Parks (APSTI). APSTI is born *to act as* infrastructure, promoting Italian scientific and technological centres as integrators between the companies' need for innovative growth and the wealth of knowledge embodied by the centres of technological and scientific excellence, universities and research centres. However, this does not mean that APSTI *is* an infrastructure, as I will show.

In order to understand whether we can talk in terms of infrastructure to describe both the work of APSTI and science parks, the chapter is structured as follows.

First, I will provide an overview of the debates and issues concerning science parks, focusing on the literature both on Italian science parks and on other international experiences. Then, I will develop an analytical framework for interpreting the qualitative data, referring to the concept of "infrastructure". This is a promising construct in order to recognize the technical, socio-organizational and institutional components that are practically involved in Italian SPs' history. I will further discuss this main concept focusing on "convergence" as an important feature of infrastructuring: it enables the investigation of the mutual constitution of different actors and their social worlds. To this end, I will stress the relevance of an "ecological perspective" in my theoretical framework. I would like to specify in advance that the word "ecological" is used here to indicate an epistemological perspective concerned with the co-existence of differences. So, the unit of analysis is not a partial viewpoint but all points of view. This approach will enable us to take into account the ambivalences of a complex phenomenon such as science parks.

Then, after a focus on the research design and methodology, I will give voice to some actors in order to articulate different viewpoints on science parks. Discussion is meant to be suggestive rather than conclusive, outlining the possibility of applying the concept of infrastructure to the ecology of science parks. I conclude with a discussion on how parks and related organizations could become drivers for innovation.

1. An overview of science parks

Since the early 1980s, the subject of science parks has generated a vast amount of both direct and indirect literature. Much of the former literature touched on science parks within the broader context of technology and economic development, whereas the latter is focused on particular themes or aspects. Furthermore, there is no single and uniformly accepted definition of Science Park (Amirahmadi and Saff 1993; Löfsten and Lindelöf 2002).

There are several terms used to describe similar developments, such as *Research Park*, *Technology Park*, *Business Park*, *Innovation Centre*. Allesch (1985) focuses on the difference between research parks, innovation centres and science parks. According to Allesch, a research park is one in which young firms of detached sections of large companies carry out research and development in relatively close cooperation with a nearby university or research establishment and where the development of prototypes, but not mass production, is allowed. An innovation centre provides New Technology-Based firms (henceforth abbreviated to NTBFs) with an optimum chance of survival and development by offering an extensive range of services, proximity to universities, and the possibility of integration into the local and regional innovation network. Finally, Allesch defines science parks as a way of locating industries: existing firms in innovative technology areas are offered attractive surroundings and proximity to research establishments. The main limitation of Allesch's distinction is that, in reality, many parks include a combination of these three categories.

MacDonald (1987) says that each of the above-mentioned terms describe: (a) a property-based initiative close to a place of learning, and (b) one that provides high quality units in a pleasant environment.

If we consider the characteristics and definitions of parks according to their main associations–such as the International Association of Science Parks and Areas of Innovation (IASP), the Association of University Research Parks (AURP) and the United Kingdom Science Parks Associations (UKSPA)–we might identify several common concepts for a SP (Durão *et al.* 2005): (1) it should be economically sustainable; (2) it should have operational links with universities, R&D centres and/or other institutions of higher education; (3) it should encourage and support start-ups and incubation of innovative, high-growth and technology based companies; (4) it should stimulate the transfer of technology and business knowledge; (5) it should be a property-based initiative. This last point agrees with MacDonald's first statement. However, Amirahmadi and Saff (1993) stress that the term *property-based initiative* makes the definition of science park particularly vague and opens the door for confusion with other types of parks or economic activity locations.

The second aspect suggested by MacDonald calls to mind Felsenstein's (1994, 93-94) metaphor of science parks as seedbeds:

> "Implicit in the seedbed metaphor is the notion of the nurturing process that eventually creates an environment for growth. The science park as a seedbed therefore refers to the conditions created to promote innovation (…) As such, they are said to create a supportive environment for the development of innovation, creativity and entrepreneurship."

Leaving for a moment this metaphor–that I will further discuss–I would like to point out that parks have not always been a *supportive environment*. Unfortunately, there are many cases of partial failure of parks. For instance, Kihlgren (2003) analysing the science parks in St. Petersburg, maintains that those parks are lacking in collaboration with local industry and their technological offer is *not aligned* with the Russian economy and its very low demand for high-technology products. Similar results are given by Ratinho and Henriques (2010) that study Portuguese science parks, concluding that their contribution to job creation and economic growth is modest. Watkins-Mathys and Foster (2006) examining the performance of hi-tech companies situated on science and technology industry parks in Beijing and Shanghai, highlight that, in spite of the advancement in hi-tech, China lags behind in entrepreneurship and technology transfer. In Italy, Butera's (1995) analysis on science parks has identified networking as the key factor of a successful park. However, Butera maintains that most of the Italian science parks have been unable to develop from "service delivery" organizations to organizational networks. All along, lack of synergies between different actors at different levels (both local and international science parks, universities, research centres, enterprises, city, state government) has been the main problem of Italian science parks. Starting from the case of the Italian scientific and technological park of Sardinia, Mongili (1998) considers the parks as the "embodiment" of a demarcation between science, technology and production. Drawing from Science and Technology Studies (STS), he discusses this limit of science parks identifying them with an outdated linear model of innovation. It has been a long time that literature has recognized the inappropriateness of this model (Kline 1985). It views innovation as an orderly process, starting with the discovery of new knowledge, moving through various stages of development, and eventually emerging in a final viable form. However, the model that Mongili identifies with the functioning of science parks is so streamlined and inadequate that its use must seriously distort thinking about processes of innovation. It ignores the fact that science and technology continuously assist each other through multiple links, involving people, artifacts, material and symbolic infrastructures.

All these results allow us to maintain that the life of a science park depends on a complex array of factors. Unlike the previous cases, other researchers have identified some successful experiences in different national contexts, or–as in the case of Italy–some authors have offered encouraging considerations.

Starting from a longitudinal comparison between firms on and off science parks in Sweden, Löfsten and Lindelöf have found that "the park milieu appears to have a positive impact on their firms' growth as measured in terms of sales and jobs" (2002, 860). Nevertheless, starting from their results, the authors conclude that it would be better if future initiatives and policies focus on NTBFs on science parks, instead of on NTBFs in general or on conventional small firms. Similarly, Colombo and Delmastro (2002), talking about Italian SPs until the year 2000, maintain that they have been rather successful in attracting entrepreneurs with high quality human capital, playing a positive selection role. The authors stressed that in situations where there are substantial market failures as regards the provision of essential inputs to NTBFs, the presence of bridging institutions such as SPs, may be relatively more beneficial than in countries where the national innovation system is more advanced.

However, Colombo and Delmastro mitigate this optimistic view, saying that additional data need to be gathered to provide further insights into the drivers of parks' success and to draw out the consequent implications for national technology policy.

Other positive experiences are documented in non-Western countries. Vaidyanathan (2008), discussing the parks in East Asian countries such as Taiwan, South Korea, Hong Kong and Malaysia, says that they have been successful, especially in attracting foreign investment and promoting growth of knowledge-based industries. With special regard to India, most of the technology parks are specific to sectors like Information Technology (IT) and biotechnology (biotech), and some of the technology parks are highly specialized within a sector. As well, a majority of the Indian technology parks are not associated with any university. Similar to the technology parks in Taiwan and Singapore, technology parks in India are export oriented whereas the science parks in the West are focused on R&D. Vaidyanathan's research sheds light on the catalyst role of these parks for India's exports and economic growth.

Coming back to Felsenstein's metaphor, he argues that the seedbed role of the science park–in a behavioural sense–has a complementary set of expectations that relates to a spatial perspective. What may be of interest is remembering that

> "The common ground between the behavioural and spatial conceptions of the seedbed lies in the notion of the seedbed as creating an environment. This environment, while occupying dimensions in geometric space, is not exclusively spatial. It represents a milieu in both the functional and behavioural sense, as well as the geographic." (Felsenstein 1994, 94)

From a spatial perspective, the *exogenous* factors (city size, level of urbanization, institutional structures and community characteristics) nurturing and promoting innovation at a given location are integral components of the seedbed environment. However, this hypothesis underestimates the intra-firm characteristics. From a behavioural perspective, the seedbed environment is composed of a knowledge *infrastructure* that creates positive externalities that become public goods. Institutions such as universities, research centres, but also local chambers of commerce, banks and venture capital companies (Felsenstein 1994, 95)

> "are all milieu-creating organizations in that, when occur within a given area, they create an information–and transaction–intensive complex that not only provides economies in scale and scope and cost savings in transactions but also reduces risk."

However–despite public policy rhetoric and few examples of science parks as catalysts in urban and regional growth–the science park, rather than support innovation, could hinder it. This enclave/seedbed dichotomy can be related to the fact that science parks are often the result of a public-private partnership. As Phan *et al.* (2005) have stressed, this heterogeneity gives rise to a "principal-principal" agency problem:

> "the primary agency problem is not the failure of professional managers to satisfy the objectives of diffused shareholders, but rather the opportunistic behaviour of the controlling shareholders" (2005, 178).

This statement, which will be evoked in the final part of the discussion, leads now to the analysis of the organizational complexity of science parks employing the concept of *infrastructure*.

2. Understanding infrastructure

The term "infrastructure" evokes vast sets of collective equipment necessary to human activities (Bowker *et al.* 2010) so that people envision infrastructure as a system of substrates–railroad lines, pipes and plumbing, electrical power plants, and wires (Star 1999).

The field of Science and Technology Studies explores the phenomenon of "infrastructuring". Historians attribute the principal model of infrastructure development to Thomas Parker Hughes. His *Networks of Power* (1983) is a study on the evolution of electric power that has become a model adapted and extended over two decades by a group of historians and sociologists studying "large technical systems" (Hughes 1989), including telephone, air traffic control and other major infrastructures.

Combining with the term "information", the concept of "infrastructure" refers to digital facilities and services usually associated with the Internet: for instance computational services, help desks, and data repositories. Implicitly, this theoretical matching marks the transition *from homogeneous*, centrally controlled, often geographically local *systems* (the historical model) *to heterogeneous*, widely distributed *networks* in which central control may be partially or wholly replaced by coordination. In general–as Bowker *et al.* (2010) pinpoint–*infrastructures are not systems*. This is all the more true taking into account the development of the Internet and its relationship with the nature and production of knowledge over space and time (Karasti, Baker and Millerand 2010; Ribes and Finholt 2009). Internet is an example of "generative infrastructure" being able to enact new innovations, extending its overall functionality and range of services provided (Monteiro *et al.* 2013). In fact, generativity is "the essential quality animating the trajectory of information technology innovation" (Zittrain 2006, 1980).

Geoffrey Bowker and Susan Leigh Star (1999), through their famous book *Sorting Things Out. Classification and its consequences*, have notably contributed to shed further light on the "information infrastructure" concept, in its historical, practical and negotiated meaning. Drawing from the analysis of Bowker and Star, we could say that an information infrastructure is first of all the connection between classifications and standards as objects for cooperation across social worlds: they are integral to any working infrastructure. Classifications and standards are imbricated in our everyday life: they saturate our environment. We classify objects, people, data and these classifications may or may not become standardized. If they do not, they are *ad hoc*, limited to an individual or a local community, and/or of limited duration. At the same time, every successful standard imposes a classification system, at the very least between good and bad ways of organizing actions or things. According to Star and Ruhleder (1996), information infrastructure is something that emerges *in situ*, in relation to organized practices, when it is connected to some particular activity. Yet, these authors have stressed two relevant aspects of infrastructure.

Firstly, it typically exists in the background, it is invisible, and frequently taken for granted. People commonly envision infrastructure as a substrate: something upon which something else "runs" or "operates". Infrastructure is put in the background where practices and activities sink (Bowker *et al.* 2010; Star and Ruhleder 1996; Star 1999). This is the reason why information infrastructures are tricky to analyse: "the easier they are to use, the harder they are to see. As well, most of the time, the bigger they are, the harder they are to see" (Bowker and Star 1999, 33).

Secondly, and in relation to the image of "sinking into the background", infrastructure could be defined as a *relational* property, not as a thing stripped of use. Understanding the relational nature of infrastructure involves unfolding the political, ethical and social choices that have been made throughout its design and development (Clarke and Star 2008). Analytically, this exercise consists of "going backstage" (Goffman 1956; Star 1999) or "infrastructural inversion" (Bowker and Star 1999). This inversion allows recognition of the depths of interdependence of technical, socio-organizational and institutional components that are involved in the infrastructure development.

2.1 Infrastructural dynamics

The above-mentioned historical model of infrastructure development – from *homogeneous systems* to *heterogeneous networks* – enables us to pinpoint that (Edwards *et al.* 2007):

- infrastructures only begin to form when locally constructed, centrally controlled *systems* are linked into *networks* and *internetworks* governed by distributed control and coordination processes;
- infrastructure formation typically starts with *technology transfer* from one location or domain to another; adapting a system to new conditions introduces technical variations as well as social, cultural, organizational, legal and financial adjustment;
- infrastructures are consolidated by means of *gateways* that permit the linking of heterogeneous systems into networks. Gateways are technologies and standards applied across multiple social worlds.

Given the heterogeneity arising from the cooperation, infrastructure's "growing" has to face some critical moments or problems that are difficult to anticipate or plan for. They may be technical, but are also frequently social or organizational in nature, particularly in the network/internetwork formation phase. Edwards *et al.* (2007) note how these are the "points" at which *systems may fail to become infrastructures* (Monteiro *et al.* 2013). Difficulties in aligning the entrenched differences between different local systems may be resolved through the creation of gateways that allow multiple divergent *systems* to *interoperate*.

However, this interoperability deals minimally both with the relational property of infrastructures and the hard work required in order to lead information artifacts and social worlds to undergird each other.

The concept of *convergence* describes better this mutual constitution based on the layering of solutions, conventions and standards that some people share through *membership* or *socialization* (Pellegrino 2008; Star, Bowker and Neumann 2003).

At an infrastructural level, artifacts and people converge through processes of multiple translations. Translation is a triangular operation that generates ordering effects (Law 1992). It involves a translator, something that is translated, and a medium in which that translation is inscribed. In other words, translation is a multifaceted interaction in which actors construct–through the translator's action–common definitions and meanings and co-opt each other in the pursuit of individual and collective purposes.

Convergence measures the extent to which the process of translation leads to agreement. Then a successful process of translation generates a shared space (it *aligns*), while an unsuccessful translation means that the actors are unable to communicate. Through a process of *disalignment* actors reconfigure themselves in separate spaces.

An important part of the process of translation is the construction of rules or conventions that limits the number of accepted translations. All these codifying regulations are forms of co-ordination and they are integral parts of the infrastructure. Well-standardized rules of co-ordination *normalise* the relations and make a series of links predictable, limits fluctuations, align actors and lead to the *irreversibilisation* of translation. This process of standardization is absolutely necessary for an information infrastructure to exist. By standards, actors from different social worlds can cooperate sharing a common "language" or *modus operandi* (Star and Griesemer 1989). At the same time, the information infrastructure must be open to change or flexible (Hanseth, Monteiro and Hatling 1996).

I maintain that when looking at parks and their infrastructure and dynamics, the balance between standardization and flexibility cannot be understood from a single viewpoint. It is for this reason that applying an ecological approach could be a promising stance.

3. Looking at the ecology of science parks

The history of the research project at issue can be introduced quoting the words that Massey *et al.* (1992, 1) use at the beginning of their book on SPs.

> "The empirical project which eventually led to this book seemed, when we began, quite straightforward (…) What happened was precisely what academics warn their students against. The project blossomed and grew. It did so in part because of a concern with conceptualisation which led us in the end to rethink what is the real content of the term "science park". This took us in wider, and unexpected, directions. The project also grew because this apparently simple empirical object of study in fact increasingly linked in, as we explored it, to some major theoretical and political issues."

Research on Italian SPs has had the same fate and it is not coincidence but exactly the consequence of the subject matter's nature. Its complexity is neither predictable nor quantifiable so that the project aimed at studying the SPs must be flexible enough to embrace the multiplicity of issues and suggestions arising from the field.

Specifically, I have conducted a qualitative research–principally based on semi-structured interviews–on Italian SPs in order to understand if the ecology of actors to which they belong can be analysed in terms of infrastructure, drawing from a STS framework. To complement this work I have focused on the Italian Association of Science and Technology Parks.

The project involved six Italian parks: the Science and Technology Park Kilometro Rosso (Bergamo), the AREA Science Park (Trieste), The VEGA-Venice Gateway for Science and Technology (Venezia), the Toscana Life Sciences (Siena), the Technology Park of Lodi Cluster (Lodi), and the Technology Park of Navacchio (Pisa). The selection of parks has first taken into account their typology. Generally, a park may be cross-thematic or thematic: in this research I have selected three cross-thematic parks and three thematic parks. I have selected the case studies also in relation to the shareholding. In fact, a park may have a public nature (as with one public case study in this research), a private nature (there is one private case study in research at issue) or mixed nature (four mixed cases have been involved). Selecting these case studies, I have paid attention to the internal presence of academic spin-offs. As a matter of fact, looking at the academic tenants is one possible way to see the SPs as organizations that work across boundaries of different social worlds: universities, businesses, research centres, laboratories (Carlile 2004).

Then, I have identified a set of actors to interview:

- Academic experts on SPs and innovation processes (4)
- The President of the Italian Association of Science and Technology Parks (APSTI) (1)
- Coordinators[1] of three thematic APSTI's Committees (3)
- Professionals of academic Industrial Liaison Offices (5)
- Directors[2] of parks (6)
- The President of the Italian Network for the Valorisation of University Research (Netval) (1)
- Business Incubator[3] managers (5)

[1] One interviewed Coordinator is also Director of one involved park: I counted this person two times. Overall, the APSTI's Committees are seven.

[2] In two cases, I interviewed the Director's spokesperson instead of the Director.

[3] At that time, one park didn't have an internal Incubator. A basic definition of "Business Incubator" or "Incubator" could be retrieved in Wikipedia: "Business incubators are programs designed to support the successful development of entrepreneurial companies through an array of business support resources and

- Founders of the University spin-offs[4] localised into the parks (10)

Each group or individual actor has contributed to give a specific vision of the Italian SPs and their dynamics.

I have completely transcribed the interviews. The software Atlas.ti has been helpful for the codification of data, following the rationale of the Grounded Theory (GT) (Glaser and Strauss 1967): the iterative process of moving back and forth between empirical data and emerging analysis makes the collected data progressively more focused and the analysis successively more theoretical (Bryant and Charmaz 2007). More precisely, I have created the codes starting from a STS background and the personal research experience, as suggested by Susan Leigh Star (2007). In fact, she invites you to distance yourself from a strictly inductive approach, legitimising a more embedded pathway to Grounded Theory.

I have completed the interviews with a further analysis of documents about SPs (retrieved principally on the websites, magazines, and reports of related associations). Desk and field researches have enabled us to understand the positioning of Italian SPs in the ecology of innovation. Given this expansive view, the analysis of data has been oriented by an *ecological approach*.

Talking about an analogous objective to represent diverse intersecting social worlds, Star and Griesemer (1989) stressed that the advantage of an ecological perspective is that it does not presuppose an epistemological primacy for any one viewpoint. Turning to research on science parks, the viewpoint of the parks' directors is not inherently better or worse than that of experts or others interviewed. Echoing Latour, Star and Griesemer said that the important questions concern the "flow" of objects and the concepts through the network of actors and social worlds. Rephrasing and adapting their words to the research at issue, it is correct to assert that the ecological viewpoint is antireductionist in that the unit of analysis is the

---

services, developed and orchestrated by incubator management and offered both in the incubator and through its network of contacts. Incubators vary in the way they deliver their services, in their organizational structure, and in the types of clients they serve" (Wikipedia 2013a).

[4] A standard definition of "University spin-off" could be retrieved in Wikipedia: "University spin-offs transform technological inventions developed from university research that are likely to remain unexploited otherwise" (Wikipedia 2013b). The number of University spin-offs' founders that I have interviewed at that time corresponded to the total number of spin-offs localized into the case studies, taking into account one "unattainable" spin-off.

infrastructure to which the park refers, not the unitary point of view of a given actor, for instance the park's management. However, it does entail understanding the management across worlds involved in the innovation processes. The analysis proposed here still contains a managerial bias, as the stories of parks' managers and their spokesperson–the President of APSTI–are much more fully fleshed out than those of the tenants. However, it is a many-to-many relational mapping, where several negotiations occur among several actors, including manager-to-manager. To be precise, the research process has devoted more attention to the viewpoints about the infrastructural dynamics related to SPs because the relevance of SPs as innovation intermediaries is much discussed.

In the following part, it will not be possible to consider the visions of all actors, so I am forced to consider most fully those that contribute to give a general sense of reference and guidance. However, all voices echo in the background.

4. A thumbnail sketch of Italian science parks

The literature review enables us to understand that there is no single definition of science park. This is not surprising given the diverse forms and characteristics that SPs exhibit. This heterogeneity is also recognisable across Italian SPs in terms of size, nature of the activities performed, and ability to attract external knowledge-intensive unit.

Despite these differences, APSTI (2012) has developed its own definition of park. First, it has introduced the "technology" in the general label, so that the "science park(s)" becomes "scientific *and technological* park(s)" (they will henceforth be abbreviated in STPs). More precisely, APSTI defines its 28[5] associated parks as

> "integrators between the companies' needs for innovative growth, with special reference to small and very small-sized businesses, and the wealth of knowledge embodied by the centres of technological and scientific excellence, universities and research centres. The role of the scientific and technological parks must be regarded as a junction between the market and the production of knowledge, as a means to promote, speed up and reduce the cost of the process, from the need for support to innovation and possible solutions, with a view to the actual improvement of dialogue and a 'cross-fertilisation' between scientific research and the production of goods and services. The majority of STPs also offer services and incubation facilities for the birth and development of new, innovative companies,

[5] This data is updated to January, 2014.

which are functionally and structurally integrated with the park, in which any innovative idea coming from the centres of scientific excellence can find its natural habitat in order to develop into a business. The park is engaged in activities, which are often key to the functions needed by the technological districts, specifically acting as "instigators" of the demand for innovation from the SMEs [small and medium enterprises] and the clustering of available technology to fulfil such demand."

This is an extension of the previous definition: in 2005, APSTI defined (Balconi and Passannanti 2006, 17, *my transl.*) the STPs as

"organizations engaged in activities for the competitiveness of local communities, promoting and managing projects for research and development, knowledge transfer and business, connecting companies, universities and research centres, public administrations and lending institutions."

According to these definitions, parks should support innovation, "with special reference to small and very small-sized businesses" or "local communities" that are typical of the Italian context. Work of each park is associated with its "connecting" role as "integrator", "as a mean to promote (…) a cross fertilisation". Parks are depicted as "instigators of the demand for innovation from the SMEs", also offering "services and incubation facilities". A further understanding of what a park is or how it is perceived, arises from the viewpoints of several actors. They help to deepen the applicability of the information infrastructure perspective.

4.1 President's vision

The interview with the President of APSTI supports the vision diffused by the promotional documents of the national association of the Italian Science and Technology Parks: in particular, I refer to both a report (APSTI 2010) and an institutional video. The report is about parks as "integrators" of the "Italian innovation chain". Similar concepts and words recur both in the interview with the President and in the institutional video, which he gave me at the end of the interview.

"Only a relational system matters and the integration is the core of this system (…) A park is an integrator between the companies' needs for innovative growth, with special reference to small and very small-sized businesses that are typical of the Italian context, and the technological know-how that is shared, even if badly, by the research centres, universities, laboratories and market." (Interview to President)

Similarly, in the institutional video the President states:

> "APSTI is becoming in a small way an increasingly important tool in this country. It's a union of the most diverse, complex, well-structured businesses but it is showing that although linking on the network, they really can lead to an increased competent growth of new generation of businesses, centres of expertise, centres of excellence, universities. In short, we can really begin to think about constructing a new system, a chain of innovation in our country. This is our mission and we are hard working with strong conviction."

In the preface of the above-mentioned report, the President highlights that the construction of a national network of parks corresponds to a project aimed at supporting Italian innovation. He states that this project "arises mainly from a group of parks, administrators and managers (…) the parks do not have to be isolated models, but they have to become a network, a *flexible system* able to take into account the economic and structural *differences* of the country" (APSTI 2010, Preface, *my transl. and emphasis*).

This statement could be interpreted as the rhetoric of focal actors ("parks, administrators and managers") and it is extremely aligned with the discourse about the information infrastructure, because it stresses the requirements of standardization and flexibility. On the one hand, the reference to "standardization" is recognizable in the President's Preface of the Report (APSTI 2010), when he talks about the "systematization" and "measure" as solutions to the fragmentary nature of the Italian parks. On the other hand, the reference to flexibility is recognizable in the above-emphasized quotation ("Parks (…) have to become a network, a *flexible system*").

Yet, the focal actors have established the creation of this flexible network of parks through which common interests from multiple and different contexts could be satisfied. By this induced process of becoming a member, each involved actor has become attached to the extant infrastructure: APSTI. Here, the membership is fostered in order to produce convergence, recognizing differences in needs and coordinating the development of solutions.

To this aim, focal actors have created seven thematic Committees, enrolling experts who deal with the various sectors and general issues and appointing them as Committees' Coordinators:

1. Food and Agriculture Committee
2. Biotechnology Committee

3. Incubator Committee
4. Internationalization Committee
5. Laboratories Committee
6. Renewable Energy and Energy Efficiency Committee
7. ICT Committee

As we read on the APSTI's website, these Committees are designed:

- to systematically promote instruments and policies in support and for the qualification of work for the development of high-tech business systems;
- to boost and support the innovative evolution of the productive sectors that have a strategic value, which together can give new competitiveness to the national economic system.

Through these "infrastructural devices", focal actors have tried to generate shared spaces of heterogeneous actors, knowledge and resources from not only parks but also local and national organizations and institutions involved in the processes of innovation.

The Food and Agriculture Committee–acting as APSTI's device–has been the first (and the only one, at the time of research) to have concretely applied in building some information system related to its specific social worlds and needs. The Agrifood Platform is the output of this effort. It is productive to know it because its development has displayed–in the end–a lack of convergence. Thus, the history of the Agrifood Committee and Platform enables the "infrastructural inversion" sensitizing the reader to traces of social and political work that can compromise the growth of infrastructures.

4.2 Agrifood Committee and its platform: the Coordinator's vision

The Agrifood Platform is an information system that supports the Agrifood parks of APSTI in their relationship with the territory and distributed actors (producers and users) interested in innovation in this field.

On first look, it seems that this web-based service has been tailored to meet the needs of national end users. If so, the Platform's operation could be described as follows: an end user looking for a service in the sector of Food and Agricolture (molecular biology, phytosanitary assessment, microbiological analysis, chemical analysis, applied research, other services such as fuel saving, zootechny, knowledge transfer, business planning, patenting, management of lab-networks) may pick from the classificatory principles on the online platform in order to have information about the response time for a given service, the description of this service, the subject of requested scientific study, and the list of STPs. This classification system has different meanings for different social worlds, it refers to knowledge and practices, but its structure is common enough to more than one world in the Agrifood field to make them recognizable, a means of translation. However, the potentiality of this technological system remains largely unexpressed.

The platform looks like a remarkable attempt to align and coordinate human and non-human actors in the field of Agrifood. The rationale is promising because it recognises the need for a highly interactive–not linear–model capable of capturing the complexity of the innovation processes, providing the conditions for the convergence of different actors. Remembering her initial effort to construct the Agrifood Committee, the Coordinator explained:

> "I said to myself: if I could convince the parks to put together their specific expertise (...) if I could convince the parks to share into a network their specific expertise following a business oriented model that is the quality of services with all certifications, pricing, so as to transform these parks into a network by which to share experiences (…) they would make a quantum leap (...) The idea, which was promising on paper, was precisely to build a network that would transform the parks as branches of one big Agrifood Park, the Agrifood APSTI, capable of gathering the need for innovation throughout the national *territory* and to answer drawing from all parks' experiences across the national *territory*. So [this would allow] multiplication of (...) the capacity to respond to the need for Italian Agrifood innovation. On paper, it was a fantastic thing."

At first glance, the emphasis on territory marks a will of change inscribed into the construction of all the APSTI Committees. However, it is especially evident in the work of the Agrifood Committee that I have studied more in depth, which focuses on its Platform as a first and unique case of information system connected to the APSTI infrastructure.

Talking about innovation, the relationship with the territory and its different actors is essential, because innovation is first of all a social process related to a vast array of individuals and organizations. Then, if they wish to cooperate for innovation production and diffusion, everyone has to play his/her/its part. This implies not only the attention to produce the conditions that could determine the innovation *per se*, but also a transformation of individualistic and self-referred trajectories into an ecology of loosely bounded relations. Yet, it cannot be a top-down process. It should be participatory.

However, as the Italian parks' history shows, several problems emerge in such convergent environments and socio-political questions arise. Conflict and multiplicity are often buried beneath layers of a common rhetoric.

From a methodological viewpoint, it is essential to be concerned with what these questions mean. Pragmatically speaking, it is fundamental to see how they work as scaffolding of the infrastructures at issue.

Stressing the "structural diversity" of Italian parks, their localism and political dependence, the Coordinator of the Agrifood Committee has specified that

> "the model proposed by the Agrifood Committee was very innovative (...) because, for the first time, the parks' system was not something that was used to communicate with the institutions but it was *aimed at communicating with the end users* of innovation promoted by the parks so it was really a reverse gear (...) The major *limitation* of this type of initiative is the structural *diversity* of the parks. As the President of APSTI has probably told you, *there is not a park like any other*; there is not a shared business model related to the parks. All come from *local* peculiarities, and, you know, from several local political mandates. For example, in many cases, they are very *dependent* on politics or they certainly depend on the vision of those who wanted the park."

This viewpoint allows discussion of the process of development of an infrastructure not subject to political and/or economic regimes, but capable of scaling up from local to social. This point will be argued drawing from a final set of viewpoints. They enable going further backstage.

4.3 Multiple-heterogeneous visions

Science parks are by definition "territorial agglomerations". In the case of Italy, they have suffered a heavy relationship with the political powers. The general instability of local and national governments and their lack of foresight about innovation have produced a fragmentary and weak development of Italian parks, hindering the construction of forms of coordination. Traditional unitary infrastructure cannot be the panacea for political and social struggles around science parks, but a sociotechnical infrastructure which is dealing with heterogeneous networks could work better. The overarching character of an infrastructure as *point of reference and displacement* targeted at the ecology of all actors interested in innovation production and diffusion could be a promising way to serve multiple and different needs both at a local and national level.

The Agrifood Committee and Platform were two promising infrastructural tools, unfortunately lacking in an appropriate network to support their enhancement. In others words, the Committee and Platform still represent an amazing project on paper, but in practice a failed opportunity "to do things together" (Becker 1982). Actually–as the Coordinator of the Agrifood Committee has explained–these organizational and technological devices have been designed first of all to help marginal parks without a planning continuity, but exactly in these cases the political partners of these parks did not approve a process of technological transfer through the Agrifood Platform.

These stories suggest that the historical boundary between science and politics is hard to cross, but also that innovation requires an infrastructure which enacts a process of membership. This process could lead to convergence, even if this does not mean the erasure of pressures and tensions. In fact, convergence is fully situated and it is a situation neither natural nor unproblematic. Different interests, points of view and meanings are at work when multiple social worlds and artifacts interact so as the degree of convergence achieved becomes more subject to contention.

As an expert on STPs remarks:

> "Many of these parks are born as an opportunity for building enhancement or speculation, depending on the term we want to use, and with the idea that (...) several material or technological infrastructures would help research centres to grow. This didn't happen... almost never (...) Certainly the location is significant, but when (...) it has a certain strength (...) There is a problem of critical mass (...) Synergies: (...) this is the point. So the first point is: what kind of goals, what kind of results in terms of entrepreneurship development, prosperity of the area does a park have? And how are the territories networked to each other? And businesses between them, universities between them? These parks must be a network not only with other subjects of the territory, but also with the other subjects of different sectors. So it does not mean cancelling the identity, it does not mean cancelling the specificity or the situated character of these projects, but [it means] increasing the level of interaction."

This account stresses again the crucial issue of convergence, emphasizing the need to preserve the specificity and situatedness of everyone consolidating the relationships, but scaling up from local to social. As indirectly referred to by the President of the Italian Network for the Valorisation of University Research, the work of parks is a matter of identity, a struggle between openness and closure, flexibility and stabilisation.

> "The value of the park is the attractiveness (…) this has an impact on the territory in economic terms, in social terms, in cultural terms. Now, "attractive" means attractive to capital, and money, and attractive to people, where both the money and people should be in some way pre-selected in the sense that it is the park's business model which should identify what type of capital and what kind of people you want to take there, playing a part in the development of the territory. But this is not a logic of closure, not at all. When it becomes attractive, it's opening the doors and the open doors, however, have to allow a global role. From this point of view it is clear that the specificity of the park becomes its identity (…) parks [were] born with the idea of the supermarket, but I think they should become refined shops that sell only a given stuff."

A Park should be able to identify what (human and nonhuman) actors attract inside, in order to be competitive outside in terms of production and diffusion of technological innovation. However, the trajectory of a park is strongly influenced by its partners, as this manager of an academic Industrial Liaison Office says:

> "Parks are heterogeneous (...) In different territories they arise from very different organizations and institutions. And I know that the institution or institutions that have created the parks have determined their structure, activities and goals for their own interests. So, if in a certain city a park has been created by the Chamber of Commerce with a few other partners, the Chamber of Commerce uses and finances this park for their own interests that maybe are not those of the university or research centres."

Then, the problematic identity of parks also damages the innovative power of the territory. This calls for an independent system of classification of parks that makes them recognizable as actively involved in the transformation of knowledge and production of innovation. According the manager of another academic Industrial Liaison Office, Italian parks have to become a hub of innovation:

> "[in the past] the science park served the local firms (...) as a local guest house, but no, the park now has to be the springboard which re-launches on the territory that it has collected."

This opinion is shared by the parks' tenants and it is well-expressed by the words of this spin-off founder that, finally, quotes the membership issue as the process by which a park could aggregate, becoming a generative infrastructure. Demarcation between the science and technology worlds seems to be erased; they should be intertwined in practice, not only theoretically:

> "[the expectation is] to enter into [...] small short nets and long nets. A short network definitely is the network of (...) this group of new start-ups (...) that are definitely characterized by (...) the idea of interconnection, and [by the idea] of creating a system which is one of the basic conditions. And the other [expectation] is to enter still more in long networks [related to] the parks' system, the system of subjects (...) with a strong bias, a strong innovative connotation. The park itself is a structure that aggregates and allows you to recognize one another [as members of] a propulsive, innovative initiative, related to science and technology."

I have to specify that their Parks have already satisfied this expectation, according to the Directors and Business Incubator managers. However, we might ask how this conviction is the result of a managerial rhetoric.

I maintain that science-and-technology-based innovation requires not only a more or less equipped site, but a working infrastructure in the service of multiple contexts and communities of practices. Such an infrastructure combines the focus on local needs (park and its tenants) with those of the variety of communities it serves (parks and interacting organizations). From this viewpoint, the work of a single park has to be articulated with the work of other actors, generating an infrastructure dealing with the innovation processes. Such infrastructure seems more appropriate not only to favour the innovation at large, but more precisely "to keep things moving along" (Bowker and Star 1999, 313), that is to keep alive the idea of innovation as a process of continuous negotiation and transformation of actors and resources at stake, stretched across space and time.

5. Final remarks

In this chapter, I analysed the infrastructural dynamics related to Italian SPs, taking into account the interplay between their installed infrastructure–the Italian Association of Science and Technology Parks (APSTI)–and entangled devices such as the Agrifood Committee and Platform. Then, I did not discuss the relationships between parks' tenants–as literature has usually invited–but I have instead focused the attention on processes that lead backstage and enable the unfolding of structural, political and socio-organizational questions. This "inversion" has stressed the problematic intersection between technological innovation and politics that I want to further discuss given its relationship to convergence issues and infrastructure's growth.

The boundary between technoscience and the political arena is quite controversial because of both the interstitial nature of innovation and the subjection of Italian parks to political purposes. If parks are territorial instruments of politics, their work is greatly limited by interests that typically are scarcely prone to being reconciled. This condition confirms that convergence is a negotiated, not obvious achievement, which implies a dialectical process of co-creation.

Yet, the political precariousness of local and national governments and their rhetoric commitments to science and technological innovation could be considered the main causes of the fragmentary nature and weakness of Italian parks[6].

These premises alert that work of an extant infrastructure, such as APSTI, or the potentialities of a developing infrastructure, such as the Agrifood Committee, risk vanishing. In particular, the mentioned Agrifood Platform could be appreciated as an attempt to construct a classification system of value-added services tailored to end users' needs, benefiting the parks–mainly those in a marginal position–and supporting the processes of production and innovation diffusion. Unfortunately, as the voices from the field have noted, political interference has stopped the development of this Platform and the consolidation of alliances strongly committed to innovation.

In order to rectify the fragmentary nature of the Italian parks, a promising solution is the development of a working infrastructure that serves intersecting social worlds–parks and others–responding to the separate as well as the combined agendas of heterogeneous actors. Such infrastructure could enable the management of multiplicity, maintaining the hybrid regime of innovation. However, an infrastructure calls for solutions simultaneously concrete and abstract, specific and general, conventionalized and customized. Drawing from the scientific knowledge, technologies, competencies and practical experiences of Italian parks, such a tool allows the imagining of the innovative power of a more synergistic model of interactions.

The construction of an infrastructure that values the work of Parks avoiding a redundant structure calls for a cultural change that frees the innovation from a hierarchical and politically-driven vision. For this

[6] It would be interesting to discuss how much this criticality is shared by other countries. However, I want to limit here my comment to Italy, starting from my research fieldwork.

reason it is fundamental that institutions at different levels adopt an ecological perspective, taking into account all viewpoints.

Clearly, as Bowker and Star (1999) have already stressed, infrastructure is not a perfect construction. It has not become a "chimera of a totally unified and universally applicable information system" (Bowker and Star 1999, 313). Rather, it represents a way of scaling up from the local to the social, from the traditional linear model of innovation to an interactive network. The design and the study of parks as infrastructures call for further analysis. This contribution is a first step in this direction.

References

Allesch, Jürgen. 1985. "Innovation centers and science parks in the Federal Republic of Germany: Current situation and ingredients for success." In *Science parks and innovation centers: Their economic and social impact*, edited by John M. Gibb, 58-68, Amsterdan: Elsevier.

Amirahmadi, Hooshang and Saff Grant. 1993. "Science Parks: A Critical Assessment." In *Journal of Planning Literature*, 8: 107-23.

Balconi, Margherita and Alessandro Passannanti. 2006. *I Parchi Scientifici e Tecnologici nel nord Italia*, Milano: Franco Angeli.

Becker, Howard S. 1982. *Art Worlds*. Berkeley, CA: University of California Press.

Bowker, Geoffrey C., Karen Baker, Florence Millerand and David Ribes. 2010. "Toward Information Infrastructure Studies: Ways of Knowing in a Networked Environment." In *International Handbook of Internet Research*, edited by Jeremy Hinsinger, Lisbeth Klastrup and Matthew Allen, 97-117, London, New York: Springer.

Bowker, Geoffrey C. and Susan L. Star. 1999. *Sorting Things Out. Classification and Its Consequences*. Cambridge, MA: The MIT Press.

Bryant, Antony and Kathy Charmaz (edited by). 2007. *The Sage Handbook of Grounded Theory*. London: Sage.

Butera, Federico. 1995. *Bachi, crisalidi e farfalle. L'evoluzione dei parchi scientifici e tecnologici verso reti organizzative autoregolate*. Milano: Franco Angeli.

Carlile, Paul R. 2004. "Transferring, Translating, and Transforming: An Integrative Framework for Managing Knowledge across Boundaries." In *Organization Science*, 15: 555-568.

Clarke, Adele E. and Susan L. Star. 2008. "The Social World Framework: A Theory/Methods Package". In *The Handbook of Science and Technology Studies*, edited by Edward J. Hackett, Olga Amsterdamska,

Michael Lynch, and Judy Wajcman, 113-137, Cambridge, MA: The MIT Press.

Colombo, Massimo and Marco Delmastro. 2002. "How effective are technology incubators? Evidence from Italy." In *Research Policy*, 31: 1103-1122.

Edwards, Paul N., Steven J. Jackson, Geoffrey C. Bowker and Cory P. Knobel. 2007. *Understanding Infrastructure: Dynamics, tensions, and design*. NSF Report of a Workshop on "History & Theory of Infrastructure: Lessons for New Scientific Cyberinfrastructures." January. http://deepblue.lib.umich.edu/handle/2027.42/49353. Accessed July 3, 2013.

Felsenstein, Daniel. 1994. "University-related science parks - "seedbeds" or "enclaves" of innovation?" In *Technovation*, 14: 93-110.

Glaser, Barney G. and Anselm L. Strauss. 1967. *The Discovery of Grounded Theory: Strategy for Qualitative Research*. Chicago: Aldine.

Goffman, Erving. 1956. *The Presentation of Self in Everyday Life*. New York: Doubleday.

Hanseth, Ole, Eric Monteiro and Morten Hatling. 1996. "Developing Information Infrastructure: The Tension between Standardization and Flexibility." In *Science, Technology, & Human Values*, 21: 407-426.

Hughes, Thomas P. 1983. *Networks of Power: Electrification in Western Society, 1880-1930*. Baltimore: Johns Hopkins University Press.

—. 1989. "Evolution of large technological systems." In *The Social Construction of Technological Systems: New Directions in the Sociology and History of Technology*, edited by Wiebe E. Bijker, Thomas P. Hughes and Trevor J. Pinch, 51-82, Cambridge, MA, London: The MIT Press.

Kihlgren, Alessandro. 2003. "Promotion of innovation activity in Russia through the creation of science parks: the case of St. Petersburg." In *Technovation*, 23: 65-76.

Kimble, Chris, Corinne Grenier and Karine Goglio-Primard. 2010. "Innovation and knowledge sharing across professional boundaries: Political interplay between boundary objects and brokers." In *International Journal of Information Management*, 20: 437-444.

Irwin, Alan. 2008. "STS Perspectives on Scientific Governance." In *The Handbook of Science and Technology Studies*, edited by Edward J. Hackett, Olga Amsterdamska, Michael Lynch and Judy Wajcman, 583-607, Cambridge, MA: The MIT Press.

Karasti, Helena, Karen S. Baker and Florence Millerand. 2010. "Infrastructure Time: Long-term Matters in Collaborative Development." In *Computer Supported Cooperative Work*, 19: 377-415.

Kline, Stephen J. 1985. "Innovation is not a linear model." In *Research Management*, 28: 36-45.

Law, John. 1992. "Notes on the Theory of the Actor-Network: Ordering, Strategy and Heterogeneity." In *Systems Practice*, 5: 379-393.

Löfsten, Hans and Peter Lindelöf. 2002. "Science Parks and the growth of new technology-based firms. Academic-industry links, innovation and markets." In *Research Policy*, 31: 859-76.

MacDonald, Stuart. 1987. "British Science Parks: Reflections on the Politics of High Technology." In *R&D Management*, 17: 25-37.

Massey, Doreen, Paul Quintas and David Wield. 1992. *High Tech Fantasies. Science Parks in Society, Science and Space*. London, New York: Routledge.

Miller, Clark. 2001. "Hybrid Management: Boundary Organizations, Science Policy, and Environmental Governance in the Climate Regime." In *Science, Technology, & Human Values*, 26: 478-500.

Mongili, Alessandro. 1998. "Modelli di innovazione e politiche dell'innovazione. Alcune considerazioni sul caso del parco scientifico e tecnologico della Sardegna." In *Quaderni del Dipartimento di Ricerche Economiche e Sociali-Sezione Sociologia*.

Monteiro, Eric, Neil Pollock, Ole Hanseth and Robin Williams. 2013. "From Artifacts to Infrastructures." In *Computer Supported Cooperative Work*, 22: 575-607.

Phan, Phillip H., Donald S. Siegel and Mike Wright. 2005. "Science parks and incubators: observations, synthesis and future research." In *Journal of Business Venturing*, 20: 165-182.

Pellegrino, Giuseppina. 2008. "Convergence and Saturation. Ecologies of Artefacts in Mobile and Ubiquitous Interaction." In *Integration and Ubiquity. Towards a philosophy of telecommunications convergence*, edited by Kristóf Nyíri, 75-82, Vienna: Passagen Verlag.

Ratinho, Tiago and Elsa Henriques. 2010. "The role of science parks and business incubators in converging countries: Evidence from Portugal." In *Technovation*, 30: 278-290.

Ribes, David and Thomas A. Finholt. 2009. "The Long Now of Technology Infrastructure: Articulating Tensions in Development." In *Journal of the Association for Information Systems*, 10: 375-398.

Star, Susan L. 1999. "The Ethnography of Infrastructure." In *American Behavioral Scientist*, 43: 377-391.

—. 2007. "Living Grounded Theory: Cognitive and Emotional Forms of Pragmatism." In *The Sage Handbook of Grounded Theory*, edited by Antony Bryant and Kathy Charmaz, 75-93, London: Sage.

Star, Susan L., Geoffrey C. Bowker and Laura J. Neumann. 2003. "Transparency beyond the Individual Level of Scale: Convergence between Information Artifacts and Communities of Practice." In *Digital Library Use. Social Practice in Design and Evaluation*, edited by Ann Peterson Bishop, Nancy A. Van House and Barbara P. Buttenfield, 241-269, Cambridge, MA: The MIT Press.

Star, Susan L. and James R. Griesemer. 1989. "Institutional Ecology, "Translation" and Boundary Objects: Amateurs and Professionals in Berkeley's Museum of Vertebrate Zoology." In *Social Studies of Science*, 19: 387-420.

Star, Susan L. and Karen Ruhleder. 1996. "Steps Toward an Ecology of Infrastructure: De

sign and Access for Large Information Spaces." In *Information Systems Research*, 7: 111-134.

Vaidyanathan, Geetha. 2008. "Technology parks in a developing country: the case of India." In *Journal of Technology Transfer*, 33: 285-299.

Watkins-Mathys, Lorraine and John M. Foster. 2006. "Entrepreneurship: the missing ingredient in China's STIPs?" In *Entrepreneurship & Regional Development*, 18: 249-274.

Wikipedia. "Business Incubator." Accessed August 3 (2013a). http://en.wikipedia.org/wiki/Business_incubator

Wikipedia. "University spin-off." Accessed August 3 (2013b). http://en.wikipedia.org/wiki/Business_incubator

Zittrain, Jonathan L. 2006. "The generative Internet." In *Harvard Law Review*, 119: 1974-2040.

Acknowledgements

For his comments and suggestions I express my gratitude to Liam Bannon.

CHAPTER FOURTEEN

CONVERGENCE: HISTORY, MEANINGS AND IMPLICATIONS FOR THE DESIGN OF COMMUNITY INTERACTIONS

GIUSEPPE LUGANO

Introduction

The open standards of the Internet represent the glue of today's global information infrastructure, which consists of digitally converged information and communication technologies (ICTs) daily used by billions of people via different devices and in all kinds of human activities and processes. Convergence and its opposite concept, divergence, are powerful conceptual tools to analyze, evaluate, compare and improve the design and use of information infrastructures towards a specific goal. Indeed, the integration of different technologies that were originally designed for a specific use (e.g. television, radio, telephone) has both economic and societal relevance: on the one hand, it offers new sources of potential revenues, and on the other it plays an important role in the perception, understanding and experience of social interaction contexts.

The implications and consequences of these types of interactions are often unpredictable. In this respect, convergence represents a conceptual framework through which to identify guiding principles to enable citizens, designers, policy-makers and entrepreneurs to prepare a common vision and a strategy to collaboratively develop tomorrow's information society.

This kind of approach may seem too naïve and utopian; however, it is not too far from well-known and ambitious plans, such as Europe's Digital Agenda for 2020. This plan aims at achieving sustainable and inclusive growth for European citizens and businesses by means of "smart" use of integrated digital technologies (Toporkoff 2012).

Although the completion of Europe's Digital Agenda relies much on citizens' participation, it is essentially a top-down political initiative; an alternative and potentially disruptive approach is connected to grass-roots bottom-up initiatives launched locally and rapidly spreading to broader areas thanks to online and mobile social networks (Lugano 2008). Much has been written, for instance, on the role of information and communication technology in mobilizing the crowds during the Arab Spring in 2011 (Stepanova 2011; Allagui and Kuebler 2011). It is worth underlining how, in this case, converged digital networks facilitated the citizens' physical convergence and coordination of collective action in Tahrir square. These and other forms of interaction are supported by Internet-based communication, and in particular by Mobile Social Software (MoSoSo), a class of mobile applications whose scope is to support social interaction among interconnected mobile users (Lugano 2007). Examples of popular instances of MoSoSo, often linked to location-based services (LBS) or geosocial networking, are smartphone applications of popular social networking sites such as Facebook or dedicated mobile social "apps" such as WhatsApp or WeChat.

Being always carried by their owners, mobile devices–smartphones and tablets in particular-represent the "last mile" of converged digital networks and are therefore the focus of the analysis in this chapter. Almost all applications installed on mobile devices have "social" features, allowing users for instance not only to read a book or look at a photo or watch video, but also to share them with others and to note one's comments or appreciation for them (e.g. via the famous "like" button in Facebook). When performed collectively, these simple behaviors may have enormous impact, for instance influencing the career of local artists like the Korean PSY, who earned rapid global success with his music video "Gangnam Style", one of the most watched of all time on YouTube.

Cases like the Arab Spring or the success of PSY may give the impression that citizens are truly empowered by access to Internet-based platforms for participation. However, it is worth noting that the services "empowering" them are run by powerful companies listed on the stock exchange such as Google and Facebook. These companies make their profits in large part thanks to new forms of contextual advertisement, enabled by the real-time analysis of huge quantities of personal data, which are stored on the company servers. The collection and use of "big data" (Manyika *et al.* 2011) represents the last frontier on the debate on converged digital technologies and markets, as well as on human rights. This trend has been emphasized on a global scale in particular following the revelations of Edward Snowden, former contractor of the US National Security Agency (NSA). This case has reinforced the already rather negative perception of users for "big data" (Craig and Ludloff 2011), seen as part of a global surveillance system in the hands of governments and multinationals.

As will be illustrated in this chapter, the controversial concept of "big data" can be associated with greater user benefit. Indeed, convergence and divergence may inform the design of future and emerging technologies by allowing users not only to "connect" to or "disconnect" from a network or service, but also to achieve a more fine-grained control over incoming and outgoing information flows by making use of "big data" (Manyika *et al.* 2011) stored on their personal clouds or devices rather than on company services. In other words, a design approach inspired by convergence and divergence concepts could enable, among others, an enhanced personal privacy management and limitation of cognitive overload. It goes without saying that even an optimal design approach would not be effective without converging policies agreed by all stakeholders. The policy aspect, which is a relevant one, is however beyond the scope of this chapter.

The chapter presents the history, meanings and implications of convergence, a phenomenon that is far from being only technological. Notably, convergence is discussed from the perspective of community and in relation to the emergence of digital communities, the contemporary form of community based on the opportunity to connect, communicate, share and use tangible and intangible resources via digital networks. As technologies become more social and communities more technological, novel approaches to the design of technologies for digital community interaction are necessary. The chapter introduces a design approach based on the convergence and divergence concepts, which would enable us to address the challenges of social convergence by allowing interconnected users to, among others, amplify and attenuate social signals based on contextual interaction needs.

1. Convergence, a multifaceted term

The analysis that follows presents an historical account of how the concepts and understanding of convergence, Internet and community have evolved and influenced each other in the last few decades. This analysis does not claim to be comprehensive, but it is included because it is functional to the objective of the chapter; specifically, to illustrate how an understanding of convergence/divergence can affect, in one way or in another, the design and practice of both contemporary communities and the technological infrastructure of Internet-based converged digital networks.

The term convergence was introduced for the first time in the seventeenth century by an English scientist, and later adopted by Darwin (1968) in describing the evolution of the species (Gordon 2003). Hence, for historical reasons convergence possesses a strong technological connotation since its etymology comes from the world of science and mathematics. It is no surprise that still today convergence is often regarded as a synonym of technological convergence. In the area of information and communication technologies, technological convergence is associated with the coming together of telecommunication systems, computer networks and media industry (Fig. 14-1) (Yoffie 1997; Covell 1999; Mueller 1999).

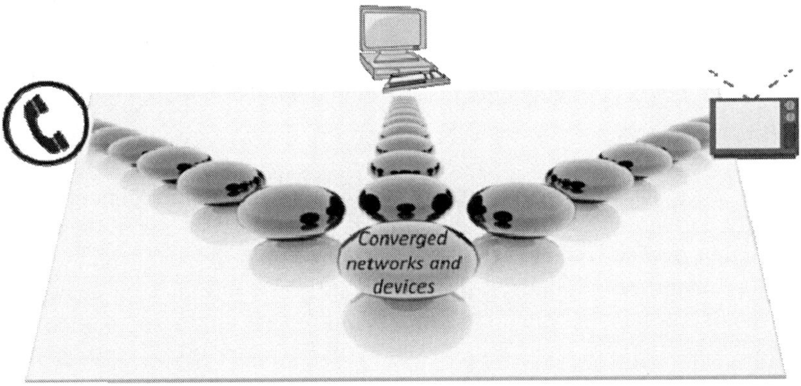

Fig. 14-1: technological view of Convergence.

The history of the Internet shows how heterogeneous computer networks were interconnected thanks to a common communication protocol, the Transmission Control Protocol (TCP). Interestingly, already in the Sixties a computer scientist, J.C.R. Licklider (1968) conceived the ARPANET, the Internet's ancestor, as a powerful communication tool. Indeed, early computer networks in American universities were used both to exchange information and to communicate at a distance. Although these practices were in their infancy, they show that technological convergence already entailed many other forms of convergence, in particular affecting the social and cultural dimensions.

In the Eighties, De Sola Pool (1983, 38) conducted groundbreaking work in this area by proposing the vision of a democratic "integrated digital network serving all purposes". His message was partly embedded in the promotional slogans on the "electronic superhighway" (Gates *et al.* 1995), the information infrastructure that would lead to a new era of progress and prosperity enabled by the digitalization process (Elmer-Dewitt and Jackson 1993; Negroponte 1996). The potential of technological convergence was described, not without emphasis, with global societal implications and impacts on politics and the economy.

There was an idea of society rapidly entering into a new promising era thank to digital networks. This vision is clearly illustrated in a *Time* article by journalists Elmer-Dewitt and Jackson (1993), who described digitalization as the main factor driving the implementation of the "electronic superhighway": indeed, the digitalization process would have facilitated access to textual and audiovisual content, allowing service providers to offer a variety of services supporting work, learning, entertainment and socialization. The same authors also pointed out that digital convergence was an industry-driven technological process:

> "Welcome to the information highway. [...] Already the major cable operators and telephone companies are competing – and collaborating – to bring this communicopia to your neighborhood [...] now the only questions are whether the public wants it and how much it is willing to pay."

In the mid-Nineties social scientists significantly contributed to develop the discourse around convergence. The area was mature and attracted widespread interest; as a concrete demonstration, the inaugural issue of *Convergence*, a journal addressing the interdisciplinary issues raised by the advent of new media technologies, was launched. In the opening issue, Silverstone published the highly thought-provocative article titled "Convergence is a dangerous word"; specifically, it was observed that throughout the Nineties a technologically deterministic view of convergence was largely dominant and the discourse on its impact on professional and everyday life remained mostly at a speculative level. Based on this, Silverstone (1995, 11) called for a more critical and broad analysis of convergence, essential for being able to truly profit from its potential and understanding its challenges:

> "If we are to make any headway in understanding not just the process of convergence but also the barriers to convergence, we need to address the social dimensions of technology in their multiple and complex arrays and we need to do so in ways that address the process and dynamics of innovation, both in development of hardware and software and in their use."

After the turn of the Millennium, digitally converged technologies became mainstream and started to be investigated from a broader perspective;

the approach called by Silverstone started to be increasingly adopted in a series of interesting studies on the topic. This contributed to associate convergence not only with the technological and economic dimensions, but also to the cognitive, social and cultural ones. Fortunati (2007, 110) provided a convincing description of convergence as a complex and multifaceted phenomenon associated with parallel trends of divergence. In particular, convergence was described as

> "a strategy to rationalize the physical and technological space or, to put it more adequately, to overcome the distribution of technologies in space, and to rationalize the time and convenience of their use."

She also underlined the implications that technological convergence has on the time and space dimensions by observing that

> "while being a spatial advantage, [...] the rationalization of space might entail a loss of effectiveness in functionality, definition of languages, modalities and services and an increase, at the same time, of the functional complexity of technology, which is scarcely tolerated by users. [...] Convergence may imply also a model of control that deserves to be discussed."

These observations can be also interpreted as a call for a new approach in designing technologies, which embeds the notion of convergence/divergence to address common interaction challenges.

Taking a media-centric approach, Jenkins (2006) introduced a conceptual framework for convergence described as a process made up of several dimensions: technological, economic, global, cultural and social or organic convergence. Technological convergence refers to the digitalization of media content (i.e. atoms to bits) that is tightly linked to economic convergence, the horizontal integration of the media entertainment industry. In other words, the discourse of the relationships between these two forms of convergence, particularly active in the Nineties, concentrates on the convergence of markets as a result of the convergence of technologies. A third type of convergence is global convergence, a direct consequence of the globalization process; this refers to the cultural hybridity resulting from the global circulation of digital media content. This process has profound consequences also on social and cognitive levels; indeed, Jenkins (2006) describes cultural convergence and social convergence as two other notable facets of convergence. Both have been particularly emphasized by the massive diffusion of online and mobile social networks within the overall landscape of Web 2.0 applications

(O'Reilly 2007). Cultural convergence describes the emergence and multiplication of new forms of creativity at the intersection between technology, media and user communities. At an individual level, these new forms of participation trigger cognitive processes related to the multi-tasking strategies for navigating and processing the rich and rapidly changing information environment. This latter form is what Jenkins (2006) defines as social convergence.

Although not part of Jenkins' analysis, the concept of social convergence can be also considered in relation to two different processes, namely social coordination and virtual co-presence. The former process has been widely described in relation to the role mobile devices have in facilitating rendezvous in a physical location, or micro-coordination (Ling and Yttri 2002; Farnham and Keyani 2006). In brief, in this context social convergence refers to the physical convergence to a commonly agreed location. Unlike micro-coordination, virtual co-presence is linked to the virtual convergence to a specific online space in a given period of time. By means of social awareness applications (Oulasvirta *et al.* 2005), users are made aware of who is online and can typically contact other users through synchronous and/or asynchronous communication channels (e.g. instant messaging, comments to posts).

Micro-coordination has been therefore traditionally associated with mobile communication, while virtual co-presence to online communication. Digitally converged technologies have brought together the offline, online and mobile dimensions of interaction, thus making possible the two phenomena of micro-coordination and virtual co-presence to occur concurrently. This is common in the use of smartphone-based mobile social software (MoSoSo) applications (Lugano 2007), specific social computing platforms often referred in the media as social "apps". For instance, applications such as Foursquare can both support "serendipitous" micro-coordination and allow users to leave comments and interact around the specific locations that have been "checked-in" to (Lindqvist *et al.* 2011).

At a cognitive level, boyd (2008) underlines that the new modalities of efficient and immediate interaction also make users increasingly exposed to the phenomena of exposure and invasion, particularly in social networking applications like Facebook. These issues are even more evident in the rapidly changing and cognitively challenging mobile context (Oulasvirta *et al.* 2005). For this reason, the design of technologies for social interaction should allow users to better cope with these challenges by minimizing risks, while enjoying the benefits of such interactions.

While technological, economic and global convergence seem to be mostly driven by political and economic actors in a top-down manner, cultural and social convergence are instead driven by single individuals and communities of users interacting in rich-media environments in which the online, offline and mobile interaction contexts are overlapping. For this reason, changes in the design of technologies for digital community interaction play a role in shaping future outlooks and consequences of cultural and social convergence.

Research on trends associated with cultural and particularly social convergence has received some scholarly interest (Morris Paris and Rubin 2013), but it is still in its infancy and therefore not coherently organized. In this respect, a recent contribution has been brought by Lugano (2010a), who has proposed the framework of "Digital Community Design" for addressing social convergence issues from the perspective of MoSoSo interaction within digital communities.

2. On the emergence of digital communities

Community is a classic sociological concept that has been debated for more than a century, starting with Tönnies' (1967) well-known description of *Gemeinschaft* and *Gesellschaft*. Community, like society, is by nature an evolving concept affected in particular by social, cultural, technological and economic trends.

In the following section, I will describe how the most recent technological trend of converged digital networks has affected the nature of contemporary communities. Methodologically, Actor-network theory (Callon 1986; Law 1992; Latour 1996) has been employed to analyze virtual and online communities (Lepa and Tatnall 2006; Beekhuyzen, von Hellens and Nielsen 2012). Also, the perspective of socio-technical systems (Trist 1981) would allow analysis of the intersections between converged digital networks and community. Instead, this chapter adopts the alternative framework of digital community (Lugano 2010a), which is considered more suitable for informing technology design.

The framework of digital community builds on the concept of hybrid social space, a type of socio-technical infrastructure integrating the offline, online and mobile dimensions of interaction (de Sousa e Silva 2006; Rheingold *et al.* 2007; Crabtree and Rodden 2008; Bilandzic *et al.* 2009). The hybrid social space is characterised by an always-on, ubiquitous and integrated digital layer allowing individual users to form digital communities (Fig. 14-2), to share content within them and to communicate in a variety of manners with one or more members of the digital community (named, according to the application, as e.g. "friends" or "followers"). In this space, the disparate social contexts are collapsed into one (boyd 2008), making the design of interaction inherently challenging.

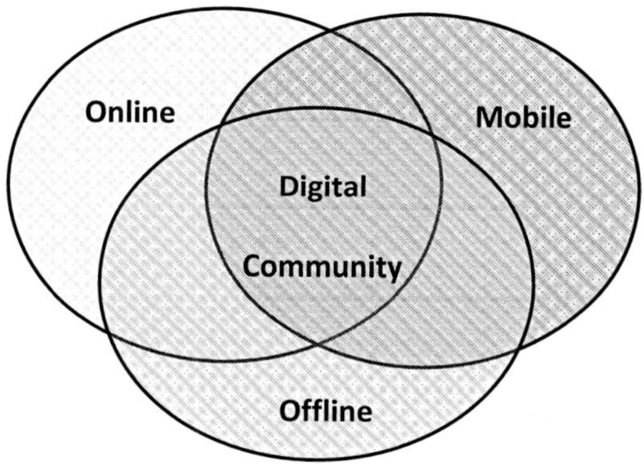

Fig. 14-2: digital Communities emerging from the Integration of the Offline, Online and Mobile Spaces of Social Interaction

At a sociological level, this view of community, which is fluid, ephemeral, typically egocentered and rapidly evolving extends the concept of personal community (Wellman *et al.* 1988) by assigning new meanings to the nature of the social tie, to the form and duration of a social relationship, and to the scale of community interaction. While personal communities were introduced to overcome the limits of the search for local solidarity, digital communities update previous community models by acknowledging users' integrated approach to social communication.

The perception that contemporary communities show very peculiar features is illustrated by the multiple labels associated with the description of community in recent scholarly research, such as hybrid (Schubert and Hampe 2005), cross-media (Trappeniers *et al.* 2008) and content-based (Turpeinen and Kuikkaniemi 2007) communities. Although there is not much agreement on the terms, the emerging form of community seems to be grounded on the convergence between the offline, online and mobile dimensions, which blend into each other.

Although not widely used, the concept of digital community introduced by Sillence and Baber (2004) is particularly suitable for my analysis, as it does not confine interactions to a single technological realm, but rather embeds the idea of technology bundles (Ling 2004). The study of community, as well as the design of technology for community, needs to acknowledge the current trends connected to the combined usage of offline, online and mobile communications. When multiple technologies are contemporarily used, the study of the social structure and dynamics of communities requires collecting and aggregating data from multiple sources. This is at the same time a challenge and an opportunity, as the Internet and mobile devices offer an enhanced capability to collect, analyze and interpret “big data”, large amounts of data on complex social sytems. This knowledge can be used both by users in goal-directed contextual social interaction (Eagle and Pentland 2005; Raento *et al.* 2005) and by researchers to develop a better understanding of human and social behavior (Barabasi 2002; Gilbert and Karahalios 2009; Raento *et al.* 2009).

From a community perspective, such advances allow the acquiring of a comprehensive understanding of contemporary communities. In particular, as the Internet and mobile technologies are typically used as personal communication devices, they allow the enhancement of how converged digital networks affect the structure, perception and experience of personal communities (Wellman 1988). Digital convergence assigns new meanings to the already multifaceted nature of personal communities, for instance to the concept of interpersonal social ties. In addition to weak or strong ties (Granovetter 1973), concepts that were at the core of the model of personal community, digital communities add also latent ties (Haythornthwaite 2002). These are not associated with established social relationships, but can be turned into weak or strong ties through the support of social computing platforms, such as MoSoSo.

Contrarily to personal communites, in digital communities not all social ties have a long-lasting nature; instead, they are often ephemeral as they may be established in *ad hoc* manner. *Ad hoc* social ties are becoming increasingly significant in contemporary information societies, characterizing entertainment, politics and the economy. In fact, smartmobs (Rheingold 2002) are a form of *ad hoc* community enabled by mobile technologies for the purpose of collective action. At the basis of smartmobs there is not a network of established social relationships, but rather a common value, ideology or objective. Therefore, in digital communities the notion of social tie is not synonymous with social relationship in the traditional sense; instead, it is closer to a logical association, either *homophilus* or *heterophilus* (Rogers and Bhowmik 1970; Lin 2001), which connects two or more persons. When the digital community is regarded from a mobile perspective, it corresponds to the notion of mobile social network (Lugano 2008).

The discussion on contemporary communities shows that digital convergence increases the overlapping between online and offline dimensions. This key trend was already recognized by Rheingold (2002), who observed that the offline and online interactions had initially different purposes, respectively related to the maintenance of local acquaintances and the creation of new social ties based on affinity. As the process of the digital convergence developed, the ubiquitous access to the social world made the two dimensions blend into each other, unifying and enriching the concept of community. For Agre, such enrichement implies an always-on condition, described as a world of freedom containing "anonymous global forces that ceaselessly rearrange all relationships to their liking" (Agre 2001, 13). From a social perspective, the always-on condition might be analyzed as a phenomenon of social convergence.

A second important and related trend concerns the overlap between the online and mobile dimensions. The first sign was recognized when laptop computers became more popular than desktop computers for their portability; a similar destiny occured with traditional mobile phones, which were overtaken in sales by smartphones. The emergence of new portable devices, such as tablets, and the acquisition of online connectivity by other devices (e.g e-book readers) and physical environments (e.g. WiFi hot-spots) contribute to blur the boundary between the online and mobile dimensions even more.

As a consequence, personal communication devices can typically access both computer and mobile networks; as such, the online world has become an extension of the mobile and vice versa. This clearly implies that digital communities are both online and mobile. Therefore, traditional community dichotomies like "online" *vs* "mobile", or "real" *vs* virtual" sound quite obsolete; rather than the nature of the technology used, it is the interaction context and the level of mediation played by such technology that define the nature of the interaction.

The third trend concerns the way mobile communication and use of contextual information on mobile devices affect face-to-face interactions and the experience of place. It was noted that the mobile and offline dimensions were already complementary to each other before digital convergence (Katz and Aakhus 2002; Ling 2004). The new developments, however, strenghten this complementary role by enriching these experiences, for instance through location-based services and social proximity interactions (Eagle and Pentland 2005).

Taken together, these trends suggest that digital convergence has lead to the emergence of digital communities. The online, offline and mobile dimensions of community do not compete and should not be regarded as distinct entities; rather, they just represent the complementary dimensions of everyday life. The integration of the online, offline and mobile dimensions opens a novel opportunity to overcome the traditional dichotomic approach to community that has remained, in different forms, since Tönnies' description of *Gemeinschaft* and *Gesellschaft* (Tönnies, 1967). Similarly, this approach applied to design allows the introduction of a complementary view of the concepts of convergence and divergence, which may have a particular importance in improving technologies for digital community interaction, notably those used in mobile contexts.

3. Function of convergence and divergence

Digital communities have recently emerged as a sociological entity characterising the era of constant connectivity to technological and social networks. On the basis of this understanding, the next step is to gain the knowledge necessary to design digital communities that allow users to minimise the risks of such interactions, while enjoying their benefits.

Considering the on-going discussion, on the one hand at a policy and business level there are demands for a stricter regulation of the Internet (Mueller 2010; Wagner 2011; Klein 2012), but also for empowering citizens through ICT solutions (Toporkoff 2012). At user level, there is desire of a more fine-grained control over personal data (boyd 2008) and also for "moments of disconnection" (Agre 2001; Birnholtz *et al.* 2012). In other words, there are different positions to be reconciled and still much research needed in order to find general design principles satifying all these requirements.

To proceed towards a solution, I propose an approach based on the idea that convergence and divergence have a complementary function determined on the basis of the interaction context, which is in turn associated with individual or community needs and goals. In other words, "smart" technologies for digital community interaction should embed both ideas of convergence and divergence and contain mechanisms to apply one or the other concept as needed by the interaction context (Fig. 14-3).

I will illustrate this idea by using the case of MoSoSo interaction. The challenges of exposure and invasion described by boyd (2008) are associated with outgoing and incoming information flows. They respectively deal with the personal privacy management and social information overload. It is not obvious to distinguish relevant information from noise, as the same information could be classified in either manners depending on the context and moment. This is particularly challenging in a mobile context and when such streams are continuous, as in the case of Twitter or Facebook feeds. Ideally, a "social algorithm" embedded in the application should be able to smartly identify what is relevant in a specific context; what is then left to the user is to specify how much support to the human processing and decision-making is needed. By default, the level of mediation could be somewhere *in between* the two extremes of technology-driven and human-driven interaction.

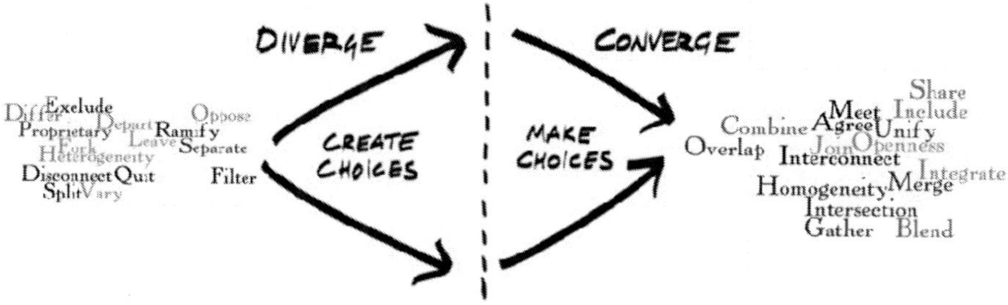

Fig. 14-3: integration of Convergence and Divergence in the Design of technologies for Social Interaction.

Technology-driven interaction is based on the user's high confidence on the reliability of the technological support, while in the human-driven approach the user should have full control because automated or semi-automated procedures would just add interferences to the interaction, making it more complex rather than simpler. Although the design should allow both solutions, it should also guide the user towards an optimal trade-off between the two opposites of technology-driven and human-driven interaction.

Implementing a balanced solution between the need for automation and the human desire to be in control is one of the major challenges of ICT design (Sheridan and Parasuraman 2006). On the one hand, there is general agreement that ICT can support users in overcoming their limited cognitive capacity and biological limits. On the other, there is an increasing awareness of the challenges of automation, especially in highly dynamic mobile contexts (Vihavainen *et al.* 2009).

Social algorithms could play an important role in making MoSoSo interaction more effective. For instance, part of the user profile could be updated on the basis of acquisition of contextual information, such as a user's current location. In addition to that, MoSoSo could also perform higher-level operations on behavioral data by inferring and displaying the structural elements of the personal community (e.g. the mobile social network). Furthermore, semi-automated support of a user's decision in sharing digital content could be implemented on the basis of behavioral and contextual data (Lugano and Saariluoma 2007). Finally, in some cases decisions could be taken by the application on behalf of the user through intelligent agents.

Let us consider a typology of social algorithms, namely those allowing MoSoSo to behave in some cases as an amplifier and in others as an attenuator of social information. These two terms, typically used in electronics, refer to tools that are used to transform the signals of a technical system by increasing or decreasing them. Similar to the concept of scalability (Lugano 2010b), these terms have been generally employed in the context of technical systems, but their meaning is relevant for social systems as well.

Until now, far more attention has been paid to the role of ICT as amplifier of social network information. For instance, adding friends in a social networking site has typically been easier than removing them. Similarly, tools for removing information from the Internet have not had the same pace of development of those supporting online publishing. Indeed, the Internet has been created to constantly grow through continuous transformations. The amount of information is growing because more digital content is added than removed, following the myth that more is better and underestimating the problem that more means more complex to manage. The increasing complexity seems to correspond to an increasing incapacity of users to efficiently deal with it. Ultimately, information might become just interference rather than knowledge, negatively affecting its usefulness.

It is therefore time to systematically address attenuation of social network signals in digital community design. In context-aware computing, this task has been discussed as the challenge of selecting the most relevant resource for the current context. MoSoSo provides a new perspective to this problem by emphasizing the structure and dynamics of the mobile social network as primary variables of context (Schilit *et al.* 1994; Dey 2001; Raento *et al.* 2005; Ahern *et al.* 2006; Tang 2007; Beach *et al.* 2008; Gay 2009).

As a final remark, it must be noted that the case of amplification and attenuation of social network signals is not the only one in which the concepts of convergence and divergence could inform design. As illustrated in Figure 14-3, a whole set of operations could be envisaged by applying the convergence principle in the context of "combining", "aggregating" or "connecting" information. Similarly, divergence would lead to a user-friendly toolbox of features for "clustering", "separating" and "filtering" information.

4. Applying convergence and divergence to design

Digitally converged ICT are already part of our everyday life. While it is of general agreement that they involve more positive than negative aspects, at the same time we sometimes feel overwhelmed and not fully in control of them. This is particularly strong when referred to social computing platforms for digital community interaction (i.e. the "Web 2.0" applications) such as Facebook. How well are key ICT companies equipping users to efficiently manage converged social spaces, social groups, roles and interaction contexts? To what extent do business and policy actors understand "digital communities" and related interaction needs? What has driven the Web 2.0 technology and business until now? What's next?

By considering the most recent trends, the complexity of dealing with such information spaces has been recently acknowledged by the main business actors. For instance, the launch of Google Plus aimed at offering a platform more advanced than Facebook in the management of "social circles". In this case, the aim was to allow the user to share information in a well-defined and clear interaction context (e.g. family, best friends, colleagues). In parallel, Facebook has been constantly making an effort to try to provide an easier and more fine-grained control over personal data. Other forms of filtering are also becoming widespread: in particular, smartphone applications make use of geo-referencing to filter information based on the user location and other parameters of interest (Lindqvist *et al.* 2011).

There seem to be clues indicating that the design approach introduced in the previous section is being increasingly followed. Indeed, a greater attention is being paid not only to aggregation, sharing and interlinking of information, places and people, but also to their clustering, filtering and protection. In this regard, the optimal trade-off between convergence and divergence is not universal because it is inherently subjective and constantly evolving.

The effort to improve design and address the challenges of interaction in working and leisure contexts is significant in consideration of the relatively young, rapidly evolving and highly competitive market of mobile "apps" (Fig. 14-4), which has rapidly grown thanks to the widespread diffusion of smartphones and the affordable rates of mobile Internet access.

Fig. 14-4: the mobile (social) web as the new frontier.

In this respect, the objective for companies is to grow not only in the quantity of registered users, but also in the quality of the user experience perceived in using their services. This can be achieved only by offering a more fine-grained level of control over incoming/outgoing social streams. In other words, it becomes relevant for the companies to tackle what boyd (2008) defined as the main challenges of social convergence, namely exposure and invasion.

Certainly there is much place for improvement, as even the most advanced devices such as the Nokia Lumia 800, launched with the slogan "People in one live stream" (Fig. 14-5), do not present optimal tools for the management of "big data" in one's device.

The Nokia Lumia 800 succeeds in the aggregation of contacts (and digital content) from different platforms. However, it does not provide any functionality to search and filter the information stream (e.g. by location, interest). The lack of these features significantly reduces the usefulness of the "one live stream". In short, the Nokia Lumia 800 embeds social algorithms acknowledging convergence, but quite neglects the importance of divergence in digital community interactions.

Although some social "apps" have started to be quite advanced in this respect, we are far from reaching maturity in this area. Indeed, the design of the basic building blocks of the smartphone, such as the address book, is generally still quite disappointing. An operation like "display all the photos that my brother or me took in this place in the last 5 years" cannot yet be easily performed.

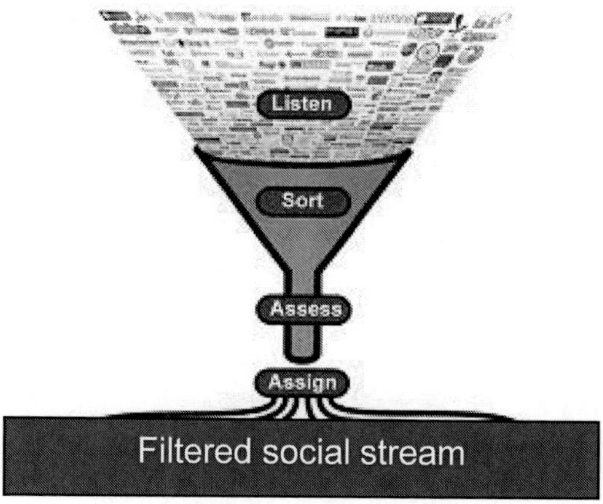

Fig. 14-5: Social Media funnel.

In order to fully enter what Fuchs *et al.* (2010) call Web 3.0-enabling human cooperation-it is necessary to effectively address social convergence issues in the design. The inclusion of an innovative social media "funnel" may represent for companies a key differentiating factor providing a significant competitive advantage. To proceed in this direction there are still challenges to overcome; for instance, it is not trivial to integrate an articulated conceptual model of the user in the design of current technologies, and even more complex to match some of its features with a given interaction context. Understanding and modelling human behavior is a key challenge which, solved, may open many new possibilities; as a matter of fact, the largest EU project is "The Human Brain Project" (Markram 2012), which aims to provide new insight on the functioning of the human brain to, among others, design revolutionary new ICT tools.

5. Conclusions

Convergence and divergence are important terms characterizing our times. Although historically linked with science and mathematics, these terms assume today a broader scope due to the multi-faceted implications of technological convergence on society and the economy. In particular, it has been illustrated how the sociological concept of digital communities has emerged in parallel to the spreading of converged digital networks.

Platforms for digital community interaction like Facebook, Twitter or WhatsApp have been very successful in supporting the seamless integration of the offline, online and mobile dimensions of interaction. Nevertheless, the design of such platforms has traditionally emphasized the "convergence" aspects of aggregation, interconnection, sharing and connectivity. As the amount of available data, devices, relations and contexts grows, it has become increasingly challenging to alleviate the phenomena of "invasion" and "exposure". In particular, as the case of Nokia Lumia shows, few are the cases in which divergence inspires the implementation of features like "social media funnels" to attenuate social signals.

What is still needed to move forward and address the challenges of social convergence, as well as the ambitious objectives of initiatives like the Digital Agenda 2020? Two areas for urgent development should be highlighted: first, the use of "big data" (Manyika *et al.* 2011) should be globally regulated, not only by agreeing on the right trade-off between citizens' privacy and nations' security, but also by allowing citizens to make smart use of "big data" in digital community interactions. Until now, while being of potential usefulness to the users, "big data" has not been used to the benefit to the user but rather kept within the realm of company marketing needs and government surveillance policies. Instead, it is particularly by making use of the possibility of collecting and analyzing "big data" almost in real-time that devices could embed a degree of intelligence to support user decision-making in rapidly evolving contexts and situations. In other words, these large datasets would allow the implementation of essential components such as social media funnels through which information flows could be "amplified or "attenuated". This is certainly an objective that requires adequate and converging international supporting policies. As previously mentioned, policy convergence is beyond the scope of this chapter, but it represents a key future extension of this study.

A second important direction concerns acknowledging a general objective of ICT design. Specifically, ICT tools should be conceived to support users in exploiting two different but complementary thinking strategies, namely convergence thinking and divergence thinking (Fig. 14-7).

Fig. 14-6: ICT tools supporting convergence and divergence Thinking.

Convergence thinking consists in the ability to recognise similarities, patterns and connections among various data and events, while divergence thinking is about recognising differences and separating individual items from a whole. These activities are performed on a regular basis by the human brain in relation to what we are able to sense. In the hybrid social space of digital communities, which evolves very rapidly and which receives contributions from thousands of invidivuals distributed globally, the human brain cannot consistently apply convergence and divergence thinking if not adequately supported by appropriate technologies. These concepts could be well integrated and extend the tradition of cognitive ergonomics and human factors in design (Salvendy 2012). It is important, however, to recognize that additional interdisciplinary research efforts are certainly needed to gain a better understanding of what it means to live constantly connected social lives. Although Silverstone's call for a better understanding of the social aspects of convergence has resulted in an advancement of the knowledge in the field, areas like computational sociology, user psychology and interaction design, just to name a few, would benefit a closer collaboration in deriving behavioral and contextual patterns and models of evolving social relationships to be tested in the field. While we are still far from understanding the socio-cultural implications of today's technology, new technological trends such as human-robot interaction and social robotics (Kanda and Ishiguro 2012; Fortunati 2013) will add new meanings and dimensions to convergence.

References

Agre, Philip E. 2001. “Welcome to the always-on world.” In *IEEE Spectrum*, 38(1), 12-13.

Ahern, Shane, Mark Davis, Dean Eckles, Simon King, Mor Naaman, Raul Nair, Miriana Spasojevic and Jeannie Yang. 2006. “Zonetag: Designing context-aware mobile media capture to increase participation.” In *Adjunct Proceedings of the Workshop on Pervasive Image Capture and Sharing (PICS 2006) at UbiComp 2006*.

Allagui, Ilhem and Johanne Kuebler. 2011. “The Arab spring and the role of ICTs: editorial introduction.” In *International Journal of Communication*, 5: 1435-1442.

Barabasi, Albert L. 2002. *Linked, the New Science of Networks*. Cambridge, MA: Perseus.

Beach, Aaron, Mike Gartrell, Sirisha Akkala, Jack Elston, John Kelley, Keisuke Nishimoto, Baishakhi Ray, Sergei Razgulin, Karthik Sundaresan, Bonnie Surendar, Michael Terada and Richard Han. 2008. “WhozThat? Evolving an ecosystem for context-aware mobile social networks.” In *IEEE Network*, 22(4): 50-55.

Beekhuyzen, Jenine, Lisa von Hellens and Sue Nielsen. 2012. “Insights from the underground: Using ANT to understand practices and motivations for file sharing in online communities.” In *ECIS 2012 Proceedings*, Paper 49.

Bilandzic, Mark, Daniel Filonik, Michael Gross, Andreas Hackel, Herbert Mangesius and Helmut Krcmar. 2009. “A Mobile Application to Support Phatic Communication in the Hybrid Space.” In *Proceedings of the Sixth International Conference on Information Technology: New Generations*: 1517-1521.

Birnholtz, Jeremy, Jeffrey Hancock, Madeline Smith and Lindsay Reynolds. 2012. “Understanding unavailability in a world of constant connection.” In *Interactions*, 19(5): 32-35.

boyd, Danah M. 2008. “Facebook’s Privacy Trainwreck: Exposure, Invasion, and Social Convergence.” In *The International Journal of Research into New Media Technologies*, 14(1): 13-20.

Callon, Michel. 1986. “Some Elements of a Sociology of Translation: Domestication of the Scallops and Fishermen of St. Brieuc Bay.” In *Power, Action and Belief: A New Sociology of Knowledge?*, edited by John Law, 196-233, London: Routledge.

Covell, Andy. 1999. *Digital Convergence: How the Merging of Computers, Communications, and Multimedia is transforming our lives*. Newport: Aegis.

Crabtree, Andy and Tom Rodden. 2008. "Hybrid ecologies: understanding cooperative interaction in emerging physical-digital environments." In *Personal and Ubiquitous Computing*, 12(7): 481-493.

Craig, Terence and Mary E. Ludloff. 2011. *Privacy and Big Data*. Sebastopol, CA: O'Reilly Media.

Darwin, Charles. 1968. *On the origin of species by means of natural selection*, 1st edition, London: Murray, 1859.

De Sola Pool, Ithiel. 1983. *Technologies of Freedom*. Cambridge, MA: Harvard University Press.

De Souza e Silva, Adriana. 2006. "From Cyber to Hybrid - Mobile Technologies as Interfaces of Hybrid Spaces." In *Space and Culture*, 9: 261-278.

Dey, Anind K. 2001. "Understanding and Using Context." In *Personal and Ubiquitous Computing*, 5(1): 4-7.

Eagle, Nathan and Alex Pentland. 2005. "Social serendipity: mobilizing social software." In *IEEE Pervasive Computing*, 4(2).

Elmer-Dewitt, Philip and David S. Jackson. 1993. "Take a trip into the future on the electronic superhighway." In *Time*, 141(15).

Farnham, Shelly and Pedram Keyani. 2006. "Swarm: hyper awareness, micro-coordination and smart convergence through mobile group text messaging." In *Proceedings of the 39th Hawaii International Conference on System Sciences (HICSS-39 2006)*.

Fortunati, Leopoldina. 2007. "Discussing a possible research agenda for the convergence of the mobile and the Internet." In *Proceedings of the conference on Communications in the XXI Century*: 109-114.

—. 2013. "Afterword: robot conceptualisations between continuity and innovation." In *Intervalla*, 1.

Fuchs, Christian, Wolfgang Hofkirchner, Matthias Schafranekl, Celina Raffl, Marisol Sandoval and Robert Bichler. 2010. "Theoretical foundations of the web: cognition, communication, and co-operation. Towards an understanding of Web 1.0, 2.0, 3.0." In *Future Internet*, 2(1): 41-59.

Gates, Bill, Nathan Myhrvold and Peter Rinearson. 1995. *The Road Ahead*. New York: Penguin.

Gay, Geri. 2009. "Context-aware mobile computing: affordances of space, social awareness, and social influence." In *Synthesis Lectures on Human-centered Informatics*, 2(1): 1-62.

Gilbert, Eric and Karrie Karahalios. 2009. "Predicting tie strength with social media." In *Proceedings of the 27th international conference on Human Factors in Computing Systems*: 211-220.

Gordon, Rich. 2003. "The meanings and implications of convergence." In *Digital Journalism: Emerging Media and the Changing Horizons of Journalism*, edited by Kevin Kawamoto, 57-75, Lanham, MD: Rowman and Littlefield.
Granovetter, Mark S. 1973. "The strength of weak ties." In *American Journal of Sociology*, 78(6): 1360-1380.
Haythornthwaite, Caroline. 2002. "Strong, weak and latent ties and the impact of new media." In *The Information Society*, 18(5): 385-401.
Jenkins, Henry. 2006. *Convergence Culture: Where Old and New Media Collide*. New York: New York University Press.
Kanda, Takayuki and Ishiguro Hiroshi. 2012. *Human-Robot Interaction in Social Robotics*. Boca Rosa, FL: CRC Press.
Katz, James E. and Mark Aakhus. 2002. *Perpetual Contact: Mobile Communication, Private Talk, Public Performance*. Cambridge: Cambridge University Press.
Klein, June R. 2012. "Wikileaks, Arab Uprisings, English Riots and Occupy Wall Street: Implications for Internet Policy and Practice from a Business and Industry Outcome Perspective." In *Information, Communication and Society Journal*, 14(6).
Law, John. 1992. "Notes on the Theory of the Actor-Network: Ordering, Strategy and Heterogeneity." In *Systems Practice*, 5: 379-393.
Latour, Bruno. 1987. *Science in Action: How to Follow Scientists and Engineers through Society*. Cambridge, MA: Harvard University Press.
Lepa, Jerzy and Arthur Tatnall. 2006. "Using actor-network theory to understanding virtual community networks of older people using the Internet." In *Journal of Business Systems, Governance and Ethics*, 1(4): 1-14.
Licklider, Joseph C.R. 1968. "The computer as a communication device." In *Science and Technology*, 76(2): 1-3.
Lin, Nan. 2001. *Social Capital: A Theory of Structure and Social Action*. Cambridge: Cambridge University Press.
Lindqvist, Janne, Justin Cranshaw, Jason Wiese, Jason Hong and John Zimmerman. 2011. "I'm the mayor of my house: examining why people use foursquare-a social-driven location sharing application." In *Proceedings of the 2011 ACM annual conference on Human factors in computing systems (CHI 2011)*: 2409-2418.
Ling, Richard and Birgitte Yttri. 2002. "Hyper-coordination via mobile phones in Norway." In *Perpetual Contact – Mobile Communication, Private Talk, Public Performance*, edited by James E. Jatz and Mark A. Aakhus, 139-169, Cambridge: Cambridge University Press.

Ling, Richard. 2004. *The Mobile Connection; The Cell Phone's Impact on Society*. San Francisco: Morgan Kaufmann Pub.

Lugano, Giuseppe and Pertti Saariluoma. 2007. "To share or not to share: Supporting the user decision in Mobile Social Software applications." In User Modeling (pp. 440-444). Berlin Heidelberg: Springer.

Lugano, Giuseppe. 2007. "Mobile Social Software: definition, scope and applications." In EU/IST eChallenges Conference, The Hague (The Netherlands), 1434-1441.

—. 2008. "Mobile social networking in theory and practice." In *FirstMonday*, 13(11).

—. 2010a. *Digital community design: exploring the role of mobile social software in the process of digital convergence*. Jyväskylä: University of Jyväskylä.

—. 2010b. "Do only computers scale? On the cognitive and social aspects of scalability." In *Encyclopaideia*, 14(28): 89-110.

Manyika, James, Michael Chui, Brad Brown, Jacques Bughin, Richard Dobbs, Charles Roxburgh and Angela Hung Byers. 2011. *Big Data: The Next Frontier for Innovation, Competition, and Productivity*. Washington: McKinsey Global Institute.

Markram, Henry. 2012. "The human brain project." In *Scientific American*, 306(6): 50-55.

Mueller, Milton L. 1999. "Digital convergence and its consequences." In *The Public*, 6(3): 11-28.

—. 2010. *Networks and States: The Global Politics of Internet Governance*. Cambridge, MA: The MIT Press.

Negroponte, Nicholas. 1996. *Being Digital*. New York: Random House.

O'Reilly, Tim. 2007. "What is Web 2.0: Design Patterns and business models for the next generation of software." In *Communications & Strategies*, 1, First Quarter.

Oulasvirta, Antti, Mika Raento and Sauli Tiitta. 2005. "ContextContacts: re-designing SmartPhone's contact book to support mobile awareness and collaboration." In *Proceedings of the 7th international conference on Human computer interaction with mobile devices & services (Mobile HCI2005)*: 167-174.

Paris, Cody M. and Simon Rubin. 2013. "Backpacking, Social Media, and Crises: A Discussion of Online Social Convergence." In *Information and Communication Technologies in Tourism*, edited by Lorenzo Cantoni and Zheng Xiang, 207-217, Berlin: Springer.

Raento, Mika, Antti Oulasvirta, Renaoud Petit and Hannu Toivonen. 2005. "ContextPhone: A prototyping platform for context-aware mobile applications." In *IEEE Pervasive Computing*, 4(2): 51-59.

Raento, Mika, Antti Oulasvirta and Nathan Eagle. 2009. "Smartphones. An emerging tool for social scientists." In *Sociological Methods and Research*, 37(3): 426-454.

Rheingold, Howard. 2002. *Smartmobs: The Next Social Revolution*. Cambridge, MA: Basic Books.

Rheingold, Howard, Saskia Sassen, Franz Vogelaar, Elizabeth Sikiardi, Noortje Marres, Koen Brams, Dirk Pültau, Marion Hamm, Kristina Andersen, Arie Altena and others. 2007. *Open11: Hybrid Space. How wireless media mobilize public space*. Rotterdam: Nai Uitgevers.

Rogers, Everett M. and Dilip K. Bhowmik. 1970. "Homophily-heterophily: relational concepts for communication research." In *Public Opinion Quarterly*, 34(4): 523-538.

Salvendy, Gavriel. 2012. *Handbook of Human Factors and Ergonomics*. New York: Wiley.

Schilit, Bill N., Norman Adams and Roy Want. 1994. "Context-aware computing applications." In *Proceedings of IEEE Workshop on Mobile Computing Systems and Applications*: 85-90.

Schubert, Petra and J. Felix Hampe. 2005. "Business Models for mobile communities." In *Proceedings of the 38th Annual Hawaii International Conference on System Sciences*: 172-183.

Sheridan, Thomas B. and Raja Parasuraman. 2006. "Human-automation interaction." In *Review of Human Factors and Ergonomics*, 1(1): 89-129.

Sillence, Elizabeth and Chris Baber. 2004. "Integrated digital communities: combining web-based interaction with text messaging to develop a system for encouraging group communication and competition." In *Interacting with computers*, 16(1): 93-113.

Silverstone, Roger. 1995. "Convergence is a dangerous word." In *Convergence*, 1(1): 11-13.

Stepanova, Ekaterina. 2011. "The role of information communication technologies in the 'Arab Spring'." In *PONARS Eurasia*, 15: 1-6.

Tang, Karen P. 2007. "Privacy Mechanisms for Context-Aware, Group-based Mobile Social Software." In *Proceedings of the 20th annual ACM Symposium on user interface software and technology (UIST'07)*.

Toporkoff, S. 2012. "ICT Convergence and Europe's Digital Agenda 2010-2020." In *The Real Issues of the Middle East and the Arab Spring: Addressing Research, Innovation and Entrepreneurship*, edited by Thomas Andersson and Abdelkader Djeflat, 315-330, New York: Springer.

Trappeniers, Lieven, Marc Godon, Laurence Claeys, Olivier Martinot and Emmanuel Marilly. 2008. "Cross-media experiences: ambient community interactions in the city." In *Bell Labs Technical Journal*, 13(2): 5-11.

Trist, Eric. 1981. "The evolution of socio-technical systems." In *Occasional paper*, 2.

Turpeinen, Marko, Kai Kuikkaniemi and Tietotekniikan Tutkimuslaitos. 2007. *Mobile Content Communities. Final Report*. Helsinki: HIIT.

Tönnies, Ferdinand. 1967. "Gemeinshaft and Gesellshaft." In *The sociology of community*, edited by Colin Bell and Howard Newby, London: Frank Cass and Co. Ltd.

Vihavainen, Sami, Antti Oulasvirta and Risto Sarvas. 2009. "I can't lie anymore!: the implications of location automation for mobile social applications." In *Proceedings of MobiQuitous 2009*.

Wagner, Ben. 2011. "Freedom of expression on the internet: Implications for foreign policy." In *Foreign Affairs*, 88(4).

Wellman, Barry, Peter J. Carrington and Alan Hall. 1988. "Networks as Personal Communities." In *Social Structures: A Network Analysis*, edited by Barry Wellman and Stephen D. Berkowitz, 130-184, Cambridge: Cambridge University Press.

Yoffie, David B. 1997. *Computing in the Age of Digital Convergence*. Boston: Harvard Business School Press.

INDEX